Lecture Notes in Economics and Mathematical Systems

378

W. Krabs J. Zowe (Eds.)

Modern Methods of Optimization

Proceedings of the Summer School "Modern Methods of Optimization", held at the Schloß Thurnau of the University of Bayreuth, Bayreuth, FRG, October 1-6, 1990

Springer-Verlag Berlin Heidelberg GmbH

Editors

Prof. Dr. Werner Krabs
Department of Mathematics
Technische Hochschule Darmstadt
Schloßgartenstr. 7
W-6100 Darmstadt, FRG

Prof. Dr. Jochem Zowe
Department of Mathematics
University of Bayreuth
Universitätsstr. 30
W-8580 Bayreuth, FRG

ISBN 978-3-540-55139-3 ISBN 978-3-662-02851-3 (eBook)
DOI 10.1007/978-3-662-02851-3

Originally published by Springer-Verlag Berlin Heidelberg New York in 1992

Typesetting: Camera ready by author

42/3140-543210 - Printed on acid-free paper

Preface

This volume contains the proceedings of the summer school "Modern Methods of Optimization", held at the Schloß Thurnau of the University of Bayreuth, October 1–6, 1990.

Like other branches of applied mathematics the area of optimization is undergoing a rapid development since the beginning of the computer age. Optimizaiton methods are of increasing importance for both, science and industry. The aim of the summer school was to present state-of-the-art knowledge by inviting 12 specialists from Optimization (and related fields) to present their areas of activity in the form of survey talks. This volume contains 10 of these presentations in slightly extended form. Most lectures started from an undergraduate level and outlined the developments up to the latest scientifique achievements. This enabled the audience, consisting of about 45 students and young researchers, to get an excellent overview of the latest trends in Optimization as well as a grasp of the breadth of its potential applications.

Equally important to the success of the summer school was the "nonmeasurable" part of the activities inherent in such a summer school. Here the inspiring atmosphere of a place like Thurnau helped to establish numerous contacts between "teachers" and "students".

The summer school was organized by the Universität Bayreuth together with the Technische Hochschule Darmstadt and was generously sponsored by the Volkswagenstiftung and the Universitätsverein Bayreuth. Their interest in the meeting and their support is hereby gratefully acknowledged.

In particular we would like to thank once more all the speakers; they were the main source for the high scientifique level of the school. Last but not least our thanks go to the students and young scientists who came to learn about modern optimization. Their active participation in the discussions following the lectures proved that the summer school fully reached its goal.

W. Krabs
(Darmstadt)

J. Zowe
(Bayreuth)

TABLE OF CONTENTS

Computability and Complexity of Polynomial Optimization Problems

Bernd Bank[(1)], Joos Heintz[(2)], Teresa Krick[(2)]
Reinhard Mandel[(1)], Pablo Solernó[(2)]

[(1)] Humboldt-Universität, FB Mathematik, PSF 1297, D-0 1086 Berlin
[(2)] Working Group Noaïota Fitchas, Universidad de Buenos Aires, Fac. de Ciencias Exactas, Depto de Matemática, Ciudad Universitaria, 1428 Buenos Aires, Argentina

Abstract. The paper is devoted to such general questions as the computability, decidability and complexity of polynomial optimization problems even for integer variables. The first part relates polynomial optimization to Tarski algebra, algorithmical semialgebraic geometry and Matijasevich's result on diophantine sets. The second part deals with the particular case of integer polynomial optimization, where all polynomials involved are quasiconvex with integer coefficients. In particular, we show that the polynomial inequality system $f_1 \leq 0, \ldots, f_s \leq 0$ admits an integer solution if and only if it admits such a solution in a ball $B(0, R)$ with $\log R = (sd)^{O(n)}\ell$, where $d \geq 2$ is a degree bound for all polynomials $f_1, \ldots, f_s \in \mathbb{Z}[x_1, \ldots, x_n]$ and ℓ is a bound for the binary length of all coefficients.

I. On Relations of Polynomial Optimization to Basic Computability and Complexity Results.

During the last two decades classes of optimization problems involving only polynomials as constraints and objective functions, respectively, have often been considered. One reason was that, on the one hand side, in several application the linear programming problem did not suffice to model real life situations appropriately, and, on the other hand, that the simplest generalizations of linear programming led to optimization problems with linear constraints and quadratic as well as convex or concave polynomial objective functions. More complicated cases arise if nonlinear constraints are considered, too. Belousov [5] was one of the first who considered convex polynomial constraints even under integer requirements on the variables. For such classes of optimization problems a comprehensive analysis of existence and stability properties is known (see e.g. Bank/Mandel [1]).
The important algorithmic-theoretical questions (i.e. decidability, computability and complexity) deeply influence also the investigations of polynomial optimization problems with or without integer requirements on the variables. In view of complexity questions, the results also yield nice contributions to the theory of complexity and stimulate even the

development of this theory (e.g. NP-completeness theory; algorithm of Khachiyan for linear programming [32], Lenstra's algorithm for integer linear programming, where the number of variables is fixed [37]).
The optimization problems considered in the following are

$$\text{(P)} \qquad \inf\{f(x) \mid f_1(x) \leq 0, \ldots, f_s(x) \leq 0,\ x \in \mathbb{R}^n\}$$
$$\text{(I)} \qquad \inf\{f(x) \mid f_1(x) \leq 0, \ldots, f_s(x) \leq 0,\ x \in \mathbb{Z}^n\},$$

where $f, f_1, \ldots, f_s \in \mathbb{Z}[x]$. As usually $\mathbb{Z}$ denotes the ring of all integer numbers, $\mathbb{R}$ the field of all reals, and $\mathbb{Z}[x]$ is the ring of all polynomials in n indeterminates, $x = (x_1, \ldots, x_n)^T$, with integer coefficients. From an algorithmic-theoretical point of view, one is immediately facing the questions whether there are algorithms for deciding the solvability of these problems or not. If the respective answer is positive, this gives rise to the problem of how one can find a solution and, finally, what kind of algorithms one can expect?
If - a priori - one does not restrict oneself to subclasses of (P) and (I) (e.g. considering polynomials with additional, nice properties) then one will investigate fairly general problems like semialgebraic sets, Tarski algebra, and Hilbert's 10-th problem.
First, we point out some relations of our problems (P) and (I) to the elementary theory of real numbers (Tarski algebra) and to Hilbert's 10-th problem.
We shortly sketch the logical backround. An elementary language $\mathcal{L}$ is given by set of constants and a finite number of operation- and relation symbols. From these symbols, together with variables $x_1, x_2, \ldots$, the symbol "=", the logical symbols $\wedge$, $\vee$, $\neg$, the quantifiers $\exists$, $\forall$ and the brackets (,), one (inductively) builds up terms and formulas. A term is formed of constants and variables by superposition of operation symbols. Two types of formulas are distinguished: atomic and arbitrary ones. Atomic formulas are equations or formal relations between terms. Arbitrary formulas are derived from atomic formulas by closure under $\wedge$, $\vee$, $\neg$ and the quantifiers $\exists$, $\forall$ (implication, equivalence and inequality symbols "$\Longrightarrow$", "$\Longleftrightarrow$" and "$\neq$", respectively, can be used als abbreviations). A formula containing no quantifier is called a quantifier-free formula. Not quantified variables in a formula are called free variables or parameters. A formula containing no free variables is called a sentence. An arbitrary formula from $\mathcal{L}$ is usually denoted by $\Phi(x_1, \ldots, x_n)$ if all free variables are contained among the symbols $x_1, \ldots, x_n$. $\mathcal{L}$ can be understood as the collection of all formulas built up in the way described before. To each formula $\Phi \in \mathcal{L}$ there corresponds its length $|\Phi|$ (the number of symbols required to write down Φ).
Tarski [62] showed that for any formula $\Phi(x_1, \ldots, x_n)$ in the elementary language with constants $\{0, 1\}$ and operation- and relation symbols $\{+, \cdot, <\}$ one effectively can associate:

(i) a quantifier-free formula $\Phi^*(x_1, \ldots, x_n)$ of the same language and,

(ii) a proof for the equivalence $\Phi \Longleftrightarrow \Phi^*$ in which only the axioms for real closed fields are used.

We recall that a real closed field is an ordered field with the intermediate value property for polynomials (e.g. $\mathbb{R}$ is such a field).

An effective procedure to obtain Φ^* from Φ such that (i) and the equivalence in (ii) hold is called a quantifier elimination procedure.
If one applies a quantifier elimination procedure to a formula of the elementary language over $\mathbb{R}$ which is a sentence, then one has a decision procedure for the elementary theory of real numbers, the so-called Tarski algebra (i.e. the set of all sentences in the considered language that hold in $\mathbb{R}$). One easily sees that

$$f(x) = 0, \quad f(x) < 0, \quad f(x) \leq 0, \quad f \in \mathbb{Z}[x],$$

are atomic formulas in the elementary language over $\mathbb{R}$.
With respect to our optimization problem (P) introduced above Tarski's result permits the following corollaries.

Corollary 1 *There is an algorithm which decides for any problem (P) whether it admits an optimal solution or not.*

This becomes easily obvious upon writing the following sentence in the elementary language over $\mathbb{R}$:

$$\exists x(f_1(x) \leq 0 \wedge \ldots \wedge f_s(x) \leq 0 \wedge (\forall y(f_1(y) \leq 0 \wedge \ldots \wedge f_s(y) \leq 0 \\ \Longrightarrow f(x) \leq f(y)))).$$

Obviously, this sentence holds if and only if (P) admits an optimal solution.
Following Grunewald/Segal [22] we say that an algorithm finds an optimal solution $x = (x_1, \ldots, x_n)^T$ of the problem (P) approximatively if

(i) x is an optimal solution and,

(ii) for each natural number m the algorithms computes a rational vector $x^m = (x_1^m, \ldots, x_n^m)^T$ such that

$$|x_i^m - x_i| < 2^{-m}, \qquad i = 1, \ldots, n,$$

(in other words, the algorithm generates a sequence of rational vectors which converges to some optimal solution within the rate given above).

Using this definition, Tarski's result permits

Corollary 2 *There is an algorithm which finds an optimal solution of (P) approximatively if there is an optimal solution for (P).*

This can be seen following the framework due to Grunewald/Segal [22]. We consider the formula

$$\Phi(x): \qquad f_1(x) \leq 0 \wedge \ldots \wedge f_s(x) \leq 0 \wedge (\forall y(f_1(y) \leq 0 \wedge \ldots \wedge f_s(y) \leq 0 \\ \Longrightarrow f(x) \leq f(y))).$$

Then the set of all optimal solutions of (P) can be written as

$$X \underset{\text{Df}}{=} \{x \in \mathbb{R}^n \mid \Phi(x) \text{ is true}\}.$$

If (P) admits an optimal solution, then X is a closed non-empty subset in $\mathbb{R}^n$. Therefore there is a natural number p such that the cube defined by $|x_i| < p$, $i = 1, \ldots, s$, intersects X, i.e. the following sentence is true

$$\exists x(x_1^2 < p^2 \wedge \ldots \wedge x_n^2 < p^2 \wedge \Phi(x)).$$

First one determines the smallest p such that this sentence becomes true. If p has been found, then, by a similar procedure, one finds a nested sequence of subcubes with edges of length $p, 2^{-1}p, \ldots, 2^{-m}p, \ldots$, all intersecting X. The sequence $\{x^m\}$ of the corresponding centers of the subcubes is a sequence of rational vectors, which converges to an x such that $|x_i^m - x_i| < 2^{-m}p$, $i = 1, \ldots, n$. Since X is closed, x must be in X.
The quantifier elimination used by Tarski is totally impracticable because of the fact that the number of steps involved tends to $2^{2^{\cdot^{\cdot^{\cdot^{2}}}}}$, where the number of exponents is given by the length $|\Phi|$ of the formula Φ.
Collins [13] proposed a substantially modified and improved quantifier elimination algorithm for the elementary theory of real closed fields using the so-called cylindrical algebraic decomposition of $\mathbb{R}^n$. The algorithm of Collins accepts any formula written in prenex from as inputs, i.e.

$$\Phi : \qquad (Q_1 x^1) \ldots (Q_r x^r)(\varphi(x^1, \ldots, x^r, y)),$$

where

- $x^i \underset{\text{Df}}{=} (x_1^i, \ldots, x_{n_i}^i)^T, \; i = 1, \ldots, r$
- $y \underset{\text{Df}}{=} (y_1, \ldots, y_m)^T$
- $n \underset{\text{Df}}{=} \sum_{i=1}^{r} n_i + m$
- φ is a quantifier-free formula with atomic formulas $f = 0$ and $f \leq 0$, $f \in \mathbb{Z}[x, y]$
- $Q_i \in \{\exists, \forall\}$ & $Q_i \neq Q_{i+1}$, $i = 1, \ldots, r-1$
- $Q_i x^i \underset{\text{Df}}{=} (Q_i x_1^i)(Q_i x_2^i) \ldots (Q_i x_{n_i}^i)$.

The number r gives the number of blocks of (alternative) quantifiers. Most of the "mathematically interesting" decision problems involve only a fixed small number r of quantifier alternations, e.g. in a prenex formulation of our optimization problem (P) only one alternation of quantifiers will appear.
To measure a formula Φ in prenex form we use $|\Phi|$, n, r and the number

$$\sigma(\Phi) \underset{\text{Df}}{=} 2 + \sum_{i=1}^{s} \deg f_i \quad \text{if } f_1, \ldots, f_s \quad \text{appear in } \varphi.$$

We note that each formula $\Phi \in \mathcal{L}$ can be brought into an equivalent prenex form in linear time with respect to $|\Phi|$. Such a procedure does not change $|\Phi|$ and $\sigma(\Phi)$, and a possible increase of n is bounded by the number of quantifiers.

The cylindrical algebraic decomposition is a process in which the $\mathbb{R}^n$ is devided into a series of regions, on each of which the members of a given family of polynomials occuring in the prenex formula are identically zero, identically positive or identically negative.
On the basis of Collins ideas, F. Müller [45] developed an algorithm to solve the optimization problem (P) where the constraint set $\{x \in \mathbb{R}^n \mid f_1(x) \leq 0, \ldots, f_s(x) \leq 0\}$ is a bounded set. In Müller's algorithm the $\mathbb{R}^{n+1}$ is divided into a finite number of disjoint connected subsets V_j, $j = 1, \ldots, p$, such that

$$\operatorname{sgn}[f_i(x_1, \ldots, x_n) \mid (x_1, \ldots, x_n, z) \in V_j] = \text{const } \forall i, j$$

and

$$\operatorname{sgn}[f(x_1, \ldots, x_n) - z \mid (x_1, \ldots, x_n, z) \in V_j] = \text{const } \forall j$$

and such that the optimal set of (P) is the projection of one of these V_j into $\mathbb{R}^n$.
In view of the optimization, Müller tries to avoid the computation of superfluous sets V_j. However, by Davenport/Heintz [15], one knows that the number of regions to be described in Collins' method is, in the worst case, doubly exponential with respect to the dimension n.
Before passing to more detailed complexity questions some general results with respect to the integer optimization problem (I) will be referred.
With respect to Hilbert's 10-th problem Matijasevich's famous theorem [42] implies that there is no algorithm to decide whether an arbitrary polynomial equation $p(x) = 0$, $p \in \mathbb{Z}[x]$ is solvable in integers or not. The immediate consequence for the problem (I) is

Corollary 3 *(Jeroslow [29])*
If all polynomials f_i, $i = 1, \ldots, s$, in (I) are of degree 2, f is linear and the problem (I) admits an optimal solution, then (even in this case) there is no algorithm which find an optimal solution.

This corollary gives immediately rise to the question for subclasses of (I) allowing the construction of a solution algorithm. Some of such classes are exposed for example in Khachiyan [33] and Mandel [40], and such expositions are also aims of our paper in the second part.
In particular, Matijasevich's result implies that such a nice result as it holds for the decidability of the theory of real numbers does not exist if the variables range in the integer numbers. Excluding the multiplication as operation one arrives at the so-called "Presburger arithmetic" (or the theory of integer numbers under addition), for which there exists a decision procedure [47]. Cooper proposed a quantifier elimination procedure that requires no more than $2^{2^{2^{O(|\ast|)}}}$ steps [1] (see Oppen [46]).
From a theoretical point of view, already Presburger's procedure shows the existence of an algorithm to solve integer linear optimization problems, see Lee [36].
While we have mainly discussed questions on the existence of algorithmic procedures so far, now we turn over to some known time-complexity results related to our problems (P) and (I).

[1] We adopt the standard notation of $O(f(n))$, $n \in \mathbb{N}$, in order to indicate that there is a constant c independent of n such that $O(f(n)) \leq c|f(n)|$ for all $n \in \mathbb{N}$.

First, we have in mind Turing-machines as algorithmic models. Then the number of steps needed in the execution of an algorithm reflects the time complexity. For cases in which the polynomials $f, f_1, \ldots, f_s \in \mathbb{Z}[x]$ appearing in the problems (P) and (I) are linear, the reader finds a nice, comprehensive analysis in the books of Schrijver [56] and Grötschel/Lovász/Schrijver [21]. By the results of Khachiyan [32] the linear optimization problem (P) is solvable by a time-polynomial algorithm; substantially modified and improved algorithms are due to Karmarkar [30], Renegar [49], Vaidya [63], and others.

Lenstra [37] has developed a time-polynomial algorithm for (I) in the linear case where the number of integer variables is fixed. The decision problem corresponding to the general linear problem (I) is NP-complete, (Karp [31], Borosh/Treybig [11]). It should be remarked that the more complicated case of Presburger arithmetic exposes even doubly-exponential lower bounds on nondeterministic time, see Fisher/Rabin [17].
Now, we turn over to complexity results for semialgebraic sets. Let us denote the constraint sets of (P) by

$$M \underset{\mathrm{Df}}{=} \{x \in \mathbb{R}^n \mid f_1(x) \leq 0, \ldots, f_s(x) \leq 0\}$$

(then M is a special case of semialgebraic set, see e.g. Bochnak/Coste/Roy [10], Benedetti/Risler [7]), and let

$$\deg f_i \leq d, \quad i = 1, \ldots, s,$$

where deg f_i means the total degree of the polynomial $f_i \in \mathbb{Z}[x_1, \ldots, x_n]$.
Further let ℓ be the maximum binary length of all coefficients appearing in $f_1, \ldots, f_s \in \mathbb{Z}[x_1, \ldots, x_n]$.

Grigoriev/Vorobjov [20] have shown that there is an algorithm with a running time bounded by $p(\ell, (sd)^{n^2})$, where p is a polynomial in the two variables ℓ and $(sd)^{n^2}$ which finds a representative subset $T \subset M$ of a cardinality not exceeding $\bar{p}((sd)^{n^2})$, where $\bar{p}$ is a univariate polynomial. For every point $x \in T$ the algorithm constructs a univariate polynomial g that is irreducible over the rational numbers $\mathbb{Q}$, and it generates expressions for each component x_i:

$$x_i = x_i(\omega) = \sum \beta_j^{(i)} \omega^j \in \mathbb{Q}[\omega], \quad i = 1, \ldots, n,$$

where $\beta_j^{(i)} \in \mathbb{Q}$, $0 \leq j < \deg g$ and $g(\omega) = 0$. Further, the algorithm yields an interval (b_1, b_2) with $b_1, b_2 \in \mathbb{Q}$ such that ω is the unique real root of g in (b_1, b_2). Finally, the degree of g admits a polynomial bound $p^*((sd)^{n^2})$, and the binary length of all coefficients of the polynomials g and $x_i(\omega)$, $i = 1, \ldots, n$, and of the constants b_1 and b_2 is bounded by $\ell p^*((sd)^{n^2})$.
Combining a result due to Heindel [23] with the algorithm of Grigoriev/Vorobjov one can find for any rational number ε, $0 < \varepsilon \leq 1$, rational ε-approximations of the points in T within a time $\tilde{p}(\log 1/\varepsilon, \ell, (sd)^{n^2})$, where $\tilde{p}$ is a trivariate polynomial.

Now, some remarks on theories of computation and complexity over real numbers are inevitable.
If we agree to code algebraic numbers ω by their (irreducible over $\mathbb{Q}$) algebraic equations $g(\omega) = 0$, $g \in \mathbb{Z}[\omega]$, and if necessary, by an interval (a, b), $a, b \in \mathbb{Q}$, not containing other roots of g than ω (as above related to the result of Grigoriev/Vorobjov), then we may consider exact algorithms to solve the optimization problem (P).
With respect to this agreement Khachiyan [33] studied exact algorithms for (P) under the additional hypothesis that all polynomials appearing in (P) are convex functions.
The various models of computation over the reals or an arbitrary (ordered) ring are discussed by Blum/Shub/Smale [9] with respect to their theory of computation and complexity over the real numbers. This theory allows to reflect the classical case of computation over $\mathbb{Z}$ definitions of universal machines, partial, recursive functions and NP-complete problems. It further reflects the properties of the reals and it provides basic settings for the study of algorithms in numerical analysis.
A common basic model for computations over real numbers consists in a so-called computation tree: an input root and internal nodes of two types

- computation nodes which transmit a program of real numbers modified by a rational operation from $\{+, -, \times, \div\}$,
- branching nodes which go right or left according to whether an inequality is true or false.

Following Smale [58] the number of branching nodes in a computation tree is called the topological complexity of the tree.
Now, considering a "problem", the topological complexity of the problem means the minimum topological complexity of all computation trees for that problem. With respect to further research it should be noted that Levine [38] succeeded in deriving a lower bound estimate for the topological complexity of the problem: Find, within ε, zeros of n complex polynomials in n indeterminates. This is a generalization of Smale's result for $n = 1$.
For all $\varepsilon < \varepsilon(d)$ the topological complexity of the problem:
"Find within ε all roots of a complex univariate polynomial of degree d and with a leading coefficient equal to 1"
is greater than $(\log d)^{2/3}$.

Recently, considerable effort has been made in view of the so-called "fast" quantifier elimination due to Solernó [59], Heintz/Roy/Solernó [25], [27] and Renegar [48], [50], which we apply to a particular case in the second part of the paper.
To formulate the general result we use, as algorithmic model, arithmetic networks (in the sense of von zur Gathen [64] introduced) with unit costs for arithmetical operations in $\mathbb{Z}$. The notion of arithmetical network may be described by a directed, acyclical graph where each vertex represents an arithmetical or boolean operation, a comparation or a selection.
To each arithmetical network there correspond two notions of complexity:

(i) the parallel complexity or depth of the arithmetical network, i.e. the length of a longest directed path in the corresponding graph;

(ii) the sequential complexity (or the number of processors) or the size of the arithmetical network, i.e. the number of vertices of the corresponding graph.

Let Φ be a prenex formula in the elementary language over $\mathbb{R}$ of length $|\Phi|$ built up by polynomials $f_1, \ldots, f_s \in \mathbb{Z}[x]$, $x = (x_1, \ldots, x_n)^T$. Further, let $d \underset{\text{Df}}{=} 2 + \sum_{i=1}^{s} \deg f_i$, and let r be the number of blocks of quantifier alternations. Then it holds:

(i) There is an algorithm which eliminates the quantifiers in Φ with
- parallel complexity $n^{O(r)}(\log d)^{O(1)} + O(|\Phi|)$
- sequential complexity $d^{n^{O(r)}}|\Phi|$.

(ii) The same bounds hold for the decision problem in the elementary language over $\mathbb{R}$.

A quantifier elimination algorithm with a complexity structure as in (i) is called a "fast" one. An analysis of the algorithm announced in (i) yields

Corollary 4 *If the formula Φ contains only one block of quantifiers and exactly one parameter, then the assertion (i) specializes to the following: there is an algorithm that generates from Φ by elimination of the quantifiers an equivalent formula ψ in the same language such that the maximum degree of the (univariate) polynomials involved in ψ is bounded by $d^{O(n)}$ and the maximum absolute value of the coefficients of all polynomials in ψ is bounded by $\sigma^{d^{O(n)}}$, where σ denotes the maximum absolute value of the coefficients of all polynomials appearing in Φ.*

For details the reader is referred to Heintz/Roy/Solernó [27]. The proof in Heintz/Roy/Solernó [27] yields $d^{O(n^2)}$ as a degree bound for the polynomials in ψ, but a refinement of the methods used and a careful checking of the estimations permit to obtain a degree bound as mentioned above. The necessary analysis would be very extensive, hence we omit it here.
We remark that these algorithmic results can be transferred mutatis mutandis to the context of Turing complexity. From the parallel complexity bounds or by a direct inspection of the algorithm one can observe that the coefficient growth of intermediate polynomial manipulations remains "under control". For technical details of the estimations used see Heintz [24] and Wüthrich [67].

Khachiyan and Tarasow [33], [34] announced that if $f_1, \ldots, f_s \in \mathbb{Z}[x]$ are convex polynomials in n indeterminates of degrees bounded by $d \geq 2$ and with coefficients of a binary length bounded by ℓ, then the inequality system $f_1 \leq 0, \ldots, f_s \leq 0$ admits an integer solution if and only if such a solution exists in a ball centered in the origin with a radius R of a binary length bounded by $d^{c(\tilde{n}+d)}n^{cd}\ell$, where $\tilde{n} = \max\{n, s\}$ and c is a universal constant independent of the considered parameters.
This result, in a direct way, can be extended to a bound for an optimal solution of the optimization problem (I) provided that all polynomials involved are convex. For the special case that a finite system of convex polynomial inequalities describes a bounded set Belousov/Shironin [6] claim (without proof) the radius bound $16^d(2^\ell n)^{2n}$, which is better than the bound for the general case given by Khachiyan/Tarasov [34] and the one

we will show in the second part. Our aim is to derive bounds of a structure better than the Khachiyan bounds. Moreover, these bounds are to be valid for the more general case of quasiconvex polynomials (e.g. the function $f : \mathbb{R} \rightarrow \mathbb{R}$ given by $f(x) = x^3$ is a quasiconvex polynomial, but it is not convex).

II. Bounds of Solutions and Complexity of Integer Quasiconvex Polynomial Optimization.

In the whole section we will use the following definitions and notations:
As usually, we call a polynomial $f \in \mathbb{R}[x]$ quasiconvex if all the lower level sets $\{x \in \mathbb{R}^n \mid f(x) \leq \alpha\}$, $\alpha \in \mathbb{R}$, are convex subsets of $\mathbb{R}^n$.
If $V \subset \mathbb{Z}^n$ is a finite set of integer vectors, then $\ell(V)$ denotes the maximum binary length of the coordinates of all vectors in V. In the same sense, $\ell(\mathcal{F})$ denotes the maximum binary length of the coefficients of all polynomials in the finite family $\mathcal{F} \subset \mathbb{Z}[x]$ of polynomials. The closed ball centered in the origin of $\mathbb{R}^n$ with a radius R is indicated by $B(0, R)$.
Now we will announce our main result.

Theorem 1 *Let $f_1, \ldots, f_s \in \mathbb{Z}[x_1, \ldots, x_n]$ be quasiconvex polynomials with integer coefficients and of degrees bounded by $d \geq 2$. Further let $\ell \underset{\text{Df}}{=} \ell(\{f_1, \ldots, f_s\})$ be a bound for the binary length of all coefficients involved. Then there is a radius $R \in \mathbb{N}$ of a binary length $\ell(R) = (sd)^{O(n)}\ell$ such that the set $\{x \in \mathbb{Z}^n \mid f_1(x) \leq 0, \ldots, f_s(x) \leq 0\}$ contains a point belonging to the ball $B(0, R)$ if the set is non-empty. The radius only depends on the parameters s, d, n, ℓ and it is independent of the special form of the polynomials.*

Remark. The exponential character of the bound obtained by Theorem 1 is intrinsic for the problem. This is shown by the following simple example in case of convex polynomials [33].
If $f_1 = -x_1 + 2^\ell$, $f_2 = x_1^2 - x_2, \ldots, f_n = x_{n-1}^2 - x_n$ are the polynomials, then all solutions of the system $f_i(x) \leq 0$, $i = 1, \ldots, n$, are located outside a ball $B(0, R)$ where $\ell(R) = 2^{n-1}\ell$. Therefore, it is clear that the bound of Theorem 1 is optimal with respect to the considered parameters, or, in other words it provides a general complexity measure.

The geometrical result of Theorem 1 allows to draw the algorithmic consequence:

Corollary 5 *There is a nondeterministic Turing machine which decides in a time bounded by $(sd)^{O(n)}\ell$, whether the set $\{x \in \mathbb{Z}^n \mid f_1(x) \leq 0, \ldots, f_s(x) \leq 0\}$ is empty or not.*

In other words, the corresponding decision problem belongs to the complexity class NEXPTIME ("nondeterministically simply exponential time").
Underlying a dense codification of the polynomials $f_1, \ldots, f_s$, i.e. the polynomials are represented by coefficient vectors of a length
$\binom{d+n+1}{n+1} = d^{O(n)}$ (possibly extended to the full length by zeros), then the input length of the family $\{f_1, \ldots, f_s\}$ is given by $L = sd^{O(n)}\ell$, which is a trivial lower bound for the time complexity. Then Corollary 5 implies

Corollary 6 *Assume a dense codification of the polynomials, then the time bound* $(sd)^{O(n)}\ell$ *on a non-deterministic Turing machine is optimal.*

Finally, the methods we use to prove Theorem 1 can be applied to obtain the following theorem on integer optimization with quasiconvex polynomial objective function and constraints, respectively.

Theorem 2 *Let* $f, f_1, \ldots, f_s \in \mathbb{Z}[x_1, \ldots, x_n]$ *be quasiconvex polynomials with integer coefficients and of degrees bounded by* $d \geq 2$*. Moreover, let*
$\ell \underset{\text{Df}}{=} \ell(\{f, f_1, \ldots, f_s\})$ *and*

$$M \underset{\text{Df}}{=} \{x \in \mathbb{R}^n \mid f_1(x) \leq 0, \ldots, f_s(x) \leq 0\}.$$

If the set $M \cap \mathbb{Z}^n$ *is non-empty and if* $\inf\{f(x) \mid x \in M \cap \mathbb{Z}^n\} = m > -\infty$*, then there is a radius* $R \in \mathbb{N}$ *of binary length* $\ell(R) = (sd)^{O(n)}\ell$ *such that the optimization problem* $\inf\{f(x) \mid x \in M \cap \mathbb{Z}^n\}$ *has an optimal point belonging to* $B(0, R)$.

Now we start proving the results.

Proof of Theorem 1.

Theorem 1 is a consequence of recent results in the field of algorithmic semialgebraic geometry, more precisely, Corollary 4, and a reduction technique for quasiconvex polynomial inequalities developed in Bank/Mandel [1], [2]. Here, we will consider the new methods used to develop the proofs in more detail. For a complete background the reader is referred to Bank/Mandel [1], [2], Bank/Heintz/Krick/Mandel, Solernó [3], Heintz/Roy/Solernó [25], [26], [27], Krick [35], Renegar [48], [50] and Solernó [59].
In order to make the proof transparent we divide it into different sections.

1 Basic properties of quasiconvex polynomials.

We restrict ourselves to showing theoretical properties that are indispensable for obtaining the announced bounds.

"Uniformity" property.

Let $f \in \mathbb{R}[x_1, \ldots, x_n]$ be a quasiconvex polynomial and let $x, u \in \mathbb{R}^n$, $u \neq 0$, be fixed. If the polynomial $f(x + tu)$ in the single indeterminate t is strictly decreasing (respectively constant), then the polynomial $f(y + tu)$ is strictly decreasing (respectively constant) for all $y \in \mathbb{R}^n$.

Proof. See Bank/Mandel [1]. #

"Linearity" property of a quasiconvex homogeneous form.

Let $p \in I\!R[x_1, \ldots, x_n]$ be a quasiconvex homogeneous form of degree $d \geq 1$. Let

$$\begin{aligned} L \ &\underset{\text{Df}}{=} \ \{x \in I\!R^n \mid p(x) = 0\}, \\ K \ &\underset{\text{Df}}{=} \ \{x \in I\!R^n \mid p(x) \leq 0\}. \end{aligned}$$

Then:

(i) L is a linear subspace in $I\!R^n$.

(ii) If d is even, then $K = L$.

(iii) If d is odd, then L is a hyperplane and K is a halfspace bounded by L.

Proof. This is an immediate consequence of the preceding properties explained above. #

Now, we give an effective characterization of L and K, exhibiting a basis $\mathcal{B}$ of L (and a system of generators $\mathcal{G}$ of K), which can be constructed from the coefficients of the quasiconvex homogeneous form p.

Lemma 1 *Let $p \in \mathbb{Z}[x_1, \ldots, x_n]$ be a quasiconvex form (with integer coefficients) of degree d, and let $\tilde{d} \underset{\text{Df}}{=} \max\{2, d\}$.*
Then the linear subspace $L = \{x \in I\!R^n \mid p(x) = 0\}$ admits an integer basis $\mathcal{B}$ such that $\ell(\mathcal{B}) = \tilde{d}^{O(r)}(\ell(p) + n)$, where $r \underset{\text{Df}}{=} n - \dim L$ and, if d is odd, then the halfspace K admits a system of integer generators $\mathcal{G}$ (i.e. $K = \{x \in I\!R^n \mid x = \sum_{v \in \mathcal{G}} \lambda_v v, \lambda_v \geq 0\}$) such that $\ell(\mathcal{G}) = \ell(\mathcal{B})$.

Proof. One observes that if p is a quasiconvex homogeneous form, then there is an indeterminate x_j such that the degree of p with respect to x_j is equal to d (this can be seen using, step by step, all unit vectors u of $I\!R^n$ to check the behaviour of $p(tu)$, $t \in I\!R$, with respect to the two properties above). Without loss of generality we assume x_1 to be this indeterminate. Now, we proceed by induction on n.
Let $n \geq 2$. If $L \neq \{0\}$, then let $u \in L \setminus \{0\}$. One has $p(tu) = t^d p(u) = 0$, $t \in I\!R$, which, by the "uniformity" property, implies that the polynomial $p(x + tu)$ in t is constant for all $x \in I\!R^n$. This means that $p(x) = p(x + u)$ for all $x \in I\!R^n$. Therefore,

$$(*) \qquad p(x) = p(x + u) \qquad \text{for all} \qquad u \in L.$$

We can write

$$\begin{aligned} p(x_1, \ldots, x_n) \ = \ & a_d(x_2, \ldots, x_n)x_1^d + a_{d-1}(x_2, \ldots, x_n)x_1^{d-1} + \\ & + \ldots + a_1(x_2, \ldots, x_n)x_1 + a_o(x_2, \ldots, x_n), \end{aligned}$$

where $a_d(x_2, \ldots, x_n) \neq 0$ and $a_i(x_2, \ldots, x_n) \in \mathbb{Z}[x_2, \ldots, x_n]$, for all $0 \leq i \leq d$, is a homogeneous form of degree $d-i$ (i.e. $a_d(x_2, \ldots, x_n) = a_d \in \mathbb{Z} \setminus \{0\}$ and $a_{d-1}(x_2, \ldots, x_n) \underset{\text{Df}}{=} \sum_{k=2}^{n} b_k x_k$

is an integer linear form).
We calculate the polynomial p at the point $x+u=(x_1+u_1, x_2+u_2, \ldots, x_n+u_n)$:

$$\begin{aligned} p(x+u) &= a_d(x_1+u_1)^d + \Big(\sum_{k=2}^{n} b_k \cdot (x_k+u_k)\Big) \cdot (x_1+u_1)^{d-1} + \cdots + \\ &\quad + a_o(x_2+u_2, \ldots, x_n+u_n) = \\ &= p(x) + x_1^{d-1}(da_d u_1 + \sum_{k=2}^{n} b_k u_k) + \cdots + p(u), \end{aligned}$$

where the degrees with respect to x_1 of the terms omitted are less than $d-1$.
By (*) one concludes that $L \subset L^1 \underset{\text{Df}}{=} \{u \in \mathbb{R}^n \mid da_d u + \sum_{k=2}^{n} b_k u_k = 0\}$.
The hyperplane L^1 admits an integer basis

$$\{(-b_2, da_d, 0, \ldots, 0)^T, \ldots, (-b_n, 0, \ldots, 0, da_d)^T\}$$

which can be completed to a basis $\mathcal{B}^1$ of $\mathbb{R}^n$ by adjoining the unit vector $(1, 0, \ldots, 0)^T$ such that $\ell(\mathcal{B}^1) = \ell(p) + \log d$. The restriction of p to L^1 can be written as

$$p^1(y_1, \ldots, y_{n-1}) \underset{\text{Df}}{=} p(-b_2 y_1 - \cdots - b_n y_{n-1}, da_d y_1, \ldots, da_d y_{n-1}),$$

which is a quasiconvex homogeneous form in $n-1$ indeterminates (or identical to the zero form, i.e. $L = L^1$) with respect to the basis $\mathcal{B}^1$. One can estimate:

$$\begin{aligned} \ell(p^1) &\le \ell(p) + d(\log d + \ell(\mathcal{B}^1)) + n\log(d+1) \\ &= (d+1)\ell(p) + 2d\log d + n\log(d+1) \\ &= \tilde{d}^c(\ell(p)+n), \end{aligned}$$

where c is a universal constant.
If $L \neq L^1$, then one can repeate the same procedure with p^1 instead of p.
Suppose that $L = L^r$ for a certain r $(0 \le r \le n)$, i.e. the restriction of p^{r-1} to L^r is the zero form, then $\dim L = n - r$ and $\ell(p^{r-1}) \le \tilde{d}^{c(r-1)}(\ell(p) + (r-1)n)$. Moreover, one has

$$\ell(\mathcal{B}^r) \le \ell(p^{r-1}) + \log d = \tilde{d}^{c(r-1)}(\ell(p) + (r-1)n),$$

where $\mathcal{B}^r$ is the basis of L written in terms of the basis $\mathcal{B}^{r-1}$ of L^{r-1}.
Finally, one canonically obtains the basis $\mathcal{B}$ of L from the basis $\mathcal{B}^r$ of L by a multiplication of r matrices with elements of binary length bounded by $\tilde{d}^{c(r-1)}(\ell(p)+(r-1)n)$, and it holds

$$\begin{aligned} \ell(\mathcal{B}) &\le (r-1)\log n + r(\tilde{d}^{c(r-1)}(\ell(p)+(r-1)n)) \\ &= \tilde{d}^{O(r)}(\ell(p)+n). \end{aligned}$$

In case d is odd, the bound for $\ell(\mathcal{G})$ is obtained easily. #

Corollary 7 *Let $f = \sum_{i=1}^{d} p_i \in \mathbb{Z}[x_1, \ldots, x_n]$ be a quasiconvex polynomial written as the sum of homogeneous forms of degree i and let $\tilde{d} \underset{\text{Df}}{=} \max\{2, d\}$.*
Then the set

$$L_i(f) \underset{\text{Df}}{=} \{x \in \mathbb{R}^n \mid p_d(x) = 0, \ldots, p_i(x) = 0\},$$

is a linear subspace in $\mathbb{R}^n$, for all $1 \le i \le d$. Every subspace $L_i(f)$ admits an integer basis B_i with $\ell(B_i) = \tilde{d}^{O(n)}\ell(f)$.
The set

$$K_i(f) \underset{\mathrm{Df}}{=} \{x \in \mathbb{R}^n \mid p_d(x) = 0, \dots, p_{i+1}(x) = 0, p_i(x) \le 0\}$$

is a halfspace in $\mathbb{R}^n$ (or coincides with $L_i(f)$), for all $1 \le i \le d$, and admits a system G_i of integer generators with $\ell(G_i) = \tilde{d}^{O(n)}\ell(f)$.

Proof. If f is a quasiconvex polynomial, then the form of highest degree p_d is quasiconvex, too. Now, the corollary is a consequence of a successive application of Lemma 1. #

Next, we exhibit a particular property of a system of quasiconvex polynomial inequalities.

2 Elimination of superfluous constraints and the reduction to a bounded set.

Let $f_1, \dots, f_s \in \mathbb{Z}[x_1, \dots, x_n]$ be quasiconvex polynomials of degrees bounded by $d \ge 2$, and let $\ell \underset{\mathrm{Df}}{=} \ell(\{f_1, \dots, f_s\})$. Let $M \subset \mathbb{R}^n$ be the convex set

$$M \underset{\mathrm{Df}}{=} \{x \in \mathbb{R}^n \mid f_1(x) \le 0, \dots, f_s(x) \le 0\}.$$

If the recession cone

$$O^+M \underset{\mathrm{Df}}{=} \{u \in \mathbb{R}^n \mid x + tu \in M \ \ \forall x \in M, \forall t \ge 0\}$$

of the convex set M is a linear subspace in $\mathbb{R}^n$, then one knows from convex analysis [54] that the closed convex set M can be represented as the sum of a compact convex set and a linear subspace, i.e.

$$M = (M \cap (O^+M)^\perp) + O^+M,$$

where $(O^+M)^\perp$ denotes the orthogonal complement of the linear subspace O^+M.
With this section we aim at showing that it is possible to eliminate inequalities from $f_1 \le 0, \dots, f_s \le 0$ such that the remaining system describes a convex set M' allowing a representation as mentioned above and satisfying

$$M \cap \mathbb{Z}^n \neq \emptyset \Longleftrightarrow M' \cap \mathbb{Z}^n \neq \emptyset.$$

The elimination procedure bases on the following considerations.
Suppose that there is a direction $u \in \mathbb{R}^n$, $u \neq 0$, such that $f_1(tu)$ is strictly decreasing and $f_2(tu), \dots, f_s(tu)$ are decreasing or constant (i.e. u is a recession direction of $f_1, \dots, f_s$, but u is not a direction of constancy for the polynomial f_1). Moreover, let $u \in \mathbb{Z}^n$. If $x^1 \in \mathbb{Z}^n$ is a point such that $f_2(x^1) \le 0, \dots, f_s(x^1) \le 0$, then the "uniformity" property of f_1 implies that $f_1(x^1 + tu)$ is strictly decreasing and one can choose a $t \in \mathbb{N}$ such that $f_1(x^1 + tu) \le 0$. Setting $x \underset{\mathrm{Df}}{=} x^1 + tu$ one has $f_i(x) = f_i(x^1 + tu) \le f_i(x^1) \le 0$, for all $2 \le i \le s$. Since $x \in \mathbb{Z}^n$ and $f_1(x) \le 0, \dots, f_s(x) \le 0$, the inequality $f_1(x) \le 0$ can be considered as a "superfluous" one. In such a way one can eliminate all "superfluous"

inequalities and one arrives at the following

Proposition. There is an index set $\{i_1, \dots, i_t\} \subset \{1, \dots, s\}$ such that, for the set

$$M' \underset{\mathrm{Df}}{=} \{x \in \mathbb{R}^n \mid f_{i_1}(x) \le 0, \dots, f_{i_t}(x) \le 0\},$$

the following holds:

(i) $M \cap \mathbb{Z}^n \ne \emptyset \Longleftrightarrow M' \cap \mathbb{Z}^n \ne \emptyset$.

(ii) If $x' \in M' \cap \mathbb{Z}^n$, then one can find a point $x \in M \cap \mathbb{Z}^n$ such that

$$\ell(x) = d^n \ell(x') + d^{O(n)} \ell$$

(If $t = s$, then one can choose $x' = 0$).

(iii) $M' = V + (M' \cap V^\perp)$, where $V = O^+ M'$ is a linear subspace in $\mathbb{R}^n$ and V has an integer basis $\mathcal{B}$ such that $\ell(\mathcal{B}) = d^{O(n)}\ell$ and $M' \cap V^\perp$ is a compact convex set.

Proof. We use the elimination procedure described before. Let

$$\begin{aligned} L(f_i) &\underset{\mathrm{Df}}{=} \{u \in \mathbb{R}^n \mid \sup_{t \in \mathbb{R}} f_i(tu) < +\infty\}, \\ K(f_i) &\underset{\mathrm{Df}}{=} \{u \in \mathbb{R}^n \mid \sup_{\tau \ge 0} f_i(tu) < +\infty\}, \end{aligned}$$

for all $1 \le i \le s$. Then it is clear that $u \in \mathbb{Z}^n$ is a recession direction of $f_1, \dots, f_s$ and not a constancy direction of f_j if and only if $u \in K(f_1) \cap \dots \cap K(f_s)$ and $u \notin L(f_j)$. In Bank/Mandel [1] it is shown that, if $f = \sum_{k=1}^{d} p_k$ is a quasiconvex polynomial of degree d written as a sum of homogeneous forms p_k of degree k, then there is an index k_o, $1 \le k_o \le d$, such that

$$\begin{aligned} L(f) &= \{u \in \mathbb{R}^n \mid p_d(u) = \dots = p_{k_o}(u) = 0\} \\ K(f) &= \{u \in \mathbb{R}^n \mid p_d(u) = \dots = p_{k_o+1}(u) = 0, p_{k_o}(x) \le 0\}, \end{aligned}$$

where $L(f)$ and $K(f)$ are defined for f in the same way as for f_i above.
If one follows the notation of Corollary 6, then $L(f) = L_{k_o}(f)$ and $K(f) = K_{k_o}(f)$. Further, $u \in \mathbb{Z}^n$, $u \ne 0$, is a recession direction of $f_1, \dots, f_s$ and not a constancy direction of f_j if and only if the vector $u \ne 0$ belongs to the polyhedral cone $K(f_1) \cap \dots \cap K(f_s)$ but not to the linear subspace $L(f_j)$. Now, we need

Lemma 2 *Under the assumptions of the proposition we suppose that* $K(f_1) \cap \dots \cap K(f_s) \not\subset L(f_j)$. *Further, let* $x^1 \in \mathbb{Z}^n$ *such that* $f_i(x^1) \le 0$, *for* $i \ne j$. *Then there is a point* $x \in M \cap \mathbb{Z}^n$ *such that* $\ell(x) = d(\ell(x^1) + d^{O(n)}\ell)$.

Proof. Since the polyhedral cone $K(f_1) \cap \dots \cap K(f_s)$ is not contained in the subspace $L(f_j)$, there is a generator $u \ne 0$ of this cone not belonging to $L(f_j)$. By Lemma 1 one can even choose $u \in \mathbb{Z}^n$ satisfying $\ell(u) = d^{O(n)}\ell$. The considerations above guarantee that there is a $\bar{t} \in \mathbb{N}$ such that $f_j(x^1 + \bar{t}u) \le 0$. Such a $\bar{t}$ only depends on the size of the coefficients of $f_j(x^1 + tu)$, i.e. $\bar{t}$ only depends on the coefficients of f_j, x^1 and u (e.g. see Mignotte [44] for such a relation). Now, one puts $x \underset{\mathrm{Df}}{=} x^1 + \bar{t}u$ and obtains $\ell(x) = d(\ell(x^1) + d^{O(n)}\ell)$. #

In order to complete the proof of (i) and (ii) in the proposition we exhibit the "superfluous" constraints step by step and eliminate them all. This yields a set

$$M' \underset{\mathrm{Df}}{=} \{x \in \mathbb{R}^n \mid f_{i_1}(x) \leq 0, \ldots, f_{i_t}(x) \leq 0\}$$

for which

$$K(f_{i_1}) \cap K(f_{i_2}) \cap \ldots \cap K(f_{i_t}) \subset L(f_{i_j}),$$

holds for all $1 \leq j \leq t$. By our construction M' must satisfy: $M \cap \mathbb{Z}^n \neq \emptyset \Longleftrightarrow M' \cap \mathbb{Z}^n \neq \emptyset$, i.e. (i) of the proposition is shown.
For the proof of (ii) we use

Lemma 3 *Let M' be as just defined and let $x' \in M' \cap \mathbb{Z}^n$. Then there is a point $x \in M \cap \mathbb{Z}^n$ such that*

$$\ell(x) = d^n \ell(x') + d^{O(n)}\ell$$

(if $M' = \mathbb{R}^n$, then we put $x' = 0$).

Proof. Without loss of generality we suppose that the "superfluous" constraints are eliminated in the order $f_s, f_{s-1}, \ldots, f_{r+1}$, i.e. the set M' is given by

$$M' = \{x \in \mathbb{R}^n \mid f_1(x) \leq 0, \ldots, f_r(x) \leq 0\}.$$

Now, if one recursively applies the result of Lemma 2, then immediately, one arrives at the estimation

$$\ell(x) \leq d^s(\ell(x') + s d^{O(n)}\ell).$$

However, the following argumentation permits to restrict the number of repeated applications of Lemma 2 by the dimension n of the space.
To simplify the notation we define

$$\begin{aligned} K_{r+1} &\underset{\mathrm{Df}}{=} K(f_1) \cap \ldots \cap K(f_r) \cap K(f_{r+1}), \\ K_{r+i} &\underset{\mathrm{Df}}{=} K_{r+i-1} \cap K(f_{r+i}), \quad \text{for all} \quad i > 1. \end{aligned}$$

It is clear that $K_s \subset K_{s-1} \subset \cdots \subset K_{r+1}$. We consider, in one single step, all convex sets K_{r+i} of the same dimension, i.e.

$$\dim K_{r+1} = \dim K_{r+2} = \cdots = \dim K_{r+j} > \dim K_{r+j+1} = \cdots.$$

Then, for all $1 \leq i \leq j$, the fact that f_{r+i} can be eliminated from $\{f_1, \ldots, f_{r+i}\}$ implies that

$$\dim(K_{r+1} \cap L(f_{r+1})) < \dim K_{r+1} = \cdots = \dim K_{r+j}.$$

One confirms that, in this case, one can choose a $u \neq 0$ such that

$$u \in K_{r+j}, \quad u \notin L(f_{r+i}), \quad \text{for all} \quad 1 \leq i \leq j,$$

(in other words, u is a common recession direction for the constraints $f_{r+j}, \ldots, f_{r+1}$).
In order to do so, we consider a system $\mathcal{G}$ of integer generators of the polyhedral cone K_{r+j} such that $\ell(\mathcal{G}) = d^{O(n)}\ell$ and put $u \underset{\mathrm{Df}}{=} v_1 + \cdots + v_e$ where $\{v_1, \ldots, v_e\} \subset \mathcal{G}$ is a maximal set of linearly independent vectors of $\mathcal{G}$. Then it is clear that $u \in K_{r+j}$ and $u \notin L(f_{r+i})$, for all $1 \leq i \leq j$, and $\ell(u) = d^{O(n)}\ell$. Proceeding in such a way one can control the number of repeated applications of Lemma 2 by the dimension n of the space, and one obtains the bound

$$\ell(x) = d^n(\ell(x') + nd^{O(n)}\ell) = d^n\ell(x') + d^{O(n)}\ell \qquad \#$$

The absence of "superfluous" inequalities in the description of M' is equivalent to the condition

$$K(f_{i_1}) \cap \ldots \cap K(f_{i_t}) = L(f_{i_1}) \cap \ldots \cap L(f_{i_t}).$$

This permits to achieve the complete proof of the proposition by

Lemma 4 *Let $f_{i_1}, \ldots, f_{i_t} \in \mathbb{Z}[x_1, \ldots, x_n]$ be quasiconvex polynomials such that*

$$K(f_{i_1}) \cap \ldots \cap K(f_{i_t}) = L(f_{i_1}) \cap \ldots \cap L(f_{i_t}).$$

Further, let $\ell \underset{\mathrm{Df}}{=} \ell(\{f_{i_1}, \ldots, f_{i_t}\})$ and

$$M' \underset{\mathrm{Df}}{=} \{x \in \mathbb{R}^n \mid f_{i_1}(x) \leq 0, \ldots, f_{i_t}(x) \leq 0\}.$$

Then

$$M' = V + (M' \cap V^{\perp}),$$

where V is a linear subspace in $\mathbb{R}^n$ with an integer basis $\mathcal{B}$ such that $\ell(\mathcal{B}) = d^{O(n)}\ell$, and $M' \cap V^{\perp}$ is a compact subset in $\mathbb{R}^n$.

Proof. We define $V \underset{\mathrm{Df}}{=} L(f_{i_1}) \cap \ldots L(f_{i_t})$. Then it is clear that V is a linear subspace in $\mathbb{R}^n$ and one easily constructs an integer basis $\mathcal{B}$ such that $\ell(\mathcal{B}) = d^{O(n)}\ell$ (using the fact that $L(f_{i_j})$ admits a basis of binary length bounded by $d^{O(n)}\ell$). Let $x \in M'$. Then there are a $u \in V$ and a $y \in V^{\perp}$ such that $x = y + u$. Then $y = x - u \in M'$ (because of $-u \in V$, the definition of V and $f_{i_j}(x + (-u)) \leq f_{i_j}(x) \leq 0$, for all $1 \leq j \leq t$), i.e. $x \in V + (M' \cap V^{\perp})$.
The converse inclusion can be verified similarily.
It suffices to show that the set $(M' \cap V^{\perp})$ is compact. If the closed convex set $(M' \cap V^{\perp})$ is not bounded, then it contains a half-line $\{x + tu \mid t \geq 0\}$ where $x \in M' \cap V^{\perp}$ and $u \in \mathbb{R}^n$, $u \neq 0$ (see e.g. Rockafellar [54]). One confirms that u is a recession direction of $f_{i_1}, \ldots, f_{i_t}$ and therefore $u \in V$. On the other hand, one obtains $u \in V^{\perp}$, i.e. $u = 0$. Hence, $M' \cap V^{\perp}$ must be bounded. #

#

3 The semialgebraic bound.

By the proposition proved in the previous section the proof of Theorem 1 is completed if one can show that M' contains an integer point x' of binary length $\ell(x') = (sd)^{O(n)}\ell$ if the set $M' \cap V^{\perp}$ is non-empty.

We can suppose that M' is represented by the sum of a linear subspace in $\mathbb{R}^n$ and a compact set. Then we can reduce our consideration to this compact set. Now, using the precise results of the algorithmic semialgebraic geometry given in Corollary 4 we can estimate the radius of a ball containing this compact set.
Without loss of generality we assume for the rest of this section that $M = \{x \in \mathbb{R}^n | f_1(x) \leq 0, \ldots, f_s(x) \leq 0\}$ does not contain superfluous inequalities. Then one has $M = (M \cap V^{\perp}) + V$, where $(M \cap V^{\perp})$ is compact and the linear subspace V admits an integer basis $\mathcal{B}$ such that $\ell(\mathcal{B}) = d^{O(n)}\ell$.

Lemma 5 *If $M \cap \mathbb{Z}^n$ is non-empty, then M contains an integer point belonging to the set $(M \cap V^{\perp}) + P$, where $P \underset{\mathrm{Df}}{=} \{w \in \mathbb{R}^n \mid w = \sum_{v \in \mathcal{B}} \beta_v v,\ 0 \leq \beta_v < 1\}$.*

Proof. Let $x \in M \cap \mathbb{Z}^n$ and let $x = y + u$, $y \in M \cap V^{\perp}$, $u \in V$, be its decompsotion corresponding to $M = (M \cap V^{\perp}) + V$. Further, let $\mathcal{B} \underset{\mathrm{Df}}{=} \{v_1, \ldots, v_m\}$ be the integer basis of V. Then we can write $u = \alpha_1 v_1 + \cdots + \alpha_m v_m$ for some reals $\alpha_1, \ldots, \alpha_m$. We put

$$\alpha_i = \lfloor \alpha_i \rfloor + \beta_i, \text{ where } \lfloor \alpha_i \rfloor \in \mathbb{Z} \text{ and } 0 \leq \beta_i < 1,$$

for all $1 \leq i \leq m$. If now $\bar{x} \underset{\mathrm{Df}}{=} y + \beta_1 v_1 + \cdot\cdot + \beta_m v_m$, then $\bar{x} = x - (\lfloor \alpha_1 \rfloor v_1 + \cdots + \lfloor \alpha_m \rfloor v_m) \in \mathbb{Z}^n$. Moreover, $\bar{x} \in M + V \subseteq M$; consequently, one has $\bar{x} \in M \cap \mathbb{Z}^n$ and $\bar{x} \in (M \cap V^{\perp}) + P$.
#

Lemma 6 *$M \cap V^{\perp} \subset B(0, R)$, where $R \in \mathbb{N} : \ell(R) = (sd)^{O(n)}\ell$.*

Proof. The semialgebraic set $M \cap V^{\perp}$ can be defined by a formula without quantifiers in the elementary language over $\mathbb{R}$ with constants in $\mathbb{Z}$, in which the coefficients of the polynomials $f_1, \ldots, f_s$ and the linear equations describing $V^{\perp}$ appear. The semialgebraic set

$$S \underset{\mathrm{Df}}{=} \{\varrho \in \mathbb{R} \mid M \cap V^{\perp} \subset B(0, \varrho)\}$$

can be described by the following formula Φ (containing one single block of quantifiers):

$$\Phi: \qquad (\forall x) \qquad (x \in M \cap V^{\perp} \Longrightarrow \|x\|^2 \leq \varrho^2),$$

where $x = (x_1, \ldots, x_n)^T$ are the quantified variables and ϱ is the only free variable.
Now, if one applies Corollary 4 to the formula Φ, then one obtains a formula Ψ without quantifiers in the variable ϱ which yields a description of the set S.
Ψ is a disjunction of conjunctions of sign conditions and certain polynomials $G_1, \ldots, G_k \in \mathbb{Z}[\varrho]$. Since all parameters in the original formula Φ are of a binary length bounded by $d^{O(n)}\ell$, and since the number and the degrees of the polynomials are not exceeding s and d, respectively, the quantifier elimination algorithm guarantees that the polynomials $G_1, \ldots, G_k$ satisfy:

$$\deg G_i = (sd)^{O(n)} \ (1 \leq i \leq k); \ \ell(\{G_1, \ldots, G_k\}) = (sd)^{O(n)}\ell.$$

Further, if $\alpha \in \mathbb{R}$ is the greatest real root of the polynomials $G_1, \ldots, G_k$, then one confirms that the formula Ψ is either true or false in the whole interval $(\alpha, +\infty)$ (that is why non of the polynomials $G_1, \ldots, G_k$ can change the sign in this interval). Since Ψ is true for a sufficiently large ϱ, Ψ must be true for the whole interval $(\alpha, +\infty)$. Hence, to obtain the desired bound for R it suffices to bound the largest real roots of the polynomials $G_1, \ldots G_k$. The bounds on the degrees and the binary length of the coefficients of the polynomials $G_1, \ldots, G_k$ can be directly used to obtain a bound $R \in \mathbb{N}$ for the real roots such that $\ell(R) = (sd)^{O(n)}\ell$ (e.g. apply Cauchy's inequality, Mignotte [44]). This completes the proof of Theorem 1. #

It should be remarked that Schimm [55] shows that the numbers in the bound of Lemma 6 can be bounded from above by $2n$.

Proof of Corollary 5.
The proof is immediately given. If the set $\{x \in \mathbb{Z}^n \mid f_1(x) \leq 0, \ldots, f_s(x) \leq 0\}$ is non-empty, then it contains a point belonging to the ball $B(0, R)$, where R is such that $\ell(R) = (sd)^{O(n)}\ell$. Therefore, it suffices to verify the following: check whether a given point $x \in \mathbb{Z}^n$ with $\ell(x) \leq (sd)^{O(n)}\ell$ satisfies the polynomial inequalities $f_1(x) \leq 0, \ldots, f_s(x) \leq 0$. This can be executed in the time $(sd)^{O(n)}\ell$. (One observes that a deterministic bound is of order $2^{(sd)^{O(n)}}\ell$, because of the fact that the polynomials $f_1, \ldots, f_s$ have to be calculated in all integer points of the ball $B(0, R)$.) #

Proof of Theorem 2.
Here, one again applies the tools used in the proof of Theorem 1: elimination of superfluous constraints and reduction to a compact set. Instead of elaborating a complete proof we can sketch the underlying ideas, only.

(i) In Bank/Mandel [1] it is shown that the optimization problem in Theorem 2 admits an optimal point if and only if $M \cap \mathbb{Z}^n \neq \emptyset$ and $\inf\{f(x) \mid x \in M\} > -\infty$.

(ii) By Theorem 1, we know that if $M \cap \mathbb{Z}^n$ is non-empty, then there is a point $x^o \in M \cap \mathbb{Z}^n \cap B(0, R)$ where $\ell(R) = (sd)^{O(n)}\ell$. Further, we have $\ell(f(x^o)) = (sd)^{O(n)}\ell$.

(iii) We consider the following set

$$N \underset{\mathrm{Df}}{=} \{x \in \mathbb{R}^n \mid f_1(x) \leq 0, \ldots, f_s(x) \leq 0,\ f(x) \leq f(x^o)\}$$

and define $\mu \underset{\mathrm{Df}}{=} \inf\{f(x) \mid x \in M \cap \mathbb{Z}^n\}$. Obviously, $N \cap \mathbb{Z}^n \neq \emptyset$ and $\mu = \inf\{f(x) \mid x \in N \cap \mathbb{Z}^n\}$.

Now, one proceeds by elimination of superfluous constraints from $f_1(x) \leq 0, \ldots, f_s(x) \leq 0$, $f(x) - f(x_o) \leq 0$ in the manner described above. One obtains a set N' corresponding to M' in the proposition above. By (i) and (iii) it is impossible that the inequality $f(x) - f(x_o) \leq 0$ is eliminated and, consequently,

$$\mu = \inf\{f(x) \mid x \in N \cap \mathbb{Z}^n\} = \inf\{f(x) \mid x \in N' \cap \mathbb{Z}^n\}.$$

As before, one obtains a decomposition

$$N' = W + (N' \cap W^{\perp});$$

where W is a linear subspace in $\mathbb{R}^n$ with an integer basis $\mathcal{B}$ satisfying $\ell(\mathcal{B}) = (sd)^{O(n)}\ell$ and $N' \cap W^\perp$ is a compact subset in $\mathbb{R}^n$. For this latter set,

$$N' \cap W^\perp \subset B(0, R) \quad \text{with} \quad \ell(R) = (sd)^{O(n)}\ell,$$

holds, too.
One observes that $f(x) - f(x^0)$ is constant on the subspace W. If $P \underset{\mathrm{Df}}{=} \{w \in \mathbb{R}^n \mid w = \sum_{v \in \mathcal{B}} \beta_v v,\ 0 \leq \beta_v < 1\}$, then the set $(N' \cap W^\perp) + P$ contains an integer point x' with $f(x') = \mu$. Starting from this point x' one can construct a point $\bar{x} \in M \cap \mathbb{Z}^n$ satisfying $\ell(\bar{x}) = (sd)^{O(n)}\ell$ and $f(\bar{x}) = \mu$. #

References

[1] Bank, B., R. Mandel (1988): Parametric integer optimization. Akademie-Verlag, Berlin.

[2] Bank, B., R. Mandel (1988): (Mixed-) Integer solutions of quasiconvex polynomial inequalities. Mathematical Research Advances in Mathematical Optimization, Akademie-Verlag, Berlin, Vol. 45, 20-34.

[3] Bank, B., J. Heintz, T. Krick, R. Mandel, P. Solernó (1990): Une borne géométrique pour la programmation entière à contraintes polynomiales. C. R. Acad. Sci. Paris t. 310, Série 1, 475-478.

[4] Bank, B., J. Heintz, T. Krick, R. Mandel, P. Solernó (1991): Une borne optimale pour la programmation entière quasiconvexe. To appear in Math. Nachr.

[5] Belousov, E. G. (1977): Introduction to convex analysis and integer programming. Edition Moscow University (in Russian).

[6] Belousov, E. G., V. M. Shironin (1990): Geometry of numbers and integer programming. Podpi. nauk.-popularnaja Ser. 3/1990 (in Russian).

[7] Benedetti, R., J.-J. Risler (1990): Real algebraic and semi-algebraic sets. Hermann, Paris.

[8] Ben-Or, M., D. Kozen, J. Reif (1986): The complexity of elementary algebra and geometry. J. of Computation and Systems Sciences 32, 251-264.

[9] Blum, L., M. Shub, S. Smale (1989): On a theory of computation and complexity over the real numbers: $\mathcal{NP}$-completeness, recursive functions and universal machines. Bull. (N.S.) of the AMS, Vol 21, 1, 1-46.

[10] Bochnak, J., M. Coste, M. F. Roy (1987): Géométrie algébrique réelle. Ergebnisse der Mathematik u. ihrer Grenzgebiete, 3. Folge, Bd. 12, Springer-Verlag, Berlin, Heidelberg.

[11] Borosh, L., L. B. Treybig (1976): Bounds on positive integral solutions of linear diophantine equations. Proceedings of the AMS, 55, 299-304.

[12] Cohen, P. J. (1969): Decision procedures for real and p-adic fields. Commu. Pure Appl. Math., 22, 131-151.

[13] Collins, G. E. (1975): Quantifier elimination for real closed fields by cylindrical algebraic decomposition. In: Lecture Notes in Computer Science 33, 134-183.

[14] Coste, M., M. F. Roy (1981): Thom's Lemma, the coding of real algebraic numbers and the computation of the topology of semi-algebraic sets. J. on Symbolic Computation 5, 121-130.

[15] Davenport, J., J. Heintz (1988): Real quantifier elimination is doubly exponential. J. Symbolic Comput. 5, 29-35.

[16] Davis, M. (1973): Hilbert's tenth problem is unsolvable. Amer. Math. Monthly 80, 233-269.

[17] Fischer, M. J., M. O. Rabin (1974): Super-exponential complexity of Presburger arithmetic. In: Complexity of Comput., ed. R. M. Karp, AMS, 27-41.

[18] Garey, M. R., D. S. Johnson (1979): Computers and Intractability. Freeman and Company, San Francisco.

[19] Grigor'ev, D. Yu. (1988): Complexity of deciding Tarski algebra. J. Symbolic Comput. 5, 65-108.

[20] Grigor'ev, D. Yu., N. N. Vorobjov (1988): Solving Systems of polynomial Inequalities in Subexponential Time. J. Symbolic Comput. 5, 37-64.

[21] Grötschel, M., L. Lovász, A. Schrijver (1988): Geometric Algorithms and Combinatorial Optimization. Springer-Verlag, Berlin, Heidelberg.

[22] Grunewald, F., D. Segal (1980): Some general algorithms. I: Arithmetic groups. Annals of Mathematics, 112, 531-583.

[23] Heindel, L. E. (1971): Integer Arithmetic Algorithms for Polynomial Real Zero Determination. J. of the Ass. for Comp. Machinery, Vol. 18, No. 4, 533-548.

[24] Heintz, J. (1983): Definability and fast quantifier elimination in algebraically closed fields. Theor. Comput. Sci. 24, 239-277.

[25] Heintz, J., M.-F. Roy, P. Solernó (1989): Complexité du principe de Tarski-Seidenberg. C. R. Acad. Sci. Paris, t. 309, Série I, 825-830.

[26] Heintz, J., M.-F. Roy, P. Solernó (1989): On the complexity of semialgebraic sets (Extended abstract). Proc. IFIP Congress'89, IX World Computer Congress, North-Holland.

[27] Heintz, J., M.-F. Roy, P. Solernó (1990): Sur la complexité du principe de Tarski-Seidenberg. Bull. Soc. Math. France. 118, 101-126.

[28] Hilbert, D. (1900): Mathematische Probleme. Vortrag gehalten auf dem internationalen Mathematiker-Kongreß zu Paris 1900. Nachr. Akad. Wiss. Göttingen, Math.-Phys., 253-297.

[29] Jeroslow, R. G. (1973): There Cannot be Any Algorithm for Integer Programming with Quadratic Constraints. Operations Research, 21, 221-224.

[30] Karmarkar, N. (1984): A new polynomial-time algorithm for linear programming. Cominatorica, 4, 373-395.

[31] Karp, R. M. (1972): Reducibility among combinatorial problems. In: Complexity of Computer Computations (R. E. Miller, J. W. Thatcher, eds.), Plenum Press, New York, 85-103.

[32] Khachiyan, L. G. (1979): A polynomial algorithm in linear programming. Doklady Akademii Nauk SSSR, 244, 1093-1096 (in Russian).

[33] Khachiyan, L. G. (1983): Convexity and complexity in polynomial programming. Proc. Int. Congress Math., Varsovic.

[34] Khachiyan, L. G., S. P. Tarasov (1980): Bounds and algorithmic complexity of convex Diophantine inequalities. Dokl. Akad. Nauk SSSR, 225 (in Russian).

[35] Krick, T. (1990): Complejidad para problemas de geometria elemental. Thèse, Université de Buenos Aires.

[36] Lee, R. D. (1976): An application on mathematical logic to the integer linear programming problems. Notre Dame Journal Formal Logic, XIII, 2, 279-282.

[37] Lenstra, Jr., H. W. (1983): Integer programming with a fixed number of variables. Mathematics of Operations Research, 8, 538-548.

[38] Levine, H. I. (1989): A lower bound for the topological complexitiy of Poly (D, n). J. of Complexity 5, no. 1, 34-44.

[39] Mandel, R. (1983): Zur Konstruktivität bei (gemischt-) ganzzahligen Optimierungsaufgaben. Math. Operationsforschung u. Statistik, Ser. Optimization, Heft 3, 343-357.

[40] Mandel, R. (1985): Beiträge zur Theorie ganzzahliger Optimierungsaufgaben. Dissertation (B), Humboldt-Universität zu Berlin.

[41] Manders, K., L. Adleman (1976): $\mathcal{NP}$-complete decision problems for quadratic polynomials. Proc. 8-th Annu. ACM Symp. on Theory of Comput., Hershey, Pennsylvania, 23-29.

[42] Matijasevich, Yu. V. (1970): Enumerable sets are Diophantine. Dokl. Akad. Nauk SSSR, 191, 279-282 (in Russian).

[43] Matijasevich, Yu. V. (1988): Diophantine complexity. Zap. Nauchn. Sem. Leningr. Otd. Steklova, 174, 122-131 (in Russian).

[44] Mignotte, M. (1982): Some useful bounds. Computer Algebra (Symbolic and Algebraic Computation). Springer-Verlag, 259-264.

[45] Müller, F. (1978): Ein exakter Algorithmus zur nichtlinearen Optimierung für beliebige Polynome mit mehreren Veränderlichen. Meisenheim, Verlag Anton Hain.

[46] Oppen, D. C. (1978): A $2^{2^{2^{pn}}}$ upper bound on the complexity of Presburger arithmetic. J. Comput. and Syst. Sci. 16, 3, 323-332.

[47] Presburger, M. (1929): Über die Vollständigkeit eines gewissen Systems der Arithmetik ganzer Zahlen, in welchem die Addition als einzige Operation hervortritt. Comptes Rendus, 1er congr. des math. des Pays Slaves, Warszawa.

[48] Renegar, J. (1989): On the computational complexity and geometry of the first order theory of the reals. Part III. Quantifier elimination. Technical Report 856, Cornell University.

[49] Renegar, J. (1988): A polynomial-time algorithm based on Newton's method for linear programming. Mathematical Programming, 40, 59-94.

[50] Renegar, J. (1988): A Faster PSPACE algorithm for decising the existential theory of the reals. Technical Report No. 792, Cornell University, Ithaca, NY.

[51] Robinson, J. (1971): Hilbert's tenth problem. Proc. Symp. Pure Math. 20, 191-194.

[52] Robinson, J. (1969): Diophantine decision problems. Studies in number theory. MAA Studies in Math. 6, 76-116.

[53] Robinson, J. (1969): Unsolvable Diophantine problem. Proc. Amer. Math. Soc. 222, 534-538.

[54] Rockafellar, R. T. (1970): Convex Analysis. Princeton Mathematical Series no. 28.

[55] Schimm, B. (1990): Über Schranken von Lösungen (gemischt-) ganzzahliger Optimierungsaufgaben mit quasikonvexen Polynomen in den Restriktionen und als Zielfunktion. Diplomarbeit, Humboldt-Universität zu Berlin.

[56] Schrijver, A. (1986): Theory of linear and integer programming. Wiley Interscience Series in Discrete Mathematics.

[57] Seidenberg, A. (1954): A new decision method for elementary algebra and geometry. Ann. Math. 60, 365-374.

[58] Smale, S. (1987): On the Topology of Algorithms, I. J. of Complexity 3, 81-89.

[59] Solernó, P. (1989): Complejidad de conjuntos semialgebraicos. Thèse, Université de Buenos Aires.

[60] Sonnevend, G. (1985): New algorithms in convex programming based on a notation of "centre" (for systems of analytic inequalities) and on rational extrapolation. Department of Numerical Analysis, Institute of Mathematics, L. Eötvös University, Budapest.

[61] Strassen, V. (1984): Algebraic complexity. Birkhäuser.

[62] Tarski, A. (1951): A decision method for elementary algebra and geometry. University of Californa Press.

[63] Vaidya, P. M. (1987): An algorithm for linear programming which requires $O(((m+n)n^2+(m+n)^{1,5n})L)$ arithmetic operations. AT& T Bell Laboratories, Murray Hill, New Jersey 07974.

[64] von zur Gathen, J. (1986): Parallel arithmetic computations: a survey. Lecture Notes in Computer Science 233, Springer-Verlag, Berlin, 93-112.

[65] Vorobjov, N. N. (1984): Bounds of real roots of algebraic equations. Notes of Sci. Seminars of Leningrad Dep. of Stehlov Inst. 137, 7-19.

[66] Weispfenning, V. (1988): The complexity of linear problems in fields. J. Symbolic Computation 5, 3-27.

[67] Wüthrich, H. R. (1976): Ein Entscheidungsverfahren für die Theorie der reell-abgeschlossenen Körper. In: Lecture Notes in Computer Science, 43, 138-162.

IDENTIFICATION BY MODEL REFERENCE ADAPTIVE SYSTEMS

Johann Baumeister
Fachbereich Mathematik, Johann Wolfgang Goethe-Universität Frankfurt/M.
Robert-Mayer-Straße 6-10, D-6000 Frankfurt/Main

Abstract: In this note, we present an introduction to the so-called model reference adaptive system method which can be used to identify a spatially varying parameter in an evolution equation. The inherent mathematical difficulties are briefly discussed and relevant results are given.

§1 Adaptive identification: Introduction

In this introductory section we will give a description of the problem field in an almost non-mathematical way.

1.1 Dynamical systems

A ***dynamical system*** is a mathematical model of a dynamical phenomenon. The evolution of such a system obeys certain laws that, in general, are given in terms of differential equations. The dimension of a system is the number of variables which is necessary to describe the configuration of the system (state of the system) at a given instant of time. Two of the problems which have attracted the interest of scientists for a long time are the problems of celestical mechanics and of the motion of fluids. The first problem is of finite dimension, the latter one has infinite dimension.

Given a dynamical system starting from a particular initial state, it is not easy to predict the evolution of the system. The mathematical problem here is to study the long-time behavior of the system corresponding to the practical problem of determining which "permanent" state will be observed after a short transient period. Interesting permanent states are equilibrium points, periodic or quasi-periodic states. To predict the long-time bevavior of a dynamical system described by a differential equation the following questions are addressed:

a) ***Existence* and *uniqueness*** of the (local) solution and ***continous dependence*** on the initial data.

b) Existence of a solution which cannot be extended in time: Existence of a ***maximal solution*** (***orbits***).

c) Existence and properties of an ***attractor***. An attractor is a set of states towards all orbits converge.

Results in a) and b) are part of the definition of the dynamical system (well-posedness) and are also needed in other parts of the study. Several adjectives are attributed to an attractor: global, strange, universal, fractal, compact,... .

1.2 Control systems

A ***control system*** is a dynamical system in which three entities called ***input***, ***state*** and ***output*** are distinguished. The distinction is such that the state is causally dependent on the input and the output is causally dependent on the state. The output is also called the observation variable.

A ***control objective*** in a control system is defined by a collection of specifications on the behavior of the system. The rules of the game are then to manipulate the input of the system in such a way that the specifications are met. Well-known examples of control objectives are ***stabilization***, ***optimal control*** and ***pole assignment***.

To be more specific, let us consider a dynamical system associated with a linear differential equation:

(1.1) $$z' = Az + Bu$$ (differential equation)

(1.2) $$z(0) = z_0$$ (initial condition)

(1.3) $$y = Cz$$ (observation equation)

We interpret u as the input, z as the state and y as the output; z_0 is the initial state.

Moreover, we assume:

$u \in U,\ z \in Z,\ y \in Y$; U,Z,Y are normed vector spaces;

$A : Z \to Z,\ B : U \to Z,\ C : Z \to Y$ are linear operators.

Of course, one has to develop a conception what should be a solution of (1.1), especially if the operator A is an unbounded operator as it is the case if (1.1) comes from a partial differential equation. For the considerations in this section it is enough to consider (1.1) as a scalar equation ($U = Z = Y = \mathbb{R}$). Then it is clear how we have to look at (1.1-3).

Now it is possible to formulate the control objectives mentioned above:

Stabilization: Find a feedback control $u = Fy = FCy$ such that y converges to an equilibrium point $\bar{z}$ as t goes to infinity.

Optimal control: Find a control u^* which minimizes a given performance criterion which may functionally depend on the control effort u and on the achieved output y.

Pole assignment: Find a feedback law F (see stabilization) such that the state converges to an equilibrium point $\bar{z}$ with a desired rate.

1.3 Adaptive control

The solution of every non-trivial control problem will depend on the laws of the system. Usually, at each time t there is only available an estimation of the laws of the system under consideration; the controller, constructed in order to meet the control objective, depends on this estimation. In the formulation (1.1-3) this means that there is a dependence on the operators A,B,C, and on z_0. If the laws of the system to be controlled are not completely known we say that the control problem is an ***adaptive control problem.*** The term adaptive stems from the fact that the control mechanism has to be adjusted according to the behavior of the system. This should be done in such a way that every adjustment leads to an improvement of the quality of the controller.
This note is exclusively concerned with systems of which the laws do not change in time; in the formulation (1.1-3) this is already taken into consideration.

Usually, adaptation of the laws of the system is based on the principles of ***recursiveness, neutrality*** and ***certainty equivalence***, respectively.

Recursiveness: The estimation of the laws of the system at each time t can be calculated on the basis of past estimations and past observations.

Neutrality: There is no adaptation if the current estimation is compatible with the observed data.

Certainty equivalence: The applied controller is calculated as if the estimation of the system were indeed the true system.

1.4 Identification

It is part of the game in adaptive control to estimate the laws of the system under consideration from the observed data. This process is called ***identification.*** In this note we want to discuss exclusively the problem of identification; in practice, control and identification have to be done simultaneously. This separation of the identification problem from the controlling part makes it easier to discuss the principal aspects of adaptive identification concepts. This is important if the dynamics in (1.1) is actually a partial differential equation since up to now the problem of adaptive control in infinite-dimensional systems is not completely understood.

Adaptive identification is a technique to determine the laws of the system from observed data simultaneously to the evolution of the true system. An important problem is to find out which input u and initial state z_0 is sufficient to identify the true system. Clearly, the experiments have to be diverse enough: If there is no experimentation at all ($z_0 = 0$, $u \equiv 0$), we learn nothing about the system, if the experiments are too poor we learn only a little. Since in adaptive control the input is generated during the adaptation process it cannot be expected that a rich enough input results. By decoupling the identification problem from the control part this difficulty is easier to discuss.

In the following we restrict ourselves to the case that B and C are already known. Moreover, we use a parametric representation of the system operator A:

$A = A(p)\ ,\ p \in P,$ where P is a set of admissible parameters in the normed vector space $\mathbb{H}$.

The identification problem consists therefore in finding out the true parameter p from observed data.

Finally, we assume that the whole state is accessible for observation, that is:

$$C = \text{identity},\ y = z.$$

The requirement of accessibility of all the state variables is a serious constraint in many practical situations, especially in infinite-dimensional systems since in this case the observer theory is not very well developed.

Now, the system we want to consider has the following formulation:

(1.4) $z' = A(p)z + f(t), \; t \geq 0;$

(1.5) $z(0) = z_0;$

(1.6) $y = z.$

Here $p \in P \subset H$, $f : [0,\infty) \to H$, $z_0 \in H$.

1.5 Model reference adaptive systems

Let us consider the system (1.4-6). A ***model reference adaptive system (MRAS)*** for the identification of the true parameter p in (1.4) from the observation z contains three parts:

a) the true system with unknown true parameter $p \in P$ and output z;

b) a model reference system with an adjustable parameter $q(t) \in P$ and output x(t); q(t) and x(t) is an estimation of p and z(t), respectively;

c) an adaptive system adjusting the parameter q(t) according to the error $e(t) := x(t) - z(t)$.

The following diagram visualizes the approach.

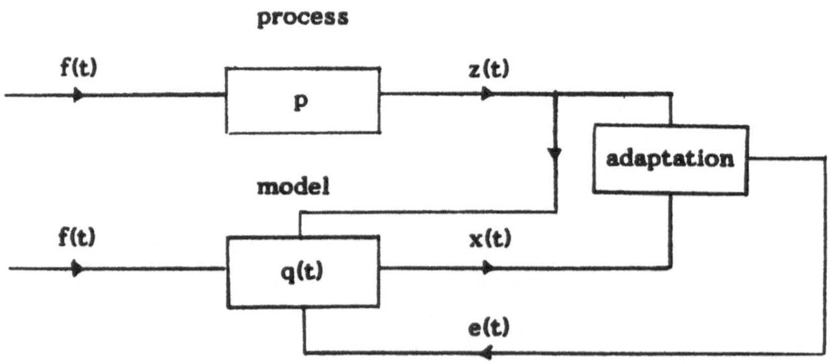

The goal is to design the model reference system and the adaptive system in such a way that the following assertions hold:

(1.7) $\lim_{t \to \infty} e(t) = 0$ (in an appropriate topology of H)

(1.8) $\lim_{t\to\infty} e(t) = 0$ implies $\lim_{t\to\infty} q(t) = p$ (in an appropriate topology of $\mathbb{H}$)

The main problem consists in finding an input f and an initial state z_0 such that (1.7),(1.8) can be verified. In the following sections we are concerned with the analysis of this problem in an abstract framework.

Bibliographical comments

There is a vast amount of literature on dynamical systems, adaptive control and identification. An almost up-to-date presentation of a modern theory on dynamical systems can be found in [13] and [31]. References on adaptive control theory are [4], [10], [21] and [26]. The design of a model reference adaptive system for distributed parameter systems is considered in [18]. The problem of richness of nputs lin finite-dimensional systems is discussed in [16] and [25].

§2 Identification problems and identifiability

In this section we present some concrete examples and a general framework for studying the identifiability question. The mathematical tools used in the following considerations can be found in Wloka [32] and / or Showalter [28].

2.1 A tutorial example

Let us consider the most simple example of an initial value problem:

(2.1) $z' = pz + f(t)\ ,\ t \geq 0;\ \ z(0) = z_0\ .$

We set

$$H := \mathbb{H} := \mathbb{R}\ ,\ P := \mathbb{H},$$

define

$$A(p) : H \to H\ ,\ A(p)h := ph\ ,$$

and assume

$$f : [0,\infty) \to H \text{ stetig.}$$

As it is well-known the solution z(p; .) of (2.1) is uniquely determined and given by the following formula (variation of constants):

(2.2) $z(p;t) = e^{pt} z_0 + \int_0^t e^{p(t-s)} f(s)\,ds\ ,\ t \geq 0.$

Let T > 0. Then we have the following well-defined mapping

(2.3) $\quad L_T : P \ni p \mapsto z(p;.) \in L_2(0,T;H).$

The identification problem as described in the last section consists in finding out the parameter $p \in P$ from the observation $L_T p$.

Let $p \in P$ and let $z := L_T p$.

a) If $q \in P$ with $L_T q = z$ then we have $(q - p)z(t) = 0$ for all $t \in [0,T]$.
   This implies:
   If there exists $t^* \in [0,T]$ with $z(t^*) \neq 0$ then $q = p$.
   If $z(t) = 0$ for all $t \in [0,T]$ we cannot decide wheter $q = p$ holds. The reason is that the experiment represented by z_0 and f is not rich enough.

b) Suppose that there exists $t^* \in [0,T]$ with $z(t^*) \neq 0$. Then the equation

(2.4) $\quad L_T q = z$

   is drastically overdetermined.

c) Suppose that $z(t) = 0$ for all $t \in [0,T]$. Then the equation (2.4) is underdetermined.

d) Notice that each experiment "$f = 0, z_0 \neq 0$" is sufficient to solve the equation (2.4) uniquely.

Let $f(t) = 0$ for all $t \geq 0$ and let us define for each $n \in \mathbb{N}$:

$$p_n := -n,\ z_n(t) := z(p_n;t) := e^{-nt},\ t \geq 0.$$

We have

$$\lim_n z_n = 0 \quad \text{in } L_2(0,T;H) \text{ (endowed with the usual topology)}$$

$$|p_n - p_{n+1}| = 1 \text{ for all } n \in \mathbb{N}.$$

This shows that the mapping L_T is not continous invertible. Thus, we see that the identification problem which is related to equation (2.4) is difficult to solve stable way.

2.2 Identifiability

We start from the following linear initial value problem (see 1.4-5):

(2.5) $\quad z' = A(p)z + f(t)\ ,\ t \geq 0;\ z(0) = z_0\ .$

In order to study the evolutionary problem (2.5) in a variational framework we have to introduce some notations and assumptions.

Let V, H and **H** be three Hilbert spaces and let **P** be a subset of **H**. Consider the following conditions:

(H1) i) $V \subset H$, V is dense in H.

ii) The dual space H^* of H is identified with H and H^* is identified with a dense subspace of the dual space V^* of V:

$$V \subset H \subset V^* \qquad \text{(Gelfand - triple)}$$

iii) The injections $V \subset H$ and $H \subset V^*$ are continous.

iv) For each $q \in P$ the operator $A(q)$ may be considered as a linear, continous and bijective mapping from V onto V^*.

v) There exist constants $\varkappa_1, \varkappa_2 > 0$ such that with

$$a(q;u,v) := \langle A(q)u,v \rangle \ , \ u,v \in V, \ q \in P,$$

the following inequalities hold:

$$-\,a(q;u,v) \geq \varkappa_1 \|u\|_V^2 \qquad , \ u \in V,$$

$$|a(q;u,v)| \leq \varkappa_2 \|u\|_V \, \|v\|_V \ , \ u,v \in P.$$

(Here and throughout $\langle .,. \rangle$ is the pairing in $V^* \times V$)

Lemma 1

Suppose that the hypothesis (H1) is satisfied and let $p \in P$. The we have:

i) $A(p)$ is a closed operator, $D(p) := \{v \in P \mid A(p) \in H\}$ is dense in H.

ii) $A(p)$ generates a C_0 - semigroup $\{S_p(t)\}_{t \geq 0}$.

iii) The initial value problem (2.5) posesses a unique solution z satisfying

$$z \in C(0,\infty;D(p)) \cap C^1(0,\infty;H),$$

$$z'(t) = A(p)z(t) + f(t), \ t \geq 0, \ z(0) = z_0 \ ,$$

$$z(t) = S_p(t)z \ + \int_0^t S_p(t-s)f(s)\,ds \ , \ t \geq 0.$$

Proof: See appendix A).

Let $T > 0$. As a consequence we have a mapping (***solution mapping***; see (2.3))

$$L_T : P \ni p \mapsto L_T p \in L_2(0,T;H),$$

$L_T p$ is a solution of (2.5) according to Lemma 1.

Now, we are ready to formulate the problems we are interested in:

Identifiability in p $\in$ P

Given $q \in P$ with $L_T q = L_T p$ prove $q = p$.

Identification problem

Given $z \in L_2(0,T;H)$, find $p \in P$ with $L_T p = z$.

Since L_T is, in general, a nonlinear mapping (see the tutorial example), identifiability should be considered also locally:

Local identifiability in p $\in$ P

Show that there exists a constant $r > 0$ such that

$$q \in P,\ L_T q = L_T p,\ \|q - p\|_H < r$$

implies $q = p$.

The identification problem is called an ***inverse problem*** since it is a problem of finding out unknown causes (p!) of known consequences (z!). In contrast, the initial value problem (2.5) is called a direct or ***forward problem*** since it is oriented along a cause - effect sequence. In a complete solution of inverse problems the questions of existence uniqueness and stability are to consider. If one of these conditions does not hold the problem is called ***ill-posed***. When solving ill-posed problems numerically, we have to expect difficulties since any error acts as a perturbation on the original equation. Because errors cannot be completely avoided there may be a range of plausible solutions and we have to find out a reasonable solution. These ambiguities in the solution of inverse problems which are ill-posed can be reduced by incorporating some sort of ***a-priori - information*** that limits the class of allowable solutions. An a - priori - information is a collection of conditions which have been obtained independently of the observed values of the data. Such conditions can be:

smoothneess, bounds on the size, convexity of functional parameters, ...

The incorporation of an a - priori - information such that the problem becomes well-posed is caled ***stabilization,*** the realization of stable computational schemes for ill-posed problems using an a - priori - information is called ***regularization***.

Illustration: Tutorial example

In this example we have

$$H = \mathbb{H} = \mathbb{R},\ V = H,\ A(p)h = ph,\ p \in H,\ h \in H.$$

To follow the ideas above we restrict the set of admissible parameters:

$$P := \{p \in \mathbb{H} \mid p_0 < p < p_1\}\ ,\ p_0 < p_1 < 0\ .$$

A straightforward calculation shows that the hypothesis (H1) holds. Moreover, the instability reported in Section 2.1 cannot occur. *

2.3 Identification in a one-dimensional heat equation

Let us consider

$$(2.6)\quad \begin{aligned} \partial_t z &= \partial_\xi(p(\xi)\partial_\xi z) && ,\ \xi \in (0,1),\ t > 0, \\ z(0,t) &= \partial_\xi z(1,t) = 0 && ,\ t > 0, \\ z(\xi,0) &= z_0(\xi) && ,\ \xi \in (0,1). \end{aligned}$$

The initial value problem above may be used to model the temperature distribution in a bar of length $l = 1$. Here ξ is the spatial variable, t is the time, and p is a function characterizing the thermal properties of the material; the temperature at the left end is maintained at temperature zero, the right end of the bar is insulated, z_0 is the distribution of the initial temperature.

Let

$$H := \mathbb{H} := L_2(0,1)\ ,\ \text{endowed with the usual norm and inner product},$$
$$P := \{p \in H^1(0,1) \mid p_0 < p(\xi) < p_1,\ \xi \in [0,1]\}\ ,\ 0 < p_0 < p_1\ ,$$
$$V := \{v \in H^1(0,1) \mid v(0) = 0\}.$$

We introduce A(p) by defining

$$D(p) := \{v \in H^1(0,1) \mid p\partial_\xi v \in H^1(0,1),\ \partial_\xi v(1) = 0\},$$
$$A(p)u := -\ \partial_\xi(p(\xi)\partial_\xi u)\ ,\ p \in P,\ u \in D(p).$$

Now, the initial value problem (2.6) is transformed into a problem of the kind (2.5). Obviously, hypothesis (H1) holds.

Let $p \in P$ and let z be the associated solution of (2.5) according to Lemma 1.

We have

$$(p\partial_\xi z)(\xi,t) - (p\partial_\xi z)(1,t) = \int_1^\xi \partial_t z(\eta.t)\, d\eta ,$$

i.e.

$$p(\xi) = (\partial_\xi z(\xi,t))^{-1} \int_1^\xi \partial_t z(\eta.t)\, d\eta \tag{2.7}$$

for all $(\xi,t) \in [0,1] \times (0,\infty)$ with $\partial_\xi z(\xi,t) \neq 0$.

Since we have the boundary condition $\partial_\xi z(1,t) = 0$, $t > 0$, the formula (2.7) cannot be used for $\xi = 1$. But for continuity reasons it is enough to reconstruct p for $\xi \in [0,1)$; in this region we can use the maximum principle to ensure that $\partial_\xi z(\xi,t) \neq 0$. The formula (2.7) shows that, in general, the identification of p is an overdetermined problem. In a variational formulation we obtain from (2.6)

$$-\int_0^1 p(\xi)\partial_\xi z(\xi,t)\, \partial_\xi v(\xi)\, d\xi = \int_0^1 \partial_t z(\xi,t)\, v(\xi)\, d\xi , \quad v \in V.$$

By integration with respect to time we obtain

$$(p, -\int_{t_2}^{t_1} \partial_\xi z(.,t)\, \partial_\xi v(.)\, dt)_{\mathbb{H}} = (z(.,t_2) - z(.,t_1),v)_{H} , \quad v \in V.$$

This implies that the following condition is sufficient for identifiability:

(2.8) rg L_{t_1,t_2} is dense in $\mathbb{H}$ for some pair (t_1,t_2) with $0 \le t_1 \le t_2$.

Here $L_{t_1,t_2} : V \ni v \mapsto -\int_{t_1}^{t_2} \partial_\xi z(.,t)\, \partial_\xi v(.)\, dt \in \mathbb{H}$.

Clearly, condition (2.8) may be interpreted as a richness property of the (observed) state z.

2.4 Identification of a matrix in an elliptic equation

Up to now, only problems of evolutionary type were considered. But by the same reasoning also the stationary case of such an problem can be investigated. To explain this we consider a specific example which is of some interest in applied sciences.

Let $\Omega \subset \mathbb{R}^n$ be an open and bounded domain with (a sufficiently smooth) boundary $\partial\Omega$. We consider the boundary value problem

(2.9) $\qquad \mathrm{div}(p\,\mathrm{grad}\,z) + f = 0\ ,\ \text{in } \Omega, \qquad z = g\ ,\ \text{in } \partial\Omega.$

Here p is a pointwise symmetric matrix which has to be identified. To simplify our considerations we assume $g \equiv 0$.

Let

$$H := L_2(\Omega)\ ,\ \mathbb{H} := L_2(\Omega)^{m,m}\ ,\ f \in H,$$

$$P := \left\{ p \in \mathbb{H} \,\middle|\, \begin{array}{l} p(\xi) \text{ symmetric for all } \xi \in \Omega, \\ \exists \varkappa_0 > 0\ \forall \xi \in \Omega\left((u,p(\xi)u) \geq \varkappa_0 \|u\|^2\ \forall u \in \mathbb{R}^m \right) \end{array} \right\},$$

$$V := H^1(\Omega)\,|v_{|\partial\Omega} = 0\}\ .$$

Integration by parts gives

(2.10) $$-\int_\Omega (p(\xi)\,\mathrm{grad}\,z(\xi),\ \mathrm{grad}\,v(\xi))\,d\xi + \int_\Omega f(\xi)\,v(\xi)\,d\xi = 0\ ,\ v \in V.$$

Using the notation

$$b(z,.) : V \ni \left(\frac{1}{2}\left(\frac{\partial z}{\partial \xi_i}\cdot\frac{\partial v}{\partial \xi_j} + \frac{\partial v}{\partial \xi_i}\,\frac{\partial z}{\partial \xi_j}\right)\right)_{1\leq i,j\leq m} \in \mathbb{R}^{m,m}\ ,$$

$$(q,\overline{q})_{\mathbb{H}} := \sum_{i,j=1}^{m} (q_{ij},\overline{q}_{ij})\ ,\ q = (q_{ij}),\ q=(\overline{q}_{ij}) \in P\ ,$$

we may reformulate (2.9) as follows:

(2.11) $$(p,b(z,v))_{\mathbb{H}} + (f,v)_H = 0 \text{ for all } v \in V.$$

Clearly, p is uniquely determined if rg b(z,.) is dense in $\mathbb{H}$; compare this with (2.8).

Bibliographical comments

In the last decade a lot of research work on parameter identification has been done. A source of references on this topic is [5] and [27]. More on identifiability aspects can be found in [17] (parabolic case) and in [1],[2],[7] and [19] (elliptic case).

The general aspects of ill-posedness in inverse problems are discussed in [6], specific problems of ill-posedness in identification problems are discussed in [11], [12] and [15].

§ 3 Model reference adaptive systems

In this section we apply the model reference adaptive systems method in evolutionary problems which are characterized by a bilinear structure.

3.1 Design of a model reference adaptive system

Let us consider again an initial value problem:

$$z' = A(p)z + f(t)\ ,\ t \geq 0;\ z(0) = z_0\ . \tag{3.1}$$

Moreover, we assume without further mention that the hypothesis (H1) is satisfied.

Let $p \in P$ be a fixed parameter and let $z \in C(0,\infty;H)$ be the associated output; p has to be identified. As a first step we want to formulate the model equation. We assume:

(H2) Let $C : V \to V^*$ be a linear and bounded operator such that there exist constants $\nu_1,\nu_2 > 0$ with

$$-\langle Cu,u\rangle \geq \nu_1 \|u\|_V^2\ ,\quad |\langle Cu,v\rangle| \leq \nu_2 \|u\|_V \|v\|_V\quad ,\ u,v \in V.$$

The initial value problem which represents the model is then given as follows:

$$x' = Cx + A(q)z(t) - Cz(t) + f(t)\ ,\ t \geq 0;\ x(0) = x_0\ . \tag{3.2}$$

Hereby, q is the parameter which has to be adjusted such that it becomes close to the unknown parameter p. The adjustment has to be done in such a way that x converges assymptotically to the known output z. For the initial value x_0 a good guess is z_0.

Suppose that $q(t) \in P$, $t \geq 0$, is the result of an adaption rule and assume that the system (3.2) has a solution $x \in C(0,\infty;H) \cap C^1(0,\infty;V^*)$ if we plug in $q(t)$ in (3.2) at time t:

$$x' = Cx(t) + A(q(t))z(t) - Cz(t) + f(t)\ ,\ t \geq 0,\ x(0) = x_0\ .$$

Notice that the evolution equation in (3.2) becomes the evolution equation in (3.1) asymptotically if we can prove

(3.3) $\lim_{t\to\infty} (x(t) - z(t)) = 0$ (in H), $\lim_{t\to\infty} q(t) = p$ (in $\mathbb{H}$).

We set $w := x - z$, $r := q - p$; w, r are error quantities.

An adaption rule which is constructed along the principles mentioned in Section 1.3 should be based on z and on the error w. Therefore we are led to an evolutionary system of the following kind:

(3.4) $q' = G(z(t),w)$, $t \geq 0$, $q(0) = q_0$

where $G : V \times V \to \mathbb{H}$ with $G(.,0) = 0$.
Clearly, the goals in (3.4) can be reformulated as follows:

(3.5) The equilibrium point (0,0) of the system (3.2),(3.4) is asymptotically stable.

A frequently used method to verify (3.5) is Ljapunov's direct method. The simplest Ljapunov-function for the error system

(3.6) $w' = Cw + A(q)z(t) - A(p)z(t)$, $r' = G(z(t),w)$

is given as follows:

(3.7) $V(w,r) := \frac{1}{2}\|w\|_H^2 + \frac{1}{2}\|r\|_{\mathbb{H}}^2$, $w \in H$, $r \in \mathbb{H}$.

Given this Ljapunov-function V we have to choose G such that V is not increasing along a solution of (3.6). As it is well-known this follows from the fact that the derivative with respect to time of V along a solution (w,r) is not positive. By formal calculations we have

$$\begin{aligned} V(w(t),r(t)) &:= \frac{d}{dt} V(w(t),r(t)) \\ &= \langle Cw(t) + A(q)z(t) - A(p)z(t),w(t)\rangle + (G(z(t),w(t)),r(t)) \\ &\leq -\nu_1 \|w(t)\|_V^2 + \langle (A(q)z(t) - A(p)z(t)),w(t)\rangle + (G(z(t),w(t)),r(t))_{\mathbb{H}}. \end{aligned}$$

To continue we use conditions collected in the following hypothesis:

(H3) i) There exists for each $q \in P$ a normed vector space $D \subset D(q)$ which is dense in $\mathbb{H}$.

ii) The mapping

$$a : P \times V \times V \ni (q,u,v) \mapsto \langle A(q)u,v\rangle = a(q;u,v) \in \mathbb{R}$$

can be extendeed to a trilinear mapping on $\mathbb{H} \times D \times V$ satisfying

$$|a(h;u,v)| \leq \varkappa_3 \|h\|_{\mathbb{H}} \|u\|_D \|v\|_V \quad , h \in \mathbb{H},\ u \in D,\ v \in V$$

where $\varkappa_3$ is a positive constant.

iii) $z \in L_\infty(0,\infty;D)$.

By Riesz's representation theorem there exists a bilinear mapping

$$b : D \times V \to \mathbb{H}$$

such that

$$(h,b(u,v))_{\mathbb{H}} = a(h;u,v) \qquad , h \in H,\ u \in D,\ v \in V; \tag{3.8}$$

$$\|b(u,v)\|_{\mathbb{H}} \leq \varkappa_3 \|u\|_D \|v\|_V \qquad , u \in D,\ v \in V. \tag{3.9}$$

From the calculations above we obtain

$$V(w(t),r(t)) \leq -\nu_1 \|w(t)\|_V^2 \ , \ t \geq 0,$$

if we define

$$G(z(t),v) := -b(z(t),v) \qquad , v \in V,\ t \geq 0.$$

Thus we are led to the following ***model reference adaptive system***:

$$\begin{aligned} x' &= Cx + A(q)z(t) - Cz(t) + f(t) \ , \ t > 0,\ x(0) = x_0 \\ q' &= -b(z(t),x - z(t)) \qquad , \ t > 0,\ q(0) = q_0 \ . \end{aligned} \tag{3.10}$$

The system for the error quantities becomes

$$\begin{aligned} w' &= Cw + A(r)z(t),\ t > 0,\ w(0) = w_0, \\ r' &= -b(z(t),w) \qquad , \ t > 0,\ r(0) = r_0 \ , \end{aligned} \tag{3.13}$$

Hereby the mapping A is extended due to hypothesis (H3) to a mapping from $\mathbb{H}$ into the space of linear bounded operators from D into V^*. Since now the mapping

$$\mathbb{H} \times D \ni (q.u) \mapsto A(q)u \in V^*$$

may be considered as a bilinear mapping we say that we are concerned with an identification problem in a bilinear system.

As a first step to establish the model reference adaptive system (3.10) we have to prove its well-posedness. As a second step we want to verify that under certain circumstances the long-time behavior (3.3) holds. Let us illustrate this later problem by an example:

Illustration: Tutorial example.

We recall (see Section 2.1 and 2.2)):

$H = \mathbb{H} = V = \mathbb{R}$; $P = (p_0,p_1) \subset (-\infty,0)$; $A(q)v = qv$, $q \in P$, $v \in V$.

Obviously, we have with $D := H$

$$b(u,v) = uv,\ u \in D,\ v \in V.$$

Therefore the model reference adaptive system (3.10) becomes a system of two ordinary differential equations; the operator C is the multiplication with a constant $c \in (-\infty,0)$ (see (H2). Well-posedness of this system follows immediately from results for ordinary differential eqations. Clearly, if $z(t) = 0$ for all $t \geq 0$ we cannot expect that (3.3) holds (see Section 2.2).

Let $\hat{z} \in \mathbb{R}$, $\hat{z} \neq 0$. Consider the error system (3.13) as a perturbation of an autonomous system in the following manner:

$$w' = cw + \hat{z}r + r(z(t) - \hat{z}) , \quad r' = - \hat{z}w + r(\hat{z} - z(t)) \quad , t \geq 0.$$

The eigenvalues of the system matrix

$$\begin{pmatrix} c & \hat{z} \\ -\hat{z} & 0 \end{pmatrix}$$

have negative real part, let's say μ. By Gronwall's lemma we obtain for the solution $y := (w,r)$:

$$|y(t)| \leq \varkappa |y(0)| \exp(\varkappa \int_0^t |\hat{z} - z(t)| ds - \mu t) , t \geq 0,$$

where $\varkappa$ is a constant depending on c, $\hat{z}$ and on an a-priori estimate of r (see Lemma 3). This shows that (3.3) holds if

$$\int_0^\infty |\hat{z} - z(t)| ds < \infty .$$

*

3.2 Well-posedness of the model reference adaptive system

We summarize the conditions of well-posedness in

Theorem 2

Suppose that (H1),(H2),(H3) hold. Then there exists a unique determined pair (x,q) satisfying

i) $(x,q) \in L_2(0,T;V \times \mathbb{H})$, $(x',q') \in L_2(0,T;V^* \times \mathbb{H})$ for all $T > 0$.

ii) $\langle x'(t),v\rangle = \langle Cx(t) + A(q(t))z(t) + f(t),v\rangle$ for all $v \in V$ a.e. in $(0,\infty)$.

$(q'(t),h)_{\mathbb{H}} = -(h,b(z(t),x(t) - z(t))_{\mathbb{H}}$ for all $h \in \mathbb{H}$ a.e. in $(0,\infty)$.

iii) $x(0) = x_0$, $q(0) = q_0$.

Proof: See appendix B).

The conditions formulated in (H3) lead to the fact that in the trilinear form a the parameter and the state can be separated by introducing b. This separation depends on the choice of $\mathbb{H}$. This illustrated by

Illustration: Heat equation

We consider again the one-dimensional heat equation already formulated in Section 2.3:

$H := L_2(0,1)$, $V := \{v \in H^1(0,1) \mid v(0) = 0\}$;

$P := \{p \in H^1(0,1) \mid p_0 < p(\xi) < p_1 \ \forall\, \xi \in [0,1]\}$, $0 < p_0 < p_1$:

$A(q)u := -\partial_\xi(q\partial_\xi u)$, $q \in P$, $u \in D(q) := \{v \in V \mid q\partial_\xi v \in H^1(0,1), \partial_\xi v(1) = 0\}$.

We present two different adaptation rules:

Case 1:

$\mathbb{H} := L_2(0,1)$, $D := \{v \in H^2(0,1) \mid v(0) = 0, \partial_\xi v(1) = 0\}$.

We have to choose

$b(u,v) := \partial_\xi u\, \partial_\xi v$, $u \in D$, $v \in V$.

Case 2:

$\mathbb{H} := H^1(0,1)$, D as above.

We have to choose

$b(u,v) := \varphi$ where φ is the solution of the following boundary value problem

$$-\varphi'' + \varphi = \partial_\xi u\, \partial_\xi v\ ,\ \varphi'(0) = \varphi'(1) = 0,$$

$u \in D$, $v \in V$.

3.3 Identifiability

Since sufficient conditions for identifiability give us hints under which assumptions the long-time behavior (3.3) can be proved we consider the problem of identifiability in the setting of Section 3.1 (see Section 2.2).

If the conditions in (H1),(H2),(H3) are true then we may define by using Th. 2

$$L_{t_1,t_2} \ni V \mapsto - \int_{t_1}^{t_2} b(z(t),v)\,dt \in \mathbb{H} \quad ; 0 \le t_1 \le t_2 .$$

Theorem 3

Suppose that (H1),(H2),(H3) hold and assume that

(3.14) $\quad\quad \text{rg } L_{t_1,t_2}$ is dense in $\mathbb{H}$ for some pair (t_1,t_2).

Then the parameter p is uniquely determined by z.

Proo: See appendix B).

Compare Th. 4 with (2.8) and (2.11).

3.4 Adaptive identifiability

Next, we investigate the question under which assumptions it is possible to find out the parameter p by the model reference adaptive system, i. e. when it is possible to prove that (3.3) holds.

Firstly, we need an a-priori estimate for the solution of the model reference adaptive system.

Lemma 4

Suppose that (H1),(H2),(H3) hold. We have:

i) $\sup_{t \ge 0} \|w(t)\|_H^2 + \sup_{t \ge 0} \|r(t)\|_{\mathbb{H}}^2 + 2\nu_1 \int_0^\infty \|w(s)\|_V^2\, ds \le \|w(0)\|_H^2 + \|r(0)\|_{\mathbb{H}}^2 .$

ii) $w \in L_\infty(0,\infty;H) \cap L_2(0,\infty;V) \cap C([0,\infty);H)$.

iii) $r \in L_\infty(0,\infty;\mathbb{H}) \cap C([0,\infty);\mathbb{H});\ r' \in L_2(0,\infty;\mathbb{H})$.

Proof: See appendix C).

Theorem 5

Suppose that (H1), (H2), (H3) hold. Then

$$\lim_{t\to\infty} w(t) = 0 \quad , \rho_0 := \lim_{t\to\infty} \|r(t)\|_{\mathbf{H}} \text{ exists.}$$

Proof: See appendix C).

The property " $\lim_{t\to\infty} (x(t) - z(t)) = 0$ " is usually called ***output-identifiability***. The main assumption which implies identifiability, i. e. $\lim_{t\to\infty} q(t) = p$, is an assumption which implies that the identifiability condition (3.14) is satisfied repeately:

(H4) $\exists\, \sigma > 0 \;\; \exists\, l > 0 \;\; \exists\, (t_n)_{n\in\mathbb{N}} : \lim t_n = \infty \;\; \forall\, h \in \mathbf{H}$
$\forall\, n \in \mathbb{N} \;\; \exists\, t_{n_1}, t_{n_2} \in [t_n, t_n+l] \;\; \exists\, v \in V : v \neq 0$
$$(\,|(h, L_{t_{n_1}, t_{n_2}} v)_{\mathbf{H}}| \geq \sigma \|h\|_{\mathbf{H}} \|v\|_V).$$

Theorem 6

Suppose that (H1),(H2),(H3) and (H4) hold. Then

$$\lim_{t\to\infty} (x(t) - z(t)) = 0, \quad \lim_{t\to\infty} q(t) = p\ .$$

Proof: See appendix C).

For practical reasons, it is desirable that the rate of convergence is high. Actually, we want to prove that the convergence is of exponential type:

$$\|w(t)\|_H \leq M \exp(-\omega t) \|w(0)\|_H\ , \quad \|r(t)\|_{\mathbf{H}} \leq M \exp(-\omega t) \|r(0)\|_{\mathbf{H}}\ , \quad t \geq 0,$$

where M and ω are positive constants.

It is not surprising that this results from an improved version of hypothesis (H4).

(H4') $\exists\, \sigma > 0 \;\; \exists\, l > 0 \;\; \forall\, h \in \mathbf{H} \;\; \forall\, \tau > 0 \;\; \exists\, t_1, t_2 \in [\tau, \tau+l] \;\; \exists\, v \in V : v \neq 0$
$$\left((h, L_{t_1,t_2} v)_{\mathbf{H}} \geq \sigma \|h\|_{\mathbf{H}} \|v\|_V \right)$$

Hypothesis (H3) implies identifiability in a sufficient large interval [0,T]. Under somewhat weaker assumptions one can prove that even in the case that p is not uniquely determined the limit $q_\infty := \lim q(t)$ exists. Moreover, q_∞ is then the parameter of minimal norm (in $\mathbf{H}$) which is in agreement with the output z.

3.5 Concluding remarks

The formulation of the adaptive systems is mainly restricted to the case that there are no errors in the data. If there are errors as it is always the case under practical circumstances two problems may arrise. Firstly, we have to differentiate the data (see system (3.12)). This is an operation which is ill-posed but the degree of ill-posedness is not very high and regularization techniques for the operation are well-studied. Secondly, adaptive identification can be considered as an iteration method for an ill-posed problem. Such a method has to be endowed with an appropriate stopping criterion.

Numerical results for the method of model reference adaptive systems are in good agreement with the theoretical results. The possibility to use different adaptation rules - see the example concerning the heat equation - is very important in getting good results.

To have data of the state z only on a finite interval [0,T] is not a very serious problem since we may apply the MRAS method recursively. That means that we restart the adaptive process at time T with the initial state z_0 and the initial guess $q_0 := q(T)$.

Bibliographical comments

The method of model reference adaptive systems has been developed originally for finite-dimensional problems by [26], [22] and [24] and combined with observer theory by [20]. For infinite-dimensional problems the method as been developed hand studied by [27] starting from [3]; see also [8], [9]. Numerical results are reported in [27]. The bilinear structure has been exploited also in [14] and [30].

Appendix A)

Proof of Lemma 1:

α) By straightforward calculations we obtain from (3.2) that D(p) is a dense subspace of H.

β) Using the Lax-Milgram lemma (see [32, Satz 17.9]) we obtain

$$\mathrm{rg}(\lambda_0 I - A(p)) = H \quad \text{for all } \lambda_0 > 0.$$

According to [P, Th. 4.3], A(p) is a closed operator and generates a C_0-semigroup $\{S_p(t)\}_{t\geq 0}$ with $\|S_p(t)\| \leq 1$ for all $t \geq 0$.

γ) A(p) is a dissipative operator, i. e.

$$\|(\lambda I - A(p))x\|_H \geq \lambda \|x\|_H \quad \text{for all } x \in D(p),\ \lambda > 0.$$

This follows from hypothesis (H2).

δ) The semigroup $\{S_p(t)\}_{t \geq 0}$ is an analytic semigroup by Th. 6A in [28]. Using Th. 3.1 in [23] we obtain the result iii). ■

Appendix B)

Proof of Theorem 2:

Let us define for $t \geq 0$, $(x,q),(\hat{x},\hat{q}) \in V \times \mathbb{H}$:

$$\mathfrak{A}(t;(x,q),(\hat{x},\hat{q})) := \langle Cx, \hat{x}\rangle + \langle A(q)z(t), \hat{x}\rangle - (b(z(t),x), \hat{q})_{\mathbb{H}},$$

$$\varphi(t) := (Cz(t) + f(t), b(z(t),z(t)) \in V^* \times \mathbb{H}.$$

Now it is easy to verify the following assertions:

α) $-\mathfrak{A}(t;(x,q),(x,q)) \geq \nu_1 \|(x,q)\|^2_{V \times \mathbb{H}}$ for all $t \geq 0$, $(x,q) \in V \times \mathbb{H}$.

β) $\mathfrak{A}(t;(x,q),(\hat{x},\hat{q})) \leq \alpha \, \|(x,q)\|_{V \times \mathbb{H}} \|(\hat{x},\hat{q})\|_{V \times \mathbb{H}}$ for all $t \geq 0$ and $(x,q),\ (\hat{x},\hat{q}) \in V \times \mathbb{H}$; here α is a constant independent of t, (x,q) and $(\hat{x},\hat{q})$.

γ) $[0,\infty) \ni t \mapsto \mathfrak{A}(t;(x,q),(\hat{x},\hat{q})) \in \mathbb{R}$ is measurable for all $(x,q),\ (\hat{x},\hat{q}) \in V \times \mathbb{H}$.

δ) $\varphi \in C([0,T]; V^* \times \mathbb{H})$ for all $T > 0$.

Using these facts we can apply Th. 5.5.1 in [29]. ■

Proof of Theorem 3:

Suppose that $q \in P$ has the same out put z as p. Then

$$0 = a(q - p; z(t), v) = (q - p, b(z(t),v))_{\mathbb{H}} \quad \text{for all } v \in V \text{ and } t \in [0,t_2].$$

Integration gives with (3.14) the result ■

Appendix C)

Proof of Lemma 4:

From the error system (3.10) follows

$$\langle w'(t), w(t)\rangle + (r'(t), r(t))_{\mathbb{H}} = \langle Cw(t), w(t)\rangle, \text{ i.e.}$$

$$\frac{d}{dt}\left(\|w(t)\|^2_H + \|r(t)\|^2_{\mathbb{H}}\right) \leq -2\nu_1 \|w(t)\|^2_V \ ;\ t \geq 0.$$

Integration implies i).

ii) and iii) are consequences of i) and Th. 2. ■

Proof of Theorem 5:

From

$$\frac{1}{2}\,\Big|\,\|w(t_2)\|_H^2 - \|w(t_1)\|_H^2\Big| = \Big|\int_{t_1}^{t_2}\langle w'(s), w(s)\rangle ds\,\Big| \quad , 0 \le t_1 \le t_2 ,$$

one deduces the following assertion:

$$\forall \rho > 0\ \forall \varepsilon > 0\ \exists t_0 \ge 0\ \forall t_1, t_2 \ge t_0 \left(|t_1 - t_2| < \rho \Rightarrow \Big|\,\|w(t_2)\|_H^2 - \|w(t_1)\|_H^2\Big| \le \varepsilon\right).$$

From this property the result " $\lim_{t\to\infty} w(t) = 0$ " follows by contradiction. Moreover, the limit $\lim_{t\to\infty} V(w(t),r(t))$ exists since the Ljapunov - function V is nonincreasing. Therefore $\lim_{t\to\infty} \|r(t)\|_H$ exists. ■

Proof of Theorem 6:

To prove by contradiction that ρ_0, defined in Th. 4, is zero one verifies the following assertion:

Let $(t_n)_{n\in\mathbb{N}}$ be a sequence with $\lim_n t_n = \infty$. Moreover, let $l > 0$, $\rho > 0$. Then

$$\lim_n\, (r(t_n), L_{t_{n,1},t_{n,2}} v)_H = 0 \quad \text{uniformly for all } t_{n,1}, t_{n,2} \in [t_n, t_n+l]\ ,\ v \in V$$

with $\|v\|_V \le \rho$. ■

References

[1] Alessandrini,G.: An identification problem for an elliptic equation in two variables. Annali di Mat. Pura ed Appl. **145** (1986), 265 - 296.

[2] Alessandrini,G.: Singular solutions of elliptic equations and the determination of conductivity by boundary measurements. J. Differential Equations **84** (1990), 252 - 272.

[3] Alt,H.W., Hoffmann, K.-H., Sprekels, J.: A numerical procedure to solve certain identification problems. In: Optimal control of differential equation, 11 - 43. Eds. Hoffman, Krabs, ISNM 68, Birkhäuser Verlag, Boston, 1984.

[4] Anderson,B.D.O., e.a.: Stability of adaptive systems: passivity and averaging anaysis. The MIT Press, Cambridge, 1986.

[5] Banks,H.T., Kunisch,K.: Estimation techniques for distributed parameter systems. Birkhäuser, Boston,1989.

[6] Baumeister,J.: Stable solution of inverse problems. Vieweg Verlag, Braunschweig, 1987.

[7] Baumeister,J., Kunisch,K.: Identifiability and stability of a two - parameter estimation problem. To appear in "Applicable Analysis".

[8] Baumeister,J., Scondo,W.: Adaptive methods for parameter identification. In: Methoden und Verfahren der Mathem. Physik, Band 34, 87 - 116. Eds. Brosowski, Martensen, Lang Verlag 1987.

[9] Baumeister,J., Scondo,W.: Asymptotic embedding methods for parameter estimation. In: Proc. of the 26th IEEE- Conference on Decision and Control, Los Angeles 1987.

[10] Böcker,J., Hartmann,I., Zwanzig,Ch.: Nichtlineare und adaptive Regelungssysteme. Springer, Berlin, 1986.

[11] Chavent,G.: Local stability of the output least square parameter estimation technique. Math. Applic. Comp. **5** (1983), 3 - 22.

[12] Colonius,F., Kunisch,K.: Output least squares stability in elliptic systems. Appl. Mat. Optim. **19** (1989), 33 - 63.

[13] Hale, J.K.: Asymptotic behavior od dissipative systems. Math. Surveys and Monographs, Amer. Math. Soc., Providence, 1988.

[14] Hsiao,G.C., Sprekels,J.: A stability result for distributed parameter identification in bilinear systems. Math. Methods Appl. Sci. **10** (1988), 447 - 456.

[15] Ito,K., Kunisch,K.: The augmented Lagrangian method for parameter estimation in elliptic systems. SIAM J. Control and Optimization **28** (1990), 113 - 136.

[16] Janecki,D.: Persistency of excitation for continous-time systems - Time-domain approach. Systems & Control Letters **8** (1987), 33 - 344.

[17] Kitamura,S., Nakagiri,S.: Identifiability of spatially - varying and constant parameters in distributed systems of parabolic type. SIAM J.Contr. Opt. **15** (1977), 785 - 802.

[18] Kobayashi, T.: Adaptive control for infinite-dimensional systems. Int. J. Systems Sci. **17** (1986), 887 - 896.

[19] Kohn,R.,V., McKenney,A.: Numerical implementation of a variational method for electrical impedance tomography. Inverse problems **6** (1990), 389 - 414.

[20] Kreisselmeier,G.: Adaptive observers with exponential rate of convergence. IEEE Trans. Aut.Contr. Vol. **22** (1977), 509 - 535.

[21] Landau,Y.D.: Adaptive control. Marcel Dekker, New York, 1979.

[22] Lion,P.M.: Rapid identification of linear and nonlinear systems. AIAA J. **5** (1987), 1835 - 1841.

[23] Lunardi,A.: Abstract quasilinear parabolic equations. Math. Ann. **267** (1984), 395 - 415.

[24] Morgan, A. P.: On the construction of nonautonomous stable systems with applications to adaptive identification and control. SIAM J.Contr.Opt. **17** (1979), 400 - 431.

[25] Narendra,K.S., Annaswamy,A.M.: Persistent excitation in adaptive systems. Int.J.Contr. **45** (1987), 147 - 160.

[26] Narendra, K. S., Kudva, P.: Stable adaptive schemes for system identification and control I,II. IEEE, **SMC - 4** (1974), 542 - 560.

[26] Polderman,J.W.: Adaptive control & identification: Conflict or conflux? CWI Tract, Amsterdam, 1989.

[27] Scondo,W.: Ein Modellabgleichsverfahren zur adaptiven Parameteridentifikation in Evolutionsgleichungen. Dissertation, Universität Frankfurt/M., 1987.

[28] Showalter,R.E.: Hilbert space methods partial differential equations. Pitman, London, 1976.

[29] Tanabe,H.: Equations of evolution. Pitman, London 1979.

[30] Tautenhahn,U., Muhs,J.: On the regularized error equation method for parameter identification in bilinear systems. Preprint Nr. 179, Techn. Universität Chemnitz.

[31] Temam, R.: Infinite-dimensional dynamical systems in mechanics and physics. Springer, Berlin, 1988.

[32] Wloka,J.: Partielle Differentialgleichungen. Teubner, Stuttgart, 1982.

Matching problems with Knapsack side constraints[1]
– A computational study –

U. Derigs and A. Metz
Lehrstuhl für Wirtschaftsinformatik, Universität zu Köln
Albertus-Magnus-Platz, D 5000 Köln 41

Abstract: We present and compare procedures for the approximate solution of the weighted matching problem with side constraints. The approaches are based on Lagrangean relaxation as well as on Lagrangean decomposition. Furthermore we develop an enumerative approach to solve this class of problems exactly. Our computational experiments investigate the efficiency of the relaxation and decompostion method for this lass of problems and are especially focusing on the comparison of different "subgradient methods" for solving the Lagrangean dual.

1. Introduction

Matchings in graphs built one of the few genuine structures studied within the field of combinatorics which are of practical relevance as well as efficiently tractable. Therefore matching problems have been analysed in detail from a theoretical as well as from a numerical point of view. However, pure matching problems are rather uncommon in practice. Nevertheless matchings are useful in solving real-world problems since several practical problems can be transformed into matching problems with side constraints. To illustrate some numerical aspects of the practical applications of matchings we discuss here some methods for solving matching problems with knapsack side constraints.

Let $G = (V, E)$ be a graph and $c : E \to \mathbb{R}_{\geq 0}$ a cost function. For any

[1] This work was partially supported by a grant from the Deutsche Forschungsgemeinschaft (DFG)

subset $F \subseteq E$ we define the costs of the subset as:

$$c(F) := \sum_{e \in F} c(e) \quad .$$

A subset $X \subseteq E$ is called a perfect matching of G, if every node of V is met by exactly one edge of X. Let $\mathcal{M}$ be the set of all perfect matchings of G. Then the (cost minimal) perfect matching problem (MP) can be stated as follows:

$$(MP) \qquad z_{MP} = min \; \{ \; c(X) \mid X \in \mathcal{M} \; \} \quad .$$

With any subset $X \in \mathcal{M}$, we can associate its incidence vector x, $x \in \{0,1\}^{|E|}$ where

$$x_e := \begin{cases} 1 & \text{if } e \in X \\ 0 & \text{else} \end{cases} \quad .$$

In the following subsets $X \in \mathcal{M}$ and their incidence vectors x are used as synonyms. This allows a simple notation and treatment of the problem.

With M the node-edge incidence matrix of G we can write (MP) in the following form

$$min \; \{c'x \mid Mx = 1, \; x \in \{0,1\}^{|E|} \; \} \quad .$$

From the results of Edmonds [1965] the linear characterization of the matching polytope $Co\{ \; x \mid Mx = 1, \; x \in \{0,1\}^{|E|} \; \}$ is well-known and (MP) can be formulated as a linear program using either the set of ***blossom constraints*** or the set of *cut constraints*. Let $W \subseteq V$ with $|W| \geq 3$, odd and $\delta(W)$ the set of edges which have exactly one node in W then the following *cut constraint*

$$\sum_{e \in \delta(W)} x_e \geq 1$$

is fulfilled by every $X \in \mathcal{M}$. With $Bx \geq 1$ the system of all cut constraints we obtain the following LP-formulation for (MP)

$$min \; \{ \; c'x \mid Mx = 1, \; Bx \geq 1, \; x \geq 0 \; \} \quad .$$

Let S be a $m \times |E|$ – matrix and $t \in \mathbb{R}^m$, $m \in \mathbb{N}$. Now the matching problem with side constraints (MPS) can be formulated as follows:

$$(MPS) \qquad v(MPS) = min \{ \, c(X) \mid X \in \mathcal{M} \, , \, Sx \leq t \, \} \qquad .$$

The theoretical relevance of (MPS) lies in the fact that set partitioning problems and thus every 0–1 optimization problem can be formulated as an (MPS) problem (see Nemhauser and Weber [1979]). Special crew and vehicle scheduling problems have been investigated in terms of (MPS) (see Ball and Benoit [1988] and Fisher [1987]), which shows the practical relevance of (MPS).

In an earlier paper (Ball et al. [1990]) we have studied the special case of (MPS) where the side-constraints were so-called GUB-conditions. The study of this class of problems was motivated by their relevance in certain scheduling and routing problems. In this paper we investigate the case where the matching problem is complicated by one (or a few) knapsack-constraint(s). The interest in this problem type stems from the following facts:

- While GUB-constraints are "easy" on their own, the knapsack-condition alone defines a NP-hard problem.
- Among all NP-hard "side constraints" the knapsack problem is still computationally tractable and a number of codes are available which promise to be able to solve even large problem instances in reasonable time.
- For the GUB-case the best bounds obtained from Lagrangean relaxation and Lagrangean decomposition coincide. Hence the knapsack-conditions are in a sense the "easiest conditions" for which decomposition may "improve" relaxation.

2. Relaxations for (MPS).

In this section we analyse several relaxations for (MPS) with respect to their relative quality and their complexity. To begin with, we state two linear programs which are immediate relaxations of (MPS):

$$(LP1) \qquad min \; \{ \; c'x \mid Mx = 1, \; Sx \leq t, \; x \geq 0 \; \}$$

and

$$(LP2) \qquad min \; \{ \; c'x \mid Mx = 1, \; Bx \geq 1, \; Sx \leq t, \; x \geq 0 \; \} \; .$$

The problem $min\{c'x \mid Mx = 1, \; x \geq 0\}$ is called *the fractional matching problem* and it can be shown that the fractional matching problem is equivalent to a bipartite matching problem (assignment problem) over a graph G_{bip} obtained from G by doubling the nodes and introducing two symmetric copies for every edge from G. Hence $(LP1)$ can be interpreted as an assignment problem with additional linear constraints. These constraints will in general destroy the integrality of the assignment problem.

$(LP2)$ is called the LP-relaxation of (MPS) which is computationally intractable in this form since the number of cut constraints, i.e. the size of the system $Bx \geq 1$, may grow expotentially with the size of the graph.

Obviously (MPS) can be reformulated as follows

$$(MPS) \qquad \begin{aligned} & min \sum_{X \in \mathcal{M}} \alpha_X c(X) \\ & \text{s.t.} \sum_{X \in \mathcal{M}} \alpha_X = 1 \\ & \sum_{X \in \mathcal{M}} \alpha_X Sx \leq t \\ & \alpha_X \in \{0,1\} \; , \; \forall X \in \mathcal{M} \end{aligned}$$

Here we have associated with each matching X a decision variable $\alpha_x \in \{0,1\}$. Now the *LP–relaxation* $(LP2)$ is obtained by replacing these conditions by

$\alpha_x \geq 0$. The LP-dual (LPD) of $(LP2)$ reads:

$$(LPD) \qquad \begin{array}{ll} & max\, \beta - t'\lambda \\ & \beta - S(x)\lambda \leq c'x \qquad \text{for all } X \in \mathcal{M} \\ & \lambda \geq 0 \qquad . \end{array}$$

For practical purpose we can assume that the number m of side-constraints is small while the number of variables/matchings is large. For LP's of this type that have many variables *column generation* may be used to increase the efficiency of the LP-simplex-method.

Here we start with a feasible basis $\mathcal{B} = \{X^1, \ldots, X^{m+1}\}$ of $(LP2)$. Then the following system of equations defines the complementary dual solution (β, λ) associated with $\mathcal{B}$:

$$\beta - S(x^i)\lambda = c'x^i \quad \text{for } i = 1, \ldots, m+1 \qquad .$$

If this solution is not dual feasible and hence the basis $\mathcal{B}$ is not optimal, we would then solve the following (pertubated matching) problem

$$min\ \{c'x + \lambda'(Sx - t) \mid X \in \mathcal{M}\}$$

using the complementary dual solution λ as penalty.

The associated optimal matching X_λ "violates the dual constraints most" and is then pivoted into the basis.

Then the whole process is repeated with the new basis until the optimal basis is constructed.

Types of relaxations which are widely used in combinatorial optimization are Lagrangean relaxations, and since more recently Lagrangean decompositions, too. In Lagrangean relaxation the set of side constraints is partitioned into two sets and one of those is dualized, i.e. introduced into the objective function with some kind of penalty-parameter. For a general discussion of Lagrangean relaxation we refer to Fisher [1981].

Applying this general idea to (MPS) there are two obvious types of relaxations. If we dualize the knapsack constraints we obtain the following Lagrangean relaxation

$$\begin{aligned} RK(\lambda) &= min\ \{c'x + \lambda(Sx - t) \mid X \in \mathcal{M}\ \} \\ &= min\ \{c'x + \lambda(Sx - t) \mid Mx = 1,\ Bx \geq 1,\ x \geq 0\ \} \quad . \end{aligned}$$

$RK(\lambda)$ is a (pure) matching problem and for any fixed $\lambda \geq 0$

$$RK(\lambda) \leq v(MPS) \quad .$$

Let $\lambda \geq 0$ given, then a matching is called λ-optimal if it is an optimal matching for $RK(\lambda)$. The best lower bound for $v(MPS)$ is obtained by optimizing the so-called Lagrangean dual

$$(DRK) \qquad v(RK) = max\ \{\ RK(\lambda) \mid \lambda \geq 0\ \} \quad .$$

Dualizing the set of matching constraints $Mx = 1$, we obtain a second relaxation

$$RM(\lambda) = min\ \{c'x + \lambda(Mx - 1) \mid Sx \leq t,\ x \in \{0,1\}^{|E|}\ \}$$

with the property

$$RM(\lambda) \leq v(MPS) \quad \text{for all } \lambda \in \mathbb{R}^{|V|} \quad .$$

Again we can formulate the Lagrangean dual

$$(DRM) \qquad v(RM) = max\ \{\ RM(\lambda) \mid \lambda \in \mathbb{R}^{|V|}\ \} \quad .$$

While the first relaxation leads to easy solvable problems, the determination of $RM(\lambda)$ is NP-hard, and thus computationally of no advantage over (MPS). Yet, we may expect $v(RK) < v(RM)$.

Applying general results from Geoffrion [1974] we can show the following primal interpretation of (DRK) and (DRM), respectively

$$\begin{aligned} v(RK) &= min\ \{\ c'x \mid Sx \leq t, x \in Co(Mx = 1,\ x \in \{0,1\}^{|E|})\ \} \\ &= v(LP2) \\ v(RM) &= min\ \{\ c'x \mid Mx = 1, x \in Co(Sx \leq t,\ x \in \{0,1\}^{|E|})\ \} \quad . \end{aligned}$$

In Lagrangean decomposition (cf. Guignard and Kim [1987]) the (combinatorial) optimization problem is split into two separate problems by dividing the set of constraints into two parts. These problems are joined by an additional restriction which will then be relaxed. The approach is also known under the label "variable splitting" (cf. Jörnsten and Näsberg [1986]). For (MPS) the decomposition or splitting into a matching problem and a knapsack-type-problem is obvious.

Let $\alpha, \beta \in [0,1]$ with $\alpha+\beta = 1$ and $Z \supseteq \{0,1\}^{|E|}$, then (MPS) is equivalent to the following three problems

$$min\ \{\alpha c'x + \beta c'y \mid Mx = 1,\ Sy \leq t,\ x \in \{0,1\}^{|E|},\ y \in Z,\ x = y\ \}$$
$$min\ \{\alpha c'x + \beta c'y \mid Mx = 1,\ Bx \geq 1,\ Sy \leq t,\ x \in Z,\ y \in \{0,1\}^{|E|},\ x = y\ \}$$
$$min\ \{\alpha c'x + \beta c'y \mid Mx = 1,\ Sy \leq t,\ x \in Z,\ y \in \{0,1\}^{|E|},\ x = y\ \}$$

Dualizing the constraints $x = y$ we obtain three (possibly different) Lagrangean decompositions

$$\begin{aligned} DK_Z(u) &= min\ \{\alpha c'x + \beta c'y + u(x-y) \mid Mx = 1,\ Sy \leq t,\ x \in \{0,1\}^{|E|},\ y \in Z\ \} \\ &= min\ \{(\alpha c + u)'x \mid X \in \mathcal{M}\ \} - max\ \{(u - \beta c)'y \mid Sy \leq t,\ y \in Z\ \} \\ DM_Z(u) &= min\ \{(\alpha c + u)'x \mid Mx = 1,\ Bx \geq 1,\ x \in Z\ \} \\ &\quad - max\ \{(u - \beta c)'y \mid Sy \leq t,\ y \in \{0,1\}^{|E|}\ \} \end{aligned}$$

and

$$\begin{aligned} \overline{DM}_Z(u) &= min\ \{(\alpha c + u)'x \mid Mx = 1,\ x \in Z\ \} \\ &\quad - max\ \{(u - \beta c)'y \mid Sy \leq t,\ y \in \{0,1\}^{|E|}\ \} \end{aligned}$$

For any $u \in \mathbb{R}^{|E|}$ and any $Z \supseteq \{0,1\}^{|E|}$ – and any feasible (α, β)-combination – the following relations hold:

$$DK_Z(u) \leq v(MPS)$$
$$DM_Z(u) \leq v(MPS)$$
$$\overline{DM}_Z(u) \leq v(MPS) \quad .$$

The best lower bounds are again obtained by solving the associated Lagrangean decomposition duals

$$(DDK_Z) \qquad v(DK_Z) = max \{ DK_Z(u) \mid u \in \mathbb{R}^{|E|} \}$$
$$(DDM_Z) \qquad v(DM_Z) = max \{ DM_Z(u) \mid u \in \mathbb{R}^{|E|} \}$$
$$(\overline{DDM}_Z) \qquad v(\overline{DM}_Z) = max \{ \overline{DM}_Z(u) \mid u \in \mathbb{R}^{|E|} \} \quad .$$

A first analysis shows that the relaxation duals are dominated by their decomposition counterparts in the following sense:

$$v(RK) \leq v(DK_Z)$$
$$v(RM) \leq v(DM_Z)$$
$$v(RM) \leq v(\overline{DM}_Z)$$

for all $Z \supseteq \{0,1\}^{|E|}$ and feasible (α, β).

Varying the superset $Z \supseteq \{0,1\}^{|E|}$ not only the quality of the lower bound varies since obviously

$$v(DK_{Z'}) \leq v(DK_Z)$$
$$v(DK_{Z'}) \leq v(DM_Z) \qquad \text{for } Z' \supseteq Z \supseteq \{0,1\}^{|E|}$$
$$v(\overline{DM}_{Z'}) \leq v(\overline{DM}_Z)$$

yet also the complexity of the subproblems changes.

Here especially the case $Z = [0,1]$ is of interest, since in this case $DK_Z(u)$ is polynomially solvable – while $DM_Z(u)$ and $\overline{DM}_Z(u)$ are NP-complete for all $Z \supseteq \{0,1\}^{|E|}$.

Again a "primal interpretation" for the decomposition duals is cruical for an evaluation of the quality of the associated lower bound for $v(MPS)$. Applying a general result from Guignard and Kim [1987] the following can be shown:

(DDK_Z) is equivalent to

$$min \{ c'x \mid x \in Co(Mx = 1, x \in \{0,1\}^{|E|}) \cap Co(Sx \leq t, x \in Z) \}$$

while (DDM_Z) is equivalent to

$$min \; \{ \; c'x \mid x \in Co(Mx = 1, \; Bx \geq 1, \; x \in Z) \cap Co(Sx \leq t, \; x \in \{0,1\}^{|E|}) \; \}$$

and $(\overline{DDM}_Z)$ is equivalent to

$$min \; \{ \; c'x \mid x \in Co(Mx = 1, \; x \in Z) \cap Co(Sx \leq t, \; x \in \{0,1\}^{|E|}) \; \} \quad .$$

To be able to compare the quality of the different relaxations and decompositions we use the following definition:

$$\begin{aligned} &\text{A set } \{ \; x \mid Ax \leq b \; \} \text{ is } Z\text{-convex} : \; \Leftrightarrow \\ &\qquad Co(\{ \; Ax \leq b \; \} \cap Z) = Co(\{ \; Ax \leq b \; \}) \cap Co(Z) \quad . \end{aligned}$$

Then the following relations hold

$$\left. \begin{array}{r} \{ \; Sx \leq t \; \} \;\; Z\text{-convex} \\ \{ \; Mx = 1, \; x \in \{0,1\}^{|E|} \; \} \;\; \text{compact} \end{array} \right\} \Rightarrow v(DK_Z) = v(RK)$$

$$\left. \begin{array}{r} \{ \; Mx = 1, \; Bx \geq 1 \; \} \;\; Z\text{-convex} \\ \{ \; Sx \leq t, \; x \in \{0,1\}^{|E|} \; \} \;\; \text{compact} \end{array} \right\} \Rightarrow v(DM_Z) = v(RM)$$

$$\left. \begin{array}{r} \{ \; Mx = 1 \; \} \;\; Z\text{-convex} \\ \{ \; Sx \leq t, \; x \in \{0,1\}^{|E|} \; \} \;\; \text{compact} \end{array} \right\} \Rightarrow v(\overline{DM}_Z) = v(RM) \quad .$$

For $Z = \{0,1\}^{|E|}$ Kim [1985] has shown that the following holds:

Lemma: If one of the two sets of constraints is Z-convex then the bound from Lagrangean decomposition is equal to the stronger of the two bounds from Lagrangean relaxation.

Now the set $\{ \; Mx = 1, \; Bx \geq 1 \; \}$ is $Z = \{0,1\}$–convex and thus we have

$$v(DM_Z) = v(DK_Z) = max \; \{ \; v(RK), v(RM) \; \} \quad .$$

For all Z, $\{0,1\}^{|E|} \subseteq Z \subseteq \mathbb{R}^{|E|}$, we have

$$v(DM_Z) = v(RM) \quad ,$$

because $\{\ Mx = 1,\ Bx \geq 1\ \}$ is Z-convex.

Combining both results we obtain

$$v(RM) = v(DM_Z) = v(DK_{\{0,1\}}) \geq v(DK_Z) \geq v(RK) \quad .$$

Because of these relationships we will only study $(DDK_{\{0,1\}})$, $(DDK_{[0,1]})$ and (DRK) furtheron.

3. Optimizing the Lagrangean-Duals

In this section we shortly describe solution methodologies to optimize Lagrangean functions. First we review some well-known properties of these functions. The lagrangean functions $R(\lambda)$ and $D(u)$ have the following properties:

(i) They are the envelope of a finite familiy of linear functions, e.g. $RK(\lambda) = min_{\bar{X} \in \mathcal{M}}\{ \ r_{\bar{X}}(\lambda) = c'\bar{x} + \lambda(S\bar{x} - t) \ \}$ etc.

(ii) Both functions are concave, piecewice linear and continous.

Because of these properties we can adopt several standard methods from nonlinear programming to maximize the Lagrangean duals.

Problems of the type

$$\min_{v \in \mathbb{R}^n} f(v)$$

where the function f is a nonsmooth convex function with the gradient existing almost everywhere are well studied in nonsmooth optimization (see Zowe [1987]). In the case of non-differentiable optimization the elements of the so-called subdifferential $\partial f(.)$, which is defined as follows:

$$\partial f(v) := \{ \ w \in \mathbb{R}^n \mid w'(v - z) \leq f(z) - f(v) \ \} \quad ,$$

serves as a substitute for the gradient. The elements of $\partial f(v)$ are called *subgradients* of f at v. Now an element $v^* \in \mathbb{R}^n$ is optimal, if $0 \in \partial f(v^*)$ holds. For $\lambda \geq 0$ fixed let x be λ-optimal then $(Sx - t)$ is a subgradient of R at λ and for u fixed and an u-optimal pair (x, y) the function $(x - y)$ is a subgradient of $(DDK_{\{0,1\}})$ and $(DDK_{[0,1]})$ at u.

In the classical concept of subgradient optimization to minimize a convex function f the following steps are iteratively applied until a prespecified stopping criterion is fulfilled:

Step 1: (*search direction*)
Given v^k choose $w^k \in \partial f(v^k)$ and set $d^k \leftarrow -\frac{w^k}{\|w^k\|}$.

Step 2: *(descent step)*

set $v^{k+1} \leftarrow v^k + t^k d^k \ , \quad t^k > 0.$

Thereby the step-size t^k can be determined via an exact line-search, i.e t^k is the optimal solution of the problem

$$min\{\ f(v^k + td^k) \mid t \in \mathbb{R}_{>0}\ \} \quad ,$$

or t^k can be choosen off-line from a sequence with the following properties

$$t^i \rightarrow 0_+ \quad \text{and} \quad \sum_{i=1}^{\infty} t^i = \infty \quad .$$

Convergence of the sequence $\{v^k\}$ when using the second t^k - rule is assured due to a theorem by Poljak [1967].

The most popular rule to determine a v–sequence used in combinatorial optimization for optimizing a lagrangean function $f = R(\lambda)$ is the so-called *Fisher–Rule* (see Fisher [1981])

$$v^k := \begin{cases} 0 & \text{if } k = 0 \\ v^{k-1} + \gamma^{k-1} \frac{Sx^{k-1} - t}{\|Sx^{k-1} - t\|} & \text{else} \end{cases}$$

where γ^{k-1} is chosen in the following way:

$$\gamma^{k-1} := \mu^{k-1} \frac{F - f(v^{k-1})}{\|\ Sx^{k-1} - t\ \|}\ ,\ 0 \leq \mu^{k-1} \leq 2, \text{ and } F \text{ is an upper bound for } f(v).$$

Examples show that all these rules may generate bad search direction in practice and sometimes lead not to convergence. This behavior is due to the shortsightedness when using only the information of one subgradient.

The so–called *Kiev–Method* solves this problem by using the information of the two recent subgradients. Let $\alpha \in]0,1[$ then we define:

$$v^k := \begin{cases} 0 & \text{if } k = 0 \\ v^1 + \gamma^0 \frac{Sx^0 - t}{\|Sx^{k-1} - t\|} & \text{if } k = 1 \\ v^{k-1} + \gamma^{k-1} \frac{\alpha Sx^{k-1} + (1-\alpha) Sx^{k-2} - t}{\|\alpha Sx^{k-1} + (1-\alpha) Sx^{k-2} - t\|} & \text{else} \end{cases}$$

where γ is defined as above.

The so-called *bundle-approach* (see Zowe [1987]) accumulates at v^k the subgradient information from a neighbourhood of v^k by retaining the previously used subgradients. With the *bundle*

$$\mathcal{B}^k := \{ w^i \mid i = 1, \ldots, k, \, w^i \in \partial f(v^i) \}$$

the new search direction d^k is constructed as a convex combination of vectors of $\mathcal{B}^k$ and an exact line-search is performed to evaluate the optimal step-size.

4. Some Special Properties for the Single Side Constraint Case

In this section we discuss some special properties for the case of a single knapsack side-constraint, i.e. $s'x \leq t$, $t \in \mathbb{N}$, $s \in \mathbb{N}^{|E|}$.

When using Lagrangean decomposition $(DDK_{[0,1]})$ the linear program can easily be solved via a greedy approch. Let ϕ be a permutation of $\{1, 2, \ldots, |E|\}$, such that the ratios of

$$\frac{(u_{\phi(i)} - \beta c_{\phi(i)})}{s_{\phi(i)}} \quad , \; 1 \leq i \leq |E|$$

are ordered in decreasing order and let $r \in \mathbb{N}_0$ such that:

$$t_1 := \sum_{i=1}^{r-1} s_{\phi(i)} < t \leq \sum_{i=1}^{r} s_{\phi(i)} \quad .$$

Then y defined as below is u-optimal:

$$y_i := \begin{cases} 1 & \text{if } \phi(i) < r \text{ and } u_{\phi(i)} - \beta c_{\phi(i)} > 0 \\ \frac{t-t_1}{s_{\phi(i)}} & \text{if } \phi(i) = r \text{ and } u_{\phi(i)} - \beta c_{\phi(i)} > 0 \\ 0 & \text{else} \end{cases} \quad .$$

Optimizing the LP-relaxation of (MPS) via column generation is based on a simple calculation. Let λ^1 such that the corresponding optimal matching X^1 is infeasible and λ^2 such that the corresponding optimal matching X^2 is feasible then the complementary λ^0 is easily calculated via:

$$\lambda^0 := \frac{c'x^2 - c'x^1}{s'x^1 - s'x^2} \quad .$$

Now let X^0 be λ^0–optimal, we then define:

$$\lambda^2 := \lambda^0 \qquad \text{if } s'x^0 \leq t$$

or

$$\lambda^1 := \lambda^0 \qquad \text{if } s'x^0 > t \quad .$$

To initialize the column generation procedure we need two λ–values and their associated matchings with the above stated properties. This is accomplished by a simple "shooting technique". First we solve $min\{\, c(X) \mid X \in \mathcal{M}\,\}$. If the corresponding optimal matching fulfills $s'x \leq t$, then it is optimal for (MPS) too. In the other case we define:

$$v_i := i * (\underset{e}{max}\{c_e\} - \underset{e}{min}\{c_e\}) \qquad \text{for } 1 \leq i \leq \frac{1}{2}|V|$$

and successively solve the problems

$$min\{\,(c + v_i)'x \mid X \in \mathcal{M}\,\}$$

for $i = 1, 2, \ldots$ until a feasible optimal matching occures for the first time, for $i = k$ say. Then we define $\lambda^1 := v_{k-1}$ and $\lambda^2 := v_k$. If all generated optimal matchings are infeasible then there exists no feasible solution to (MPS). This simple procedure is motivated by the following lemma.

Lemma: If the set of feasible solutions for (MPS) is nonempty, then there exists a $\bar{\lambda}$-optimal matching which is feasible and $\bar{\lambda} \leq \frac{|V|}{2} * (max_e\{c_e\} - min_e\{c_e\})$.

Proof: If the graph only allows one perfect matching or feasible matchings only, then $\lambda = 0$ fullfills the above property. In the other case there exist two matchings X^1 and X^2 with X^1 feasible and X^2 infeasible, and a $\bar{\lambda} \geq 0$ such that:

$$c'x^1 + \bar{\lambda}(s'x^1 - t) = c'x^2 + \bar{\lambda}(s'x^2 - t) \qquad ,$$

because the lagrangean function is continous on the area $\mathbb{R}_{>0}$. Hence X^1 and X^2 are $\bar{\lambda}$-optimal with

$$\bar{\lambda} = \frac{c'x^2 - c'x^1}{s'x^1 - s'x^2} \leq \frac{|V|}{2} * (\underset{e}{max}\{c_e\} - \underset{e}{min}\{c_e\}) \qquad .$$

□

We now describe a method for bounding the duality gap for (MPS) with one side-constraint, i.e. we give a bound for the value

$$\Delta := v(MPS) - v(RK) \qquad .$$

This method is based on a fundamental property of the set of perfect matchings in a graph and it is an analogue of a result for the matroidal Knapsack problem obtained by Camerini and Vercellis [1981]. In Section 5 we will in a certain sense "extend" this idea when using the same combinatorial property to construct "good" feasible matchings.

Let M and N be two perfect matchings then we know that the set $M \Delta N := (M \setminus N) \cup (N \setminus M)$ is spanned by a set $Q_1, Q_2, \ldots, Q_r$ of disjoint even cycles in G, the edges of which are alternately in M and N. With the usual notation $M \oplus Q = (M \setminus Q) \cup (Q \setminus M)$ for M-alternating even cycles we obtain the following relationship

$$N = M \oplus Q_1 \oplus \cdots \oplus Q_r \quad .$$

Hence any perfect matching N in G is obtained from any other perfect matching M in G by exchanging the role of matching and nonmatching edges on a set of r M-alternating cycles and we define the distance $dist(M, N)$ between M and N to be the number r of such cycles.

Lemma: Let λ^* be the optimal Lagrange multiplier and define $\bar{c} := (c + \lambda^* s)$. Then two cases are possible for $RK(\lambda^*)$:

(i) there exists a λ^*-optimal matching X^* with $\lambda^*(s'x^* - t) = 0$

(ii) there exist λ^*-optimal matchings X^+ and X^- with $s'x^+ > t, s'x^- < t$ and $dist(X^+, X^-) = 1$.

Proof: Assume we have solved (DRK) by subgradient optimization. Then we either end up with case (i) and an optimal matching X^* or with two matchings X^+ and X^- fulfilling the second condition except maybe for $dist(X^+, X^-) = 1$. Thus assume $dist(X^+, X^-) = r,\ r > 1$, then there exist cycles $Q_1, \ldots, Q_r$ with

$$X^+ = X^- \oplus Q_1 \oplus \cdots \oplus Q_r \quad .$$

Now defining for $1 \leq i \leq r$:

$$\bar{c}(Q_i) := \bar{c}(X^+ \cap Q_i) - \bar{c}(X^- \cap Q_i)$$

and

$$s(Q_i) := s(X^+ \cap Q_i) - s(X^- \cap Q_i)$$

we get the following equalities

$$\bar{c}(X^+) := \bar{c}(X^-) + \sum_{i=1}^{r} \bar{c}(Q_i)$$

and

$$s(X^+) := s(X^-) + \sum_{i=1}^{r} s(Q_i) \quad .$$

Now assume $\bar{c}(Q_j) \neq 0$ for at least one j. Then the following is true

$$\bar{c}(X^- \oplus Q_j) < \bar{c}(X^-) \text{ or } \bar{c}(X^+ \oplus Q_j) < \bar{c}(X^+)$$

But this is a contradiction to the optimality of X^+ and X^-. Hence $\bar{c}(Q_i) = 0$ holds for all $1 \leq i \leq r$.

Based upon this property the following procedure starting from an initial pair (X^+, X^-) produces either an optimal matching X or an eventually different pair of matchings – also denoted by (X^+, X^-) – which fulfills the property (*ii*).

procedure **REDUCE DISTANCE**

$i := 0$

$X := X^-$

Until $dist(X^+, X^-) = 1$ do

$i \leftarrow i + 1$

$X \leftarrow X^- \oplus Q_i$

If $s(X) = t$ then STOP, X is optimal,

else if $s(X) < t$ then $X^- \leftarrow X$

else if $s(X) > t$ then $X^+ \leftarrow X$

end do □

For the following let Q be the cycle spanning $X^+ \Delta X^-$. Now it is obvious that

$$\bar{\Delta} := c(X^-) - \bar{c}(X^-)$$

may serve as an approximation of the duality gap and

$$\begin{aligned}\bar{\Delta} &= c(X^-) - (c(X^-) + \lambda^* s(X^-) - \lambda^* t) \\ &= \lambda^*(t - s(X^-)) \\ &< \lambda^*(s(X^+) - s(X^-)) = \lambda^* s(Q)\end{aligned}$$

Now λ^* is defined as the unique point of the intersection of the linear functions $r_{X^+}(\lambda)$ and $r_{X^-}(\lambda)$ i.e. for λ^* holds

$$c'x^+ + \lambda^*(s'x^+ - t) = c'x^- + \lambda^*(s'x^- - t)$$

hence

$$\lambda^* = \frac{c(X^-) - c(X^+)}{s(X^+) - s(X^-)} = -\frac{c(Q)}{s(Q)}$$

Combining these expressions we immediately obtain the following (worst-case) approximation for the duality gap.

Lemma: The difference Δ between $v(MPS)$ and $v(RK)$ is either zero or smaller than the minimal difference between the objective function values of any feasible/nonfeasible pair (X^-, X^+) of λ^*-optimal matchings, i.e.

$$\Delta = v(MPS) - v(RK) \leq min\ (c(X^-) - c(X^+))$$

where $\bar{c}(X^-) = \bar{c}(X^+) = v(RK)$, $s(X^-) \leq t$ and $s(X^+) > t$.

Corollary:

$$\Delta \leq min\ \{\ c(X^-) - c(X^+) \mid X^-, X^+ \in \mathcal{M}, X^+ \oplus X^- \text{ cycle in } G\ \}\ .$$

Thus for a graph G with no (even) cycles, (MPS) can be solved optimally by Lagrangean relaxation $RK(\lambda)$.

5. Improvement Techniques

Generally the procedures to optimize the lagrangean function or the LP-relaxation will terminate with a lower bound and the best feasible matching, that is generated during the optimization process, the costs of which exceeds the lower bound. In this chapter we will discuss two improvement techniques which will reduce this "duality-gap" by generating better feasible matchings.

When optimizing the lagrangean function $RK(\lambda)$ at each iteration we have to solve a matching problem. This sequence of matchings can be divided into a subsequence of feasible matchings and into a subsequence of infeasible matchings. Now it is possible by combining pairs of matchings from these subsequences to determine a feasible matching with lower costs. This procedure leads to an improvement approach which is due to Ball [1986].

Let X^0 and X^1 be two perfect matchings, one beeing feasible and one beeing infeasible, then the best feasible perfect matching with respect to $G' = (V, X^0 \cup X^1)$ can be found by solving a multiple knapsack problem. The components of the graph G' are:

- single edges $e \in X^0 \cup X^1$ and
- even cycles $Q_1, \ldots, Q_k$ the edges of which are alternately in X^0 and X^1.

Obviously any perfect matching $\bar{X}$ in G' can be represented by a vector $(y_1, \ldots, y_k) \in \{0, 1\}$ such that:

$$\bar{X} = (X^0 \cap X^1) \cup \bigcup_{j=1}^{k} (X^{y_j} \cap Q_j) \quad .$$

The costs of $\bar{X}$ are given by:

$$\begin{aligned} c(\bar{X}) &= \sum_{j=1}^{k} c(X^{y_j} \cap Q_j) + c(X^0 \cap X^1) \\ &= c(X^0) + \sum_{i=1}^{k} (c(X^1 \cap Q_j) - c(X^0 \cap Q_j)) y_j \quad . \end{aligned}$$

With respect to the side-constraints we have:

$$S\bar{x} = Sx^0 + \sum_{i=1}^{k}(S(X^1 \cap Q_j) - S(X^0 \cap Q_j))y_j \quad .$$

For $1 \leq j \leq k$ and $1 \leq i \leq m$ we define:

$$w_j := c(X^0 \cap Q_j) - c(X^1 \cap Q_j)$$
$$a^j := S(X^0 \cap Q_j) - S(X^1 \cap Q_j)$$

and

$$b_i := t_i - S(X^0 \cap X^1) \quad .$$

Thus the problem of determining the best feasible matching in G' leads to the following Knapsack problem:

$$\begin{aligned} max \ & \sum_{j=1}^{k} w_j y_j \\ \text{s.t.} \ & \sum_{j=1}^{k} a_{ij} y_j \leq b_i && 1 \leq i \leq m \\ & y_j \in \{0,1\} && 1 \leq j \leq k \quad . \end{aligned}$$

Combining (all) pairs of feasible and infeasible matchings obtained during the relaxation phase this way, may lead to better feasible solutions, i.e. perfect matchings fulfilling the side–constraints. Since the number k of cycles is at most $\frac{1}{4}|V|$ these problems are rather "efficiently" solvable for small m.

In the following another improvement technique is presented which is the adoption of a concept of Handler and Zang [1980] which they used to optimize the constrained shortest path problem. The basic idea is to determine a sequence of k–best matchings with respect to a penalized cost-function. We formulate this approach in one procedure for both cases – relaxation and decomposition. For this purpose we denote by v^* a given Lagrangean parameter for the relaxation dual as well as for the decomposition dual.

In case of Lagrangean relaxation we consider the following matching problem

$$min \{ \bar{c}(x) = c'x + v^*(Sx - t) \mid X \in \mathcal{M} \} \quad .$$

In the case of Lagrangean decomposition let y^* be the v^*-optimal knapsack solution, i.e. an optimal solution to

$$max\{ (v^* - \beta c)'y \mid Sy \leq t, \; y \in \{0,1\} \} \quad .$$

Then we consider the following matching problem

$$min \{ \bar{c}(x) = (\alpha c + v^*)'x - (v^* - \beta c)y^* \mid X \in \mathcal{M} \}$$

which is equvalent to

$$min\{ (\alpha c + v^*)'x \mid X \in \mathcal{M} \} \quad .$$

Using this (common) notation we can now define the improvement algorithm based on generating the sequence of k-best matchings:

Initialization : With X^0 being the best feasible perfect matching found during the relaxation resp. decomposition phase set $B_U := c(X^0)$. Let X^1 be the optimal matching with respect to the optimal Lagrangean multiplier v^* and set $B_L := \bar{c}(X^1)$. Set $k := 2$ and $K = X^1$.

Step 1 : If $\mathcal{M} \setminus K$ is empty then terminate, X^0 is the optimal solution of (MPS).

Else determine a k–best matching X^k with respect to the penalized costs, i.e. solve the following problem:

$$min \{ \bar{c}(X) \mid X \in \mathcal{M} \setminus K \} \quad .$$

Step 2 : If $\bar{c}(X^k) \geq B_U$ then terminate, X^0 is optimal, else set: $B_L := \bar{c}(X^k)$ and $K := K \cup \{X^k\}$.

Step 3 : If X^k is a feasible matching, i.e. $Sx^k \leq t$, and $c(X^k) < B_U$ then set: $B_U := c(X^k)$ and $X^0 := X^k$.

Set $k := k + 1$ and goto step1.

Since we have a finite graph G, the set $\mathcal{M}$ is finite and so this procedure terminates. It is clear that the matching X^0 is optimal, if the procedure terminates in **step 1**. If the algorithm terminates in **step 2**, then X^0 is optimal too. Assume the existence of a feasible matching $\bar{X}$ with $c(\bar{X}) < c(X^0)$. Then, in the case of Lagrangean relaxation we obtain

$$\bar{c}(\bar{X}) = c'\bar{x} + v^*(S\bar{x} - t) \leq c(\bar{X}) < c(X^0) \leq \bar{c}(X^k)$$

which means that $\bar{X}$ would have preceeded X^k in the sequence of k-best matchings. In the case of Lagrangean decomposition we obtain

$$\begin{aligned} \bar{c}(\bar{X}) &= (\alpha c + v^*)'\bar{x} - (v^* - \beta c)'y^* \\ &\leq (\alpha c + v^*)'\bar{x} - (v^* - \beta c)'\bar{x} \\ &\leq c(\bar{X}) < c(X^0) \leq \bar{c}(X^k) \end{aligned}$$

which leads to the same contradiction as for the relaxation.

The above procedure works for all parameters v^*. Choosing $v^* = 0$ leads to a simple enumeration of all perfect matchings with respect to their objective function value and hence the first feasible matching will be optimal, too. Choosing large values for v^* will penalize the infeasible matchings and thus feasible matchings will appear earlier in the enumeration, yet the optimal perfect matching may not be among the first matchings constructed. Computational experience shows that choosing the *optimal* Lagrangean parameter to penalize may be a good compromise.

As every enumerative approach for solving combinatorial optimization problems this procedure is as well time–consuming as space–consuming. Hence for practical purpose it may be necessary or economial to bound the number of matchings constructed by either specifying a maximum number of iterations or by prespecifying a maximal allowable tolerance $\epsilon = (B_U - B_L)/B_L$. In any case the approach is heavily depending on fast matching codes and an efficient strategy for constructing the sequence of k-best matchings.

For more details on the underlying theory and procedures for constructing the sequence of k-best matchings we refer to Derigs [1985] and Derigs and Metz [1990].

6. Computational Results

In this section we report computational experiences with several of the strategies for solving (MPS) which we have outlined so far. FORTRAN 77 subroutines based on the previously described approaches were implemented and tested on a large set of randomly generated matching problems with one and five knapsack side constraints on a DEC VAX-station 3200 under VMS 5.1.

Before we describe our construction scheme for MPS some remarks on the complexity of generating "typical" MPS' is adequate:
While the use of randomly generated matching problems, i.e. randomly weighted graphs, is generally acceptable for evaluating the performance of matching algorithms since for a given graph structure the performance on worst-case and average-case examples does not differ significantly, the construction scheme for random MPS–problems requires some more care. The specification of the knapsack constraint(s) is consequential since it may easily make the constrained matching problem trivially solvable or extremly hard. On top of this coordination problem, the knapsack problem on its own is rather sensitive with respect to the choice of the right hand side and correlations between the coefficients in the objective function and in the constraint(s). Here the last issue is rather cruical since during the decomposition phase the coefficients of the objective function are penalized by a multible of the knapsack–coefficients.

In the following we describe our model for randomly constructed MPS–problems.

The matching problems are characterized by three different parameters

- $n =$ the number of nodes of the graph G
- $m =$ the number of edges of the graph G
- $C =$ the highest cost coefficient; the cost coefficients were generated as uniformly distributed random integers in the interval $[1, C]$.

For each specification (n, m, C) we generated 10 examples. For each example the coefficients of the knapsack side constraints were generated as uniformly distributed random integers in the interval $[1, 1000]$. For (MPS) with one side-constraint the right hand side t of the side constraint was taken randomly from the interval $[T_1, T_2]$ where:

$T_1 := min\ \{sx \mid X \in \mathcal{M}\}$

$T_2 := sx^*$, with X^* the optimal (unconstrained) perfect matching in G

To specify the right hand side t in the case of five knapsack side-constraints we first constructed the sequence of the 500-best matchings. Then we randomly took a matching $\bar{X}$ out of the sequence $(X_{150}, \ldots, X_{500})$ and set $t = S\bar{x}$.

6.1 Results for a single side-constraint

(a) Comparison of different λ–strategies for the relaxation phase

For solving the Lagrangean relaxation dual (DRK) of (MPS) respectively the LP–relaxation we investigated the following λ-rules:

- binary search (BS)
- bundle technique (BT)
- column generation (CG)
- Kiev method (KI)
- Fisher rule (FR) .

We always used an initial phase, where starting with a value of $\lambda = 0$, the parameter λ was increased by $max_{e \in E}\{c_e\}$ until a feasible matching with respect to the knapsack side-constraint was constructed. After the relaxation phase we applied the knapsack heuristic to every pair of feasible/infeasible matchings obtained in the relaxation phase with the aim of constructing better feasible matchings. In contrast to our investigations on matching problems with generalized upper bounds (cf. Ball et al. [1990]) we found that this heuristic does not construct matchings of a comparable quality as we had expected. As stopping criterion for the different λ–strategies we used:

- the restriction to at most 201 iterations and
- $|\lambda_{i+1} - \lambda_i| < 10^{-4}$.

Table 1 gives the average CPU-time required to optimize the Lagrangean dual, including the time spent for the the initial - and the knapsack phase. The significant low CPU-time required by (CG) lies in the fact, that this procedure constructs the fewest number of matchings. The other procedures have to calculate 3–5 times more matchings until convergence is achieved.

In table 2 we present the average error (gap) for the different methods in percentage of the upper bounds. No problem could be solved optimal, i.e. with an error bound of 0, by any method. The (CG)-procedure produced the largest error bounds with all the other methods being comparable and slightly better. The reason for this behaviour is again the small number of matchings constructed by (CG). In table 3 we compare the quality of the lower bounds using the lower bound obtained from (CG) as a reference value, i.e. for all other methods we display the difference in percentage. This table shows that there are no significant improvements. Combining these results with the results from table 2 we can conclude that the smaller gaps for (BS), (BT), (KI) and $(F\dot{R})$ are due to better upper bounds, i.e. the construction of the better feasible matchings. Thus (CG) is highly efficient for solving the Lagrangean dual, yet it fails to produce enough feasible solutions to be used in the knapsack heuristic.

C	n	m	CPU-TIME CG AVERAGE	CPU-TIME BS AVERAGE	CPU-TIME BT AVERAGE	CPU-TIME KI AVERAGE	CPU-TIME FR AVERAGE	NUMBER OF MATCHINGS CG AVERAGE	NUMBER OF MATCHINGS BS AVERAGE	NUMBER OF MATCHINGS BT AVERAGE	NUMBER OF MATCHINGS KI AVERAGE	NUMBER OF MATCHINGS FR AVERAGE
1000	50	612	0.583	2.192	3.300	3.357	3.517	8.10	32.00	47.80	45.80	49.80
		306	0.362	1.491	1.392	3.280	3.200	8.00	32.00	31.60	72.10	69.60
		204	0.254	1.139	1.376	2.353	2.493	7.20	32.00	35.20	67.80	71.50
		153	0.200	1.018	0.530	2.364	2.593	6.60	32.00	17.10	73.50	80.00
	100	2475	2.524	9.095	8.694	13.263	17.176	9.10	32.00	33.20	48.80	62.90
		1237	1.427	5.465	7.194	8.748	9.549	8.70	32.00	42.10	53.80	61.20
		825	1.215	4.397	3.433	8.342	9.851	8.80	32.00	25.10	60.80	70.30
		618	1.096	3.895	2.432	7.019	6.934	8.90	32.00	19.20	56.20	57.80
10000	50	612	0.593	2.434	1.851	5.748	6.542	8.70	36.00	26.50	81.80	94.40
		306	0.342	1.613	1.118	3.304	3.299	7.90	36.00	24.80	69.40	74.20
		204	0.295	1.470	0.930	4.125	4.096	7.30	36.00	21.20	97.20	97.50
		153	0.224	1.186	0.783	2.891	3.114	6.90	36.00	23.20	85.40	91.60
	100	2475	2.918	11.603	7.097	17.977	24.437	9.20	36.00	23.30	57.30	75.30
		1237	1.472	6.265	3.698	14.967	12.164	8.60	36.00	21.30	87.20	70.60
		825	1.224	5.176	3.931	11.022	12.318	8.90	36.00	29.80	78.70	90.80
		618	1.103	4.515	4.105	8.962	9.011	8.90	36.00	33.10	70.60	71.90
100000	50	612	0.545	2.704	1.640	6.261	5.502	8.00	39.00	24.30	89.80	82.20
		306	0.313	1.571	0.632	3.489	2.578	7.60	39.00	16.00	91.80	68.30
		204	0.261	1.353	1.067	2.621	2.708	7.60	39.00	29.30	75.00	79.40
		153	0.216	1.268	0.775	2.844	3.452	6.70	39.00	21.30	83.20	104.30
	100	2475	2.747	11.870	6.912	24.781	24.207	9.40	39.00	23.10	85.80	82.70
		1237	1.869	7.748	5.690	14.869	14.621	9.40	39.00	29.60	77.70	75.80
		825	1.212	5.560	3.301	11.983	12.063	8.80	39.00	23.70	87.40	89.30
		618	1.150	5.310	3.043	11.339	10.687	8.80	39.00	22.70	83.50	79.20

Table 1: Average CPU-time (in CPU-seconds) and number of matchings for solving (DRK) with different λ-rules

C	n	m	DUALITY-GAP CG AVERAGE	BS AVERAGE	BT AVERAGE	KI AVERAGE	FR AVERAGE
1000	50	612	3.877	3.184	3.717	4.949	4.956
		306	4.721	3.963	3.995	3.945	3.905
		204	3.863	2.668	3.550	2.436	2.673
		153	3.690	3.623	3.623	3.610	3.611
	100	2475	4.926	3.445	3.832	4.661	4.717
		1237	4.570	3.947	3.789	4.685	4.796
		825	3.842	3.465	3.222	3.680	3.456
		618	1.571	1.447	1.569	1.400	1.426
10000	50	612	3.458	3.322	3.158	6.762	6.717
		306	2.153	2.148	2.208	1.952	1.954
		204	6.714	4.907	4.911	4.906	4.909
		153	6.310	4.152	4.151	4.153	4.153
	100	2475	5.562	3.964	4.077	3.529	3.006
		1237	6.849	4.489	5.075	6.443	6.385
		825	4.477	2.992	2.768	2.668	2.868
		618	1.930	1.648	1.649	1.650	1.648
100000	50	612	9.370	6.441	6.342	7.094	7.964
		306	4.938	4.374	4.375	5.892	5.467
		204	4.110	3.300	3.463	3.313	3.162
		153	5.417	3.234	3.234	3.376	3.362
	100	2475	5.586	3.051	3.027	7.398	7.398
		1237	1.542	1.356	1.470	1.635	1.163
		825	2.794	1.808	1.666	1.791	1.660
		618	1.749	1.718	1.718	1.720	1.602

Table 2: Average error (gap) for different λ-rules for solving (DRK) (in percentage of the upper bound)

c	n	m	BS AVERAGE	BT AVERAGE	KI AVERAGE	FR AVERAGE
1000	50	612	0.052	0.144	-0.674	-0.682
		306	0.022	-0.011	0.041	0.082
		204	0.049	0.040	0.041	0.045
		153	0.032	0.032	0.046	0.045
	100	2475	0.074	0.096	-0.307	-0.326
		1237	0.068	0.053	-0.451	-0.436
		825	0.027	-0.009	0.027	0.036
		618	0.036	0.001	0.084	0.057
10000	50	612	0.006	0.000	-0.368	-0.361
		306	0.005	-0.001	0.004	0.002
		204	0.003	-0.001	0.004	0.001
		153	0.001	0.002	0.000	0.000
	100	2475	0.012	-0.001	-0.013	0.015
		1237	0.011	0.003	-0.714	-0.716
		825	0.009	-0.001	-0.007	-0.001
		618	0.004	0.003	0.002	0.004
100000	50	612	0.001	0.001	-1.159	-1.152
		306	0.001	0.000	-0.292	-0.294
		204	0.000	0.000	-0.013	-0.003
		153	0.000	0.000	-0.024	-0.009
	100	2475	0.001	0.000	-1.527	-1.527
		1237	0.000	0.000	-0.005	-0.002
		825	0.001	0.000	-0.009	0.000
		618	0.001	0.000	-0.001	0.000

Table 3: Improvement of the lower bounds with respect to (CG)

(b) Comparison of different λ–strategies for the decomposition phase

For optimizing the decomposition dual $(DK_{\{0,1\}})$ we investigated three different λ–rules

- bundle technique (BT)
- kiev method (KI)
- Fisher rule (FR) .

Since we have m side constraints of the form $x = y$, binary search could not be applied to this problems. The column generation approach would require to solve a linear program with $m+1$ variables and $m+1$ rows. Due to the size of this LP we did not implement this approach for the decomposition.

C	n	m	CPU-time BT AVERAGE	CPU-time KI AVERAGE	CPU-time FR AVERAGE	number of iterations BT AV. ITER	number of iterations KI AV. ITER	number of iterations FR AV. ITER
1000	50	612	68.321	23.000	22.217	190.50	201.00	201.00
		306	49.386	15.683	15.099	191.90	201.00	201.00
		204	37.372	13.382	12.939	173.20	201.00	201.00
		153	36.184	9.817	8.846	183.50	198.20	187.60
	100	2475	277.660	89.392	83.603	192.60	201.00	201.00
		1237	166.040	62.443	59.084	199.00	201.00	201.00
		825	118.990	46.148	43.820	199.00	201.00	201.00
		618	97.994	41.794	41.362	199.00	201.00	201.00
10000	50	612	*	24.555	23.073	*	201.00	201.00
		306	*	15.590	14.663	*	201.00	201.00
		204	*	15.507	14.710	*	201.00	201.00
		153	*	12.115	11.283	*	201.00	201.00
	100	2475	*	83.252	78.274	*	201.00	201.00
		1237	*	62.315	60.064	*	201.00	201.00
		825	*	47.058	45.070	*	201.00	201.00
		618	*	39.949	36.359	*	201.00	198.60
100000	50	612	*	24.680	23.079	*	201.00	201.00
		306	*	17.569	16.676	*	201.00	201.00
		204	*	12.742	11.982	*	201.00	201.00
		153	*	12.255	12.139	*	201.00	201.00
	100	2475	*	94.273	90.752	*	201.00	201.00
		1237	*	58.023	56.386	*	201.00	201.00
		825	*	47.135	44.407	*	201.00	201.00
		618	*	42.926	41.816	*	201.00	201.00

Table 4: Average CPU-time (in CPU-seconds) and number of matchings for solving $(DDK_{\{0,1\}})$ with different λ-rules

In an initial phase we constructed the best dual multiplier, $\hat{\lambda}$ say, for the Lagrangean relaxation (using the (CG)-approach). Then all the three methods were started with the following u-parameter

$$u_e = \hat{\lambda} s_e \quad .$$

Thus the search techniques for solving the decomposition dual are started from the level of the optimum for the relaxation dual if $\alpha = 1.0$ is chosen.

All procedures use the knapsack code MT2 of Martello and Toth [1988]. As the authors described in their work run-time problems occur if correlated data is at hand. Therefore we set $\alpha = 0.9$, to avoid unacceptable CPU-times. The (BT)-rule could only be applied to examples with $C = 1000$ because of the occurance of integer overflows in MT2.

C	n	m	BT AVERAGE	KI AVERAGE	FR AVERAGE
1000	50	612	2.390	2.077	2.609
		306	3.640	3.266	3.670
		204	2.383	2.998	3.367
		153	1.714	1.712	1.763
	100	2475	1.737	1.965	2.162
		1237	1.847	1.517	1.838
		825	1.589	1.607	1.867
		618	2.497	1.271	1.457
10000	50	612	*	2.882	3.302
		306	*	3.428	4.174
		204	*	1.706	1.912
		153	*	3.090	3.558
	100	2475	*	1.688	1.943
		1237	*	1.059	1.191
		825	*	1.187	1.371
		618	*	1.818	2.173
100000	50	612	*	3.445	4.197
		306	*	3.384	3.393
		204	*	1.678	1.966
		153	*	1.733	1.967
	100	2475	*	1.290	1.355
		1237	*	1.720	2.028
		825	*	1.151	1.368
		618	*	1.056	1.199

Table 5: Average error (gap) for different λ-rules for solving $(DDK_{\{0,1\}})$ (in percentage of the upper bound)

Table 4 gives the average CPU-time required to optimize the decomposition dual of $(DK_{\{0,1\}})$. In comparison to the Lagrangean relaxation the CPU-time has increased significantly. The reason for this fact is that we now have to solve a matching problem and a knapsack problem at each iteration and we need more iterations to fulfill the stopping critera.

Table 5 presents the average error in percentage of the upper bound for the different procedures. These are smaller as for relaxation.

C	n	m	BT AVERAGE	KI AVERAGE	FR AVERAGE
1000	50	612	0.413	0.267	-0.078
		306	0.260	-0.044	-0.469
		204	0.163	-0.369	-0.755
		153	0.202	-0.025	-0.175
	100	2475	0.628	0.713	0.506
		1237	0.452	0.432	0.267
		825	0.260	0.159	-0.058
		618	0.124	0.109	-0.034
10000	50	612	*	-0.353	-0.757
		306	*	-0.407	-0.832
		204	*	-0.260	-0.470
		153	*	-0.457	-0.960
	100	2475	*	-0.152	-0.327
		1237	*	-0.050	-0.184
		825	*	-0.148	-0.287
		618	*	-0.285	-0.529
100000	50	612	*	-0.460	-0.878
		306	*	-0.440	-0.860
		204	*	-0.161	-0.373
		153	*	-0.198	-0.441
	100	2475	*	-0.265	-0.412
		1237	*	-0.251	-0.533
		825	*	-0.231	-0.371
		618	*	-0.173	-0.318

Table 6: Improvement of the lower bounds with respect to (CG)

In table 6 we compare the lower bounds from decompostion with the lower bounds from relaxation using the (CG) bound as a reference. This table shows that there is no significant improvement. This indicates that the optimal value of the Lagrangean decomposition dual is not significantly larger than the optimal value of the Lagrangean relaxation dual. However combining these results with the information from table 5 we can say that we now produce generally better upper bounds, i.e. feasible matchings.

Moreover we applied all λ-rules to solve the decomposition dual $(DK_{[0,1]})$ where the knapsack problem is relaxed to a LP. For this decomposition we used $\alpha = 0.9$ as well as $\alpha = 1.0$.

			α=0.9			α=1.0		
C	n	m	BT AVERAGE	KI AVERAGE	FR AVERAGE	BT AVERAGE	KI AVERAGE	FR AVERAGE
1000	50	612	166.583	22.123	19.891	158.834	16.703	17.500
		306	95.367	13.397	12.392	104.431	10.798	12.065
		204	45.377	10.584	9.600	42.708	7.703	7.400
		153	57.511	9.186	8.857	42.746	7.407	7.527
	100	2475	368.375	95.907	92.828	358.119	94.741	88.934
		1237	311.612	51.910	47.895	212.667	48.714	44.784
		825	293.505	47.352	44.664	247.151	37.055	35.238
		618	168.185	39.032	35.260	179.716	30.677	27.715
10000	50	612	137.130	22.269	21.094	205.757	23.240	20.427
		306	87.678	13.119	14.084	53.291	8.699	9.551
		204	100.956	13.703	12.742	98.762	12.095	11.183
		153	54.702	8.831	8.395	73.495	7.333	6.783
	100	2475	424.030	122.938	107.242	255.726	109.333	96.965
		1237	328.289	61.061	57.517	355.884	53.767	49.839
		825	386.970	43.451	40.292	316.005	34.718	32.359
		618	283.718	40.001	39.745	191.307	34.050	32.606
100000	50	612	166.229	21.422	19.869	392.749	19.554	19.398
		306	85.260	10.961	10.305	81.724	9.219	9.033
		204	69.896	12.111	11.298	104.046	9.689	10.096
		153	83.755	9.791	8.994	49.711	8.763	8.289
	100	2475	423.222	105.518	95.326	431.680	97.220	91.180
		1237	344.965	69.019	65.357	269.963	52.421	50.276
		825	287.499	48.740	45.856	282.454	42.962	40.609
		618	276.608	39.461	43.553	33.203	33.203	34.076

Table 7a: Average CPU-time (in CPU-seconds) for solving $(DDK_{[0,1]})$ with different λ-rules

In tables 7-9 we summarize the corresponding results for the three different λ-rules applied to the decomposition dual $(DK_{[0,1]})$. Table 7a displays the average CPU-time required to optimize the decomposition dual $(DK_{[0,1]})$. These times are slightly smaller than those for optimizing the decomposition dual $(DK_{\{0,1\}})$. In table 7b we present the average number of iterations used to achieve convergence. Again we have to solve a large sequence of matching problems. The rather time consuming computation of one (BT)–iteration is now visible since despite a comparable number of iterations the (BT)–rule is much more time consuming than (KI) and (FR).

C	n	m	α = 0.9 BT AV. ITER	α = 0.9 KI AV. ITER	α = 0.9 FR AV. ITER	α = 1.0 BT AV. ITER	α = 1.0 KI AV. ITER	α = 1.0 FR AV. ITER
1000	50	612	181.00	181.70	172.50	174.00	144.30	157.20
		306	172.30	184.10	186.20	185.80	165.10	184.20
		204	135.10	188.10	184.40	141.40	155.10	164.50
		153	181.00	188.90	200.20	145.60	174.40	182.00
	100	2475	200.00	201.00	201.00	200.00	201.00	201.00
		1237	188.80	184.90	187.00	144.90	189.60	187.00
		825	200.00	201.00	201.00	157.00	174.70	177.40
		618	167.10	184.50	185.10	166.40	176.10	163.30
10000	50	612	162.80	196.80	201.00	187.70	201.00	186.90
		306	164.10	164.70	186.70	125.80	128.60	142.80
		204	181.30	201.00	201.00	200.00	201.00	201.00
		153	155.90	175.60	179.50	200.00	162.60	169.30
	100	2475	200.00	201.00	201.00	147.00	201.00	201.00
		1237	184.90	201.00	201.00	186.70	201.00	201.00
		825	200.00	201.00	201.00	163.70	177.90	181.90
		618	183.70	193.20	201.00	172.20	190.90	201.00
100000	50	612	154.00	182.90	185.30	177.70	173.40	183.40
		306	154.90	156.90	156.60	161.20	144.50	149.20
		204	124.80	201.00	194.80	180.10	179.80	200.10
		153	147.60	199.00	196.40	174.90	191.70	192.20
	100	2475	194.00	201.00	201.00	186.20	201.00	201.00
		1237	173.30	201.00	201.00	153.30	181.20	181.20
		825	185.50	201.00	201.00	176.90	201.00	201.00
		618	187.20	174.50	201.00	148.30	174.50	193.50

Table 7b: Average number of iterations for solving $(DDK_{[0,1]})$ with different λ-rules

Table 8 presents the average errors (gaps) in percentage of the upper bound for the different procedures. For $\alpha = 0.9$ we achieve slightly smaller gaps than for $\alpha = 1.0$.

In table 9 we again compare the lower bound to the (CG)-bound for the relaxation. This table shows that for $\alpha = 1.0$ the lower bounds are somewhat better while for $\alpha = 0.9$ they are worse. Combining the information from table 8 and 9 we can see that we are able to produce better feasible matchings for $\alpha = 0.9$ than for $\alpha = 1.0$. Again this might be due to a slightly larger number of matchings constructed.

c	n	m	α = 0.9 BT AVERAGE	α = 0.9 KI AVERAGE	α = 0.9 FR AVERAGE	α = 1.0 BT AVERAGE	α = 1.0 KI AVERAGE	α = 1.0 FR AVERAGE
1000	50	612	2.613	2.075	2.254	2.658	2.504	2.254
		306	3.019	2.558	3.683	3.412	2.587	3.683
		204	2.085	1.675	2.500	2.547	1.702	2.500
		153	2.570	2.452	3.304	3.505	2.499	3.304
	100	2475	2.044	1.777	2.080	2.436	2.200	2.080
		1237	1.692	1.419	1.808	3.390	1.657	1.808
		825	1.607	1.426	1.738	2.648	1.685	1.738
		618	0.979	0.983	1.143	1.274	1.130	1.143
10000	50	612	2.993	2.739	4.026	3.274	2.874	4.026
		306	1.547	1.789	2.093	1.636	1.568	2.093
		204	2.499	2.850	4.173	3.125	2.545	4.173
		153	2.805	2.865	5.026	3.789	2.828	5.026
	100	2475	1.395	1.425	1.985	3.775	1.468	1.985
		1237	2.881	2.100	2.807	4.771	2.847	2.807
		825	1.276	1.267	1.680	2.519	1.224	1.680
		618	1.098	1.242	1.552	1.495	1.465	1.552
100000	50	612	2.526	2.475	5.246	5.823	2.520	5.246
		306	2.879	3.202	4.116	3.766	2.860	4.116
		204	2.011	1.984	2.627	3.197	1.827	2.627
		153	3.929	2.940	3.709	3.352	2.705	3.709
	100	2475	1.304	1.287	1.893	2.564	1.469	1.893
		1237	0.914	0.873	1.037	1.143	0.981	1.037
		825	0.999	1.083	1.318	1.835	1.086	1.318
		618	0.899	1.215	1.643	1.461	1.169	1.643

Table 8: Average error (gap) for different λ-rules for solving $(DDK_{[0,1]})$ (in percentage of the upper bound)

C	n	m	α = 0.9 BT AVERAGE	α = 0.9 KI AVERAGE	α = 0.9 FR AVERAGE	α = 1.0 BT AVERAGE	α = 1.0 KI AVERAGE	α = 1.0 FR AVERAGE
1000	50	612	0.321	0.198	0.015	0.580	0.430	0.291
		306	0.182	-0.013	-0.774	0.260	0.080	0.044
		204	0.111	-0.044	-0.425	0.173	0.061	0.049
		153	0.070	-0.127	-0.692	0.154	0.056	0.043
	100	2475	0.637	0.828	0.517	0.913	0.752	0.469
		1237	0.424	0.391	-0.124	0.461	0.333	0.261
		825	0.175	0.188	-0.134	0.338	0.192	0.108
		618	0.147	0.170	0.007	0.301	0.191	0.149
10000	50	612	-0.115	-0.292	-1.298	0.063	0.008	0.001
		306	0.029	-0.220	-0.532	0.031	0.007	0.007
		204	-0.001	-0.368	-1.707	0.018	0.001	0.001
		153	-0.012	-0.496	-2.138	0.014	0.004	0.003
	100	2475	-0.031	-0.159	-0.633	0.102	0.012	0.001
		1237	-0.075	-0.220	-0.903	0.053	0.002	0.002
		825	-0.008	-0.164	-0.455	0.036	0.009	0.008
		618	-0.027	-0.173	-0.488	0.025	0.004	0.003
100000	50	612	-0.079	-0.398	-1.739	0.005	0.000	0.000
		306	-0.020	-0.449	-1.405	0.003	0.000	0.000
		204	-0.075	-0.265	-0.719	0.002	0.000	0.000
		153	-0.025	-0.422	-1.217	0.001	0.000	0.000
	100	2475	-0.069	-0.188	-0.799	0.010	0.000	0.000
		1237	-0.011	-0.121	-0.286	0.005	0.002	0.002
		825	-0.047	-0.132	-0.370	0.004	0.000	0.000
		618	-0.020	-0.208	-0.599	0.003	0.000	0.000

Table 9: Improvement of the lower bounds with respect to (CG)

(c) Evaluation of the k-best-improvement technique

The tree-search strategy for finding the set of k-best matchings has been described in detail in Derigs and Metz [1990]. In our implementation here we have limited the candidate list to 1000 entries (matchings). Thus the k-best-improvement-phase fails to find the optimum if the number of matchings to be stored exceeds this limit. This did not occur for the relaxation but for the decomposition. In that case the whole approach stopped with only an improved feasible solution and an improved lower bound.

In table 10 we list the average CPU-time for the k–best phase after solving the relaxation dual (DRK) and the decomposition dual $(DDK_{\{0,1\}})$. Here the (CG)-rule was used to solve (DRK) while we considered (KI) and (FR) for solving the decomposition dual $(DDK_{\{0,1\}})$.

				CPU-time		number of solved matchings		
C	n	m	CG AVERAGE	KI AVERAGE	FR AVERAGE	CG AVERAGE	FR AVERAGE	KI AVERAGE
100	50	612	2.413	41.281	38.363	164.40	3168.80	2931.40
		306	2.408	22.438	25.324	234.50	2162.30	2441.00
		204	1.456	17.072	20.266	146.70	1693.40	2030.20
		153	1.064	7.940	9.070	119.50	935.00	1078.40
	100	2475	47.034	85.160	85.370	1347.40	2001.80	2003.60
		1237	28.250	88.463	79.937	1163.20	2472.00	2289.60
		825	22.736	89.930	90.226	951.00	3033.80	3156.30
		618	15.696	82.614	91.748	617.30	3081.40	3432.40
10000	50	612	2.036	31.517	33.223	168.90	2550.50	2761.00
		306	1.392	18.556	23.668	126.20	1875.40	2327.40
		204	2.519	10.831	12.324	269.20	1044.50	1221.70
		153	2.212	18.360	19.318	227.10	2029.80	2172.80
	100	2475	54.958	141.656	129.869	1491.60	4007.00	3523.10
		1237	36.317	92.979	88.757	1365.90	2886.40	2871.60
		825	24.473	118.551	90.439	1078.50	4583.80	3480.40
		618	15.244	64.903	66.850	649.20	2820.80	3086.60
100000	50	612	5.086	46.175	45.495	392.20	3447.80	3411.30
		306	2.076	27.882	31.747	200.50	2645.40	3069.60
		204	3.114	14.968	17.926	353.30	1573.60	1895.00
		153	1.701	13.928	16.060	168.60	1474.80	1753.80
	100	2475	39.002	141.773	124.894	1099.80	3627.50	3065.20
		1237	11.036	105.036	100.792	409.80	3686.20	3454.60
		825	24.800	135.139	105.811	1042.00	5079.50	4072.70
		618	30.004	83.181	76.929	1163.90	2932.40	2668.60

Table 10: Average CPU-time (in CPU-seconds) for the k-best phase

In table 11 we compare the quality of the associated k-best approaches. In the case of relaxation we were able to solve all problems optimally in a period of time which is significantly lower than the CPU-time necessary to optimize the decomposition dual $(DDK_{\{0,1\}})$. These results identify the combination of solving (DRK) by (CG) and than to perform k-best enumeration to be substantially superior over decomposition followed by k-best enumeration. With (DRK) + "k-best" we were able to solve all problems optimally with reasonable time, wheras the dimension of the candidate list was not sufficiently large when starting the k-best approach after decomposition.

C	n	m	CG AVERAGE	KI AVERAGE	FR AVERAGE
1000	50	612	0.000	0.420	1.117
		306	0.000	1.684	2.305
		204	0.000	0.687	0.793
		153	0.000	0.000	0.000
	100	2475	0.000	3.978	4.291
		1237	0.000	2.370	2.453
		825	0.000	1.837	2.097
		618	0.000	1.488	1.434
10000	50	612	0.000	0.484	0.789
		306	0.000	1.302	1.615
		204	0.000	0.000	0.000
		153	0.000	0.642	0.525
	100	2475	0.000	1.283	1.447
		1237	0.000	0.774	0.949
		825	0.000	0.672	0.688
		618	0.000	0.922	1.317
100000	50	612	0.000	0.697	0.687
		306	0.000	0.446	0.501
		204	0.000	0.000	0.117
		153	0.000	0.167	0.012
	100	2475	0.000	0.785	0.975
		1237	0.000	1.023	1.430
		825	0.000	0.468	0.841
		618	0.000	0.643	0.818

Table 11: Remaining error after the k-best phase measured in percentage of the upper bound

6.2 Results for multiple side-constraints

In tables 12-14 we display the results for matching problems with five knapsack constraints. To optimize the Lagrangean dual (DRK) we use the (KI)- and (FR)-rule. In table 12 we present the required CPU-time to optimize the problems and the average number of iterations to achieve convergence or to fulfill the stopping criteria. There is no significant difference between the two λ-rules. In table 13 we compare the quality of the different approaches. As in table 12 both λ–rules are comparable and we could only solve 2 of the 80 examples optimal, i.e. with an error bound of 0.

Table 14 displays the results of the k-best approach applied after the (KI)-rule. All problems could be solved optimally with the time requirement for the k-best-phase only beeing twice the time for the relaxation phase. This is also due to the relatively small true error between the upper bound and the optimum – which is also displayed in table 14. Thus the best feasible solution of the relaxation phase was already near optimal.

C	n	m	CPU-TIME		NUMBER OF MATCHINGS	
			KI AVERAGE	FR AVERAGE	KI AVERAGE	FR AVERAGE
1000	100	2475	19.560	16.752	43.10	38.60
		1237	17.560	18.799	64.30	69.20
		825	9.918	11.574	48.80	56.90
		618	12.499	11.641	69.90	68.00
10000	100	2475	19.937	22.003	45.20	50.50
		1237	21.655	21.448	81.40	81.70
		825	16.523	16.458	77.00	78.80
		618	12.901	13.098	76.40	80.50

Table 12: Average CPU-time and number of matchings for solving (DRK)

C	n	m	DUALITY-GAP	
			KI AVERAGE	FR AVERAGE
1000	100	2475	3.554	3.718
		1237	3.874	3.261
		825	2.045	2.059
		618	2.136	4.612
10000	100	2475	0.996	0.891
		1237	4.615	4.108
		825	1.648	1.378
		618	3.234	3.092

Table 13: Average error (gap) in percentage of the upper bound

C	n	m	CPU-TIME AVERAGE	TRUE ERROR AVERAGE	NUMBER OF MATCHINGS AVERAGE
1000	100	2475	39.791	0.018	1121.90
		1237	21.633	0.026	795.40
		825	26.451	0.009	980.40
		618	18.542	0.012	707.50
10000	100	2475	12.615	0.003	330.10
		1237	29.276	0.036	1026.40
		825	18.794	0.009	665.70
		618	11.661	0.026	440.70

Table 14: Results of the k-best approach

7. Conclusions

Although, from a theoretical point of view, Lagrangean decomposition may outperform Lagrangean relaxation for solving restricted combinatorial optimization problems, our results show that for (MPS) the Lagrangean relaxation seems to be the better method from a numerical point of view. The reason for this fact could be the size and the shape of the subgradients, which are of high dimension in the case of decompostion. The low CPU-time and the quality of the combined relaxation and k-best approach indicates that this is an effective method for solving matching problems with knapsack side constraints. Also for (MPS) the possible improvement of the lower bound via decompostion over relaxation could not be materialized. This was a motivation not to apply decomposition in an MPS–based approach for solving a class of special routing problems but to use relaxation (cf. Derigs and Metz [1991]).

References

Ball M.O., Benoit H. [1988]: A Lagrangean Relaxation Based Heuristic for the Urban Transit Crew Scheduling Problem. *Lecture Notes in Economics and Mathematical Systems 308, 54-67.*

Ball M.O., Derigs U., Hilbrand C., Metz A. [1990]: Matching Problems with Generalized Upper Bounds Side Constraints. *Networks, Vol. 20, 703-721.*

Camerini P., Vercellis C. [1981]: The Matroidal Knapsack: A Class of (often) Well-Solvable Problems. *Operations Research Letters 3, 157-162.*

Derigs U. [1985]: Some Basic Exchange Properties in Combinatorial Optimization to Construct the k–best Solutions. *Discrete Appl. Math. 11, 129-141.*

Derigs U., Metz A. [1990]: On the Construction of the Set of k-Best Matchings and their use in Solving Constrained Matching Problems. *Working Paper.*

Derigs U., Metz A. [1991]: A Matching Based Approach for Solving the Delivery/Pick-up Vehicle Routing Problem with Time Windows. *Working Paper.*

Edmonds J. [1965]: Maximum matching and a polyhedron with $0,1$ vertices. *Journal of Research of the National Bureau Standards 69B, 125-130.*

Fisher M.L. [1981]: The Lagrangean Relaxation Method for Solving Integer Programming Problems. *Management Sci. 27, 1-18.*

Fisher M. [1987]: System Design for Express Airlines. *Doctorial Dissertation in Flight Transportation, Massachusetts Institute of Technology.*

Geoffrion A. [1974]: Lagrangean Relaxation and its Use in Integer Programming. *Mathematical Programming Study 2, 82-114.*

Guignard M., Kim S. [1987]: Lagrangean Decomposition: A Model yielding stronger Lagrangean Bounds. *Mathematical Prog. 39, 215-228.*

Jörnsten K., Näsberg M. [1986]: A new Lagrangean Approach to the generalized Assignment Problem. *European Journal of Operational Research 27, 313-323.*

Kim S. [1985]: Primal Interpretation of Lagrangean Decomposition. *Department of Statistics Report 72, University of Pennsylvania.*

Martello S., Toth P. [1988]: A new Algorithm for the 0-1 Knapsack Problem. *Management Sci. 34, 633-644.*

Nemhauser G.L., Weber G. [1979]: Optimal Set Partitioning, Matching and Lagrangean Duality. *Naval Res. Logistics Q.26, 553-563.*

Poljak B.T. [1967]: A General Method for Solving Extremum Problems. *Soviet. Math. Dokl. 8, 593-597.*

Zowe J. [1987]: Optimization with Nonsmooth Data. *OR Spektrum 9, 195-201.*

Optimization under functional constraints
(semi-infinite programming) and applications

Rainer Hettich, Martin Gugat
Universität Trier, FB IV
Postfach 3825, D-5500 Trier

Abstract: In this paper, we will give an introductory review on theory, methods and applications of semi-infinite programming. In part I, linear problems are considered, which form a subclass of convex optimization. A central topic, therefore, is duality theory, which gives a relation to moment problems. Parametric problems are discussed by means of two examples (rational Chebyshev approximation and approximation of eigenvalues of the Laplacean). In Cahpter II, general nonlinear problems are considered with the emphasis on the theory of reduction to finite problems by use of techniques from parametric optimization. The last section is on numerical methods, where especially the reduction methods as a consequence of the reduction theory in part II are discussed. We finish with an example from robotics.

In this paper, we consider optimization problems of the kind.

(SIP) maximize $f(z)$ subject to $z \in Z$, with the feasible set Z set defined by

$$Z = \{z \mid g(z,t) \leq 0 \, , \, t \in B\}$$

with $B \subset \mathbb{R}^m$ typically a compact set. Thus, a point z is feasible (i.e. $z \in Z$) if $g(z,t)$ as a function of t is less or equal zero on the whole of B. If B is a finite set, then $g(z,t) \leq 0$, $t \in B$, represents a finite number of inequalities, and (SIP) is a common optimization problem. In this paper problems with finitely many constraints will be denoted as "finite optimization problems" in contrast to semi-infinite problems, where B is an infinite set. Due to a number of interesting applications (robotics, control problems, design problems, defect-minimization for boundary and eigenvalue-problems) a growing interest in semi-infinite programming can be observed.

Part I. Linear semi-infinite programming

I.1. Statement of the problem

In the first part of our presentation we consider linear problems, i.e. problems with f and g linear w.r.t. the variable z :

$$f(z) = c^T z \; , \; g(z,t) = a^T(t) z - b(t)$$

with $a : B \to \mathbb{R}^n$, $b : B \to \mathbb{R}$, at least continuous functions of t. In order to avoid technical assumptions in the sequel we tacitly assume all functions to be continuous with all required derivatives existing and continuous.

Our linear semi-infinite problem may be stated as

$$\text{(LSIP)}: \quad \text{maximize } c^T z \text{ subject to}$$
$$a^T(t)z \leq b(t) \quad \text{for all } t \in B$$

Note that the feasible set

$$Z = \{z \mid a^T(t)\, z \leq b(t) \ , \ t \in B\}$$

is defined by linear (continuous) inequalities, i.e. Z is a closed convex set. Thus (LSIP) is a convex programming problem in a special setting. Indeed, we will see that – for instance – we have a duality theory for (LSIP) quite similar to convex programming.

I.2. Applications

Before we give a sketch of the theory of linear semi-infinite programming we will mention some (fields of) applications which already give an impression of the relationship to other mathematical fields:

- Linear Chebyshev approximation (see I.2.1 below)
- Defect minimization methods for boundary value and eigenvalue problems (see I.2.2 for an example)
- Moment problems (see I.4.5 for an example)
- Design-problems (see [13])

I.2.1. Linear Chebyshev approximation

The typical problem of linear Chebyshev approximation is as follows:

Let a linear finite-dimensional subspace V of C[B], $B \subset \mathbb{R}^m$ a compact set, and an additional function $\varphi \in C[B]$ be given. Then we want to determine the element $v^* \in V$

of least distance to φ, measured in the maximum norm, i.e.

$$||v^* - \varphi||_\infty \leq ||v - \varphi||_\infty \ , \ v \in V.$$

Taking a basis $v_1,..,v_N$ of V we write the elements of V as

$$v = \sum_{i=1}^{N} z_i v_i .$$

Introducing an additional variable z_n, n = N+1, which by the requirement

$$|\varphi(t) - \sum_{i=1}^{N} z_i v_i(t)| \leq z_n \ , \ t \in B$$

has the meaning of an upper bound for the approximation error, we obtain as an equivalent formulation of thc problem

Maximize $f(z) = -z_n$ subject to

$$\left.\begin{array}{l} -\sum_{i=1}^{N} z_i v_i(t) - z_n \leq -\varphi(t) \\ \sum_{i=1}^{N} z_i v_i(t) - z_n \leq \varphi(t) \end{array}\right\} t \in B$$

a problem which is easily cast into the form (LSIP). So we see that linear Chebyshev approximation may be considered as a subclass of (LSIP).

I.2.2. Defect minimization method

For illustration let us consider the Dirichlet boundary value problem

$$\Delta u(y) = 0 \ , \ y \in \Omega$$
$$u(y) = \varphi(y) \ , \ y \in B = \partial\Omega$$

with Ω some given region in $\mathbb{R}^m$.

The defect minimization approach is as follows: Choose a finite dimensional subspace

$$V \in \{w \mid w \in C^2(\Omega) \cap C(\bar{\Omega}), \Delta w = 0 \text{ on } \Omega\} \ , \ \bar{\Omega} = \Omega \cup \partial\Omega$$

and approximate the function $\varphi : B \to \mathbb{R}$ by an element in the linear space $V|_B$ w.r.t. the maximum norm on B.

I.e. find $v^* \in V$ such that $v^*|_B$ is the best Chebyshev approximation to φ from $V|_B$

$$||v^* - \varphi||_\infty^B \le ||v - \varphi||_\infty^B.$$

According to I.2.1 this problem can be cast into the form (LSIP). The maximum principle gives for the norm of $v^* - \varphi$ on $\bar{\Omega}$ the bound

$$||v^* - \varphi||_\infty^{\bar{\Omega}} \le ||v^* - \varphi||_\infty^B$$

which follows immediately from the solution of the (LSIP).

We note that this approach (together with computable error bounds) can be applied to large classes of boundary value and eigenvalue problems (see I.5.3 for another example).

I.3 Optimality Conditions

I.3.1 Primal Conditions

To get an idea about the geometry of Z, similar as in finite-dimensional optimization it is important to identify topologically the part of the admissible set where the point in question lies: To this end we look at the active constraints. We define the set of active indices as

$$E(\bar{z}) = \{t \mid a^T(t)\,\bar{z} = b(t)\} \subset B.$$

This set is closed.

It is clear that if $\bar{z}$ is an optimal point for (LSIP), we cannot find a direction in which it is possible to enlarge our function and at the same time stay within the feasible set. The second condition is mainly a statement about the active constraints.

Theorem I.1 (i) Let $\bar{z}$ be an optimal point for (LSIP). Then there is no solution ξ of the linear inequality system

$$c^T\xi > 0,\ a^T(t)\xi < 0,\ t \in E(\bar{z}).$$

On the other hand, if we are sure that there is no feasible direction of ascent, $\bar{z}$ must be an optimal point:

(ii) $\bar{z}$ is optimal if there is no solution ξ of

$$c^T\xi > 0,\ a^T(t)\xi \leq 0\ , t \in E(\bar{z}).$$

For the proof of (i), we must cushion $E(\bar{z})$ with an open neighbourhood where $a^T(t)\xi < 0$. In the (closed) complement we have max $a(t)^T\bar{z} - b(t) < 0$, thus we can find ε such that $\bar{z} + \alpha\xi$ is feasible for $0 \leq \alpha \leq \varepsilon$.

We sec that there is a slight difference between (i) and (ii). Assuming the Managasarian Fromovitz constraint qualification

(MFCQ) : There exists a ξ such that $a^T(t)\xi < 0$ for all $t \in E(\bar{z})$

it can be shown that the conditions in (i) and (ii) are equivalent (see [10])

Theorem I.2. Let (MFCQ) be true. Then $\bar{z}$ is an optimal point if and only if the system of inequalities in (ii) has no solution.

I.3.2. Farkas–Lemma, systems of inequalities

The following theorem extends the well–known Farkas Lemma to systems with infinitely many inequalities. Its geometrical meaning is simply that a point which is not contained in a closed cone can be separated from this cone by a hyperplane.

Theorem I.3. Let $S \subset \mathbb{R}^n$ be given and $K\{S\}$ be the convex cone generated by S.

$$K\{S\} = \{\sum_{\ell=1}^{r} \lambda_\ell s_\ell \mid \lambda_\ell \geq 0\ , s_\ell \in S,\ r < \infty\}$$

Then for every $c \in \mathbb{R}^n$ exactly one of the following alternatives is true:

1. There is a solution ξ of $c^T\xi > 0$, $s^T\xi \leq 0$, $s \in S$.

2. $c \in \mathrm{cl}(K\{S\})$.

The main difference to finitely generated cones is that $K\{S\}$ is not necessarily closed. An example for this is the cone generated by a circle containing the origin in the boundary. The theorem of Caratheodory states that the elements of $K\{S\}$ can be written as linear combinations of $r \leq n$ elements. So if we know that $K\{S\}$ is closed, this is true for c.

I.3.3 Dual conditions; Kuhn–Tucker theorem

Now let $S = \{a(t) \mid t \in B\}$. A sufficient condition for the cone $K\{S\}$ to be closed is that (MFCQ) is valid. A proof can be found in [10].In this situation, the Farkas–Lemma allows us to state a Kuhn–Tucker theorem.

Theorem I.4 Let (MFCQ) be true. $\bar{z}$ is an optimal point for (LSIP) if and only if there exist $r \leq n$ points $t^j \in E(\bar{z})$ and $\lambda_j > 0$ such that

$$c - \sum_{j=1}^{r} \lambda_j a(t^j) = 0$$

I.4. Duality theory

I.4.1 The dual problem ($LSIP_D$)

Remember that our linear SIP

(LSIP) Maximize c^Tz subject to $z \in Z$

$$Z = \{z \mid a^T(t)z \leq b(t) \ , \ t \in B\}$$

is actually a convex programming problem. Therefore we could immediately apply all parts of convex duality theory which are independent of the representation of Z. We will see however, that there are some difficulties to derive a Lagrangian dual and respective strong duality theorems for (LSIP) arising from the representation of Z by an infinite set of (linear) inequalities instead of a finite set of convex constraints, which may result in duality gaps unknown in this form in convex programming (cf. I.4.4). To derive our dual, we start from the finite linear programming case:

If B is a finite set, the usual LP–dual is given by

($LSIP_D$) Find $\mu(t) \in \mathbb{R}$, $\mu(t) \geq 0$, $t \in B$, such that

$$\sum_{t \in B} \mu(t) b(t) \text{ is minimized subject to}$$

$$\sum_{t \in B} \mu(t)\, a(t) = c.$$

Formally, we can extend this to the case of an infinite (compact) B by interpreting

$$\sum_{t \in B} \mu(t) \begin{bmatrix} a(t) \\ b(t) \end{bmatrix} = \int_B \begin{bmatrix} a(t) \\ b(t) \end{bmatrix} d\mu(t)$$

with $\mu(t)$ now in the set $M^+(B)$ of nonnegative regular Borel–measures. Then ($LSIP_D$) is just the so–called moment problem (see also I.4.5). Now, in case that a, b are continuous on the compact B, a theorem of Rogosinski states that the convex cone

$$M_{n+1} = K\{S\} \ , \ S = \{ \begin{bmatrix} a(t) \\ b(t) \end{bmatrix} \mid t \in B\}$$

can also be represented as

$$M_{n+1} = \{\int_B \begin{bmatrix} a(t) \\ b(t) \end{bmatrix} d\mu(t) \mid \mu \in M^+(B)\}$$

Thus, every value of the objective function in ($LSIP_D$) with $\mu \in M^+(B)$ can be realized already with nonnegative μ with finite support supp(μ).

In the sequel, if not stated otherwise, we always consider ($LSIP_D$) with this additional restriction, i.e.

(LSIP_D) Maximize $\sum_{t\in B} \mu(t)b(t)$ subject to $\mu \in Z^D$,

$$Z^D = \{\mu \mid \mu \geq 0,\ \mathrm{supp}(\mu) < \infty;\ \text{and} \sum_{t\in B} \mu(t)a(t) = c\}$$

By v(LSIP) etc. we denote the value of the programming problem in question, i.e.

$$v(\mathrm{LSIP}) = \sup \{c^T z \mid z \in Z\}.$$

In case that $v(\mathrm{LSIP}) < \infty$, we say that (LSIP) is bounded. Note that also in the bounded case a solution $z^* \in Z$ is not guaranteed to exist. If $Z = \emptyset$, then, by definition of the supremum $v(\mathrm{LSIP}) = -\infty$.

For problems (LSIP), (LSIP_D) we have the following theorem.

<u>Theorem I.5</u> (weak duality) : For $z \in Z$, $\mu \in Z^D$

$$\sum_{t\in B} \mu(t)b(t) \geq c^T z \tag{1}$$

<u>Proof</u>. Due to $\mu \in Z^D$ we have

$$c^T z = (\sum_{t\in B} \mu(t)\, a(t))^T z = \sum_{t\in B} \mu(t)\, a^T(t) z.$$

From $\mu \geq 0$ and $z \in Z$ inequality (1) follows. ∎

From this, we immediately obtain

$$v(\mathrm{LSIP}) \leq v(\mathrm{LSIP}_D) \tag{2}$$

the weak duality relation. In case of equality in (2) the problems are said to be in strong duality. In the next section we give a sufficient condition for this. The example in I.4.4 shows that not always we can expect equality in (2).

I.4.2. A result on strong duality

We define the cone

$$MH_{n+1} = K\{M_{n+1} \cup \{e_{n+1}\}\} ,$$

$e_{n+1} = (0,..,0,1)^T \in \mathbb{R}^{n+1}$. We note that if M_{n+1} is closed, then MH_{n+1} is also closed, whereas the converse is not true.

Theorem I.6 (strong duality) Suppose $Z \neq \emptyset$. If MH_{n+1} is closed then $v(LSIP) = v(LSIP_D)$ and $(LSIP_D)$ has a solution μ^* if (LSIP) is bounded.

Proof. If $v(LSIP) = \infty$, then due to the weak duality relation I.4.1(2), $v(LSIP_D) = \infty$, i.e. $Z^D = \emptyset$. Now suppose $v(LSIP) = d$, $-\infty < d < \infty$. This implies that there is no solution of

$$c^T\xi > 0 \, , \, a^T(t)\xi \leq b(t) \, , \, t \in B \tag{1}$$

as otherwise (LSIP) would be unbounded.

The definition of $v(LSIP) = d$ further implies that the following system (2) also has no solution ξ

$$\begin{aligned} c^T\xi - d &> 0 \\ a^T(t)\xi - b(t) &\leq 0, \quad t \in B. \end{aligned} \tag{2}$$

This, however, together with the unsolvability of (1) implies that there is also no solution $\begin{bmatrix} \xi \\ \xi_{n+1} \end{bmatrix}$ of

$$\begin{aligned} c^T\xi + d\xi_{n+1} &> 0 \\ a^T(t)\xi + b(t)\,\xi_{n+1} &\leq 0 \\ \xi_{n+1} &\leq 0 . \end{aligned} \tag{3}$$

By the Farkas–Lemma, this is equivalent to

$$\binom{c}{d} \in \text{cl}(MH_{n+1}) = MH_{n+1}. \tag{4}$$

(4) means that there exists $\mu^* \in Z^D$ and an $\varepsilon^* \geq 0$ such that

$$c = \sum_{t \in B} \mu^*(t)a(t) \ , \ d = \sum_{t \in B} \mu^*(t)b(t) + \varepsilon^*$$

Due to the weak duality relation, ε^* must be zero and μ^* solves $(LSIP_D)$.■

I.4.3. Superconsistency and stability

It is easy to see that the condition that MH_{n+1} is closed is not stable w.r.t. small changes in the problem "data" a(t), b(t). I.e. if we take for instance the maximum norm on C[B], then in general we cannot be sure that MH_{n+1} closed for $\bar{a}$, $\bar{b}$ implies that MH_{n+1} is closed for problems with a,b in a neighborhood of $\bar{a}$, $\bar{b}$. Therefore, small perturbations may destroy strong duality.

Now it can be shown (cf. [10]) that the so-called Slater condition

(SL) There exists an $\bar{x}$ such that $a^T(t)\bar{x} < b(t)$, $t \in B$

is sufficient for M_{n+1} and MH_{n+1} to be closed.

Definition: We call (LSIP) superconsistent if (SL) holds.

Obviously, (SL) and thus superconsistency of (LSIP), is a stable condition in the above sense, and the same holds for the superconsistency of $(LSIP_D)$ in the next definition.

Definition. We call $(LSIP_D)$ superconsistent if $c \in \text{int } M_n$, $M_n = K\{a(t) \mid t \in B\}$.

For a proof of the following theorem see [15]).

Theorem I.7 (i) If one of the problems (LSIP), ($LSIP_D$) is superconsistent, then the level–sets of the other are bounded.
(ii) If one of the problems (LSIP), ($LSIP_D$) is superconsistent and has finite value, then the other has a solution and we have strong duality, i.e. v(LSIP) = v($LSIP_D$).

I.4.4. An example for a duality gap

With $B = [0,1] \cup \{2\}$ define

$$a(t) = \begin{bmatrix} t \\ -t^2 \end{bmatrix} \text{ for } t \in [0,1] \ , \ a(2) = \begin{bmatrix} 1 \\ 0 \end{bmatrix}$$

and

$$b(t) = 0 \text{ for } t \in [0,1] \ , \ b(2) = L \geq 0.$$

Consider the problem of maximizing z_1 subject to $a^T(t)z \leq b(t)$, $t \in B$.

We have

$$M_{n+1} = K\{\{(t, -t^2,0)^T \mid t \in [0,1]\} \cup \{(1,0,L)^T\}\}$$

which is obviously not closed and neither is MH_{n+1}. The feasible region is easily seen to be

$$Z = \{\begin{bmatrix} z_1 \\ z_2 \end{bmatrix} \mid z_1 \leq L, z_2 \geq z_1\}.$$

Thus, the value of our (LSIP) obviously is 0.

On the other hand, the dual problem becomes

$$\min \ \mu(2) \cdot L$$

subject to

$$\Sigma \mu(t) \begin{bmatrix} t \\ -t^2 \end{bmatrix} + \mu(2) \begin{bmatrix} 1 \\ 0 \end{bmatrix} = \begin{bmatrix} 1 \\ 0 \end{bmatrix}.$$

The only feasible μ is given by $\mu(t) = 0$, $t \in [0,1]$, $\mu(2) = 1$ and therefore the value of the dual is L.

I.4.5. An application in numerical integration

Let us consider the one-sided L_1-approximation problem:

Given f, $\varphi_1,..,\varphi_n \in C[0,1]$. Minimize

$$\int_0^1 \mid f(t) - \sum_{i=1}^n z_i \varphi_i(t) \mid dt \tag{1}$$

subject to

$$\sum_{i=1}^n z_i \varphi_i(t) \leq f(t) \quad , \quad t \in [0,1] \tag{2}$$

Due to (2), we can replace (1) by

$$\int_0^1 (f(t) - \sum_{i=1}^n z_i \varphi_i(t))dt = \int_o^1 f(t)dt - \sum_{i=1}^n z_i \int_o^1 \varphi_i(t)dt,$$

Thus we obtain the linear semi-infinite problem

(AP) Maximize $\sum_{i=1}^n c_i z_i \;,\; c_i = \int_o^1 \varphi_i(t)dt$

subject to (2).

The dual becomes

(AP_D) Minimize $\sum_{t \in B} \mu(t)f(t)$ subject to

$$\sum_{t \in B} \mu(t)\varphi_i(t) = c_i \,,\; \mu \geq 0, \mid \text{supp}(\mu) \mid < \infty.$$

(AP) has an interesting interpretation: Suppose that $\text{supp}(\mu) = \{t_1,.., t_p\} \subset [0,1]$. Then the constraints require

$$\sum_{j=1}^p \mu_j \varphi_i(t_j) = \int_o^1 \varphi_i(t)dt \;,\; i=1,..,n \tag{3}$$

with $\mu_j = \mu(t_j)$ and the object is to minimize

$$\sum_{j=1}^{p} \mu_j f(t_j)$$

subject to (3). I.e. we are looking for points t_j and weights μ_j such that the numerical integration formula

$$\int_0^1 f(t)\,dt \doteq \sum_{j=1}^{p} \mu_j f(t_j)$$

becomes exact on the space spanned by $\varphi_1,..,\varphi_n$. This reminds of Gauss–quadrature and indeed, for n even, $\varphi_i = t^{i-1}$, $i = 1,..,n$, and any function $f \in C^n[0,1]$, $f^{(n)}(t) > 0, t \in [0,1]$, the unique solution is given with $p = \frac{1}{2}n$ and t_j, μ_j the knots and weights of Gauß–quadrature (cf. [10]).

I.5 Parametric problems

Often in applications it is desirable to know whether the solution of an optimization problem is sensitive w.r.t. small changes in the problem data. Therefore, sensitivity analysis constitutes an important part of optimization theory. Moreover, in some cases, it is required to consider the problem solution over a whole range of some parameter. We start in I.5.1 with rational Chebyshev approximation as an example where it is required to adapt a parameter $\lambda \in \mathbb{R}$ such that the value $v(\lambda)$ of a linear (SIP) depending on λ becomes zero. In I.5.2 we give an algorithm which in some sense requires one–sided derivatives of $v(\lambda)$ and show that under reasonable assumptions these may be computed. The results can be applied to the rational approximation problem. In I.5.3, as a second example of a parametric model, we discuss the approximation of membrane eigenvalues by defect minimization. In this application the computation of all minima of a function $v(\lambda)$ in some given λ–interval is required. $v(\lambda)$ is again the value function of a parametric (SIP).

I.5.1. Example: General rational Chebyshev Approximation

Let $B \subset \mathbb{R}^m$ be a compact region, $\varphi_1,..,\varphi_{n_1}$ and $\psi_1,..,\psi_{n_2}$ two systems of linearly independent functions in C(B). With the notation ($z \in \mathbb{R}^n$, $n = n_1+n_2+1$)

$$v(z,t) = \sum_{i=1}^{n_1} z_i \varphi_i(t)$$

$$w(z,t) = \sum_{j=1}^{n_2} z_{n_1+j}\, \psi_j(t)$$

let R be the space of general rational functions given by

$$R = \{r(z,t) = \frac{v(z,t)}{w(z,t)} \mid 1 \le w(z,t) \le \gamma \text{ for } t \in B\}$$

with γ a fixed large number (for instance $\gamma = 10^6$).

Then, given another function $\varphi \in C(B)$ an $\bar{r} \in R$ is sought such that

$$||\varphi - \bar{r}||_\infty^B \le ||\varphi - r||_\infty^B, \quad r \in R.$$

In the same way as for linear Chebyshev approximation we can state this problem in terms of a (SIP):

(RAP) Maximize $f(z) = -z_n$ subject to

$$\left.\begin{array}{l} (\varphi(t) - z_n)\, w(z,t) - v(z,t) \le 0 \\ (-\varphi(t) - z_n)\, w(z,t) + v(z,t) \le 0 \\ 1 \le w(z,t) \le \gamma \end{array}\right\} \quad t \in B$$

Note, that (RAP) would be a linear SIP if z_n would be a constant rather than a variable.

The well-known differential correction algorithms can be derived via the linearization (RAP_ℓ) of (RAP) in a given z^k and variable $\Delta z = z - z^k$:

(RAP_ℓ) Maximize $-\Delta z_n$ subject to

$$\left.\begin{array}{r} (\varphi(t) - z_n^k)\, w(z,t) - v(z,t) \le \Delta z_n\, w(z^k,t) \\ (-\varphi(t) - z_n^k) w(z,t) + v(z,t) \le \Delta z_n\, w(z^k,t) \\ 1 \le w(z,t) \le \gamma \end{array}\right\} \quad t \in B$$

To get rid of the dependence on $r(z^k,\cdot)$ a simplified problem (RAP_ℓ') is considered, obtained by replacing $w(z^k,t)$ in the right hand sides by their lower bound 1.

Then the so-called Original Differential Correction Algorithm (ODC) proceeds as follows

Step k. Given z^k. Compute a solution z^* of (RAP_ℓ) and let

$$z_i^{k+1} = z_i^*,\ i=1,..,n_1 + n_2\ ;\ z_n^{k+1} = ||\varphi - r(z^*,\cdot)||_\infty^B$$

and the Differential Correction Algorithm (DC) is obtained by considering (RAP_ℓ') instead of (RAP_ℓ). It is known that (DC) converges globally with a linear rate. Convergence of (ODC) can be proved only locally but with a superlinear rate in case of a unique solution of (RAP), as has been shown in 1988 by Cheney and Powell.

The following algorithm not only combines these convergence properties but can be shown to converge superlinearly under a much weaker assumption.

Replacing z_n^k, Δz_n in (RAP_ℓ') by λ, z_n we obtain

(LRAP(λ)) Maximize $-z_n$ subject to

$$\left.\begin{array}{r} (\varphi(t) - \lambda)\, w(z,t) - v(z,t) \le z_n \\ (-\varphi(t) - \lambda) w(z,t) - v(z,t) \le z_n \\ 1 \le w(z,t) \le \gamma \end{array}\right\} \quad t \in B$$

a parametric linear SIP. For γ sufficiently large the following can be shown by elementary calculations (cf [11]).

Theorem I.8 (i) LRAP(λ) has a solution for every $\lambda \in \mathbb{R}$.

(ii) For $\bar{\lambda} \in \mathbb{R}$ arbitrary let (with $v(\lambda)$ the value of (LRAP(λ)))

$$\ell_1(\lambda) = v(\bar{\lambda}) - (\lambda - \bar{\lambda})$$

$$\ell_2(\lambda) = v(\bar{\lambda}) - \gamma(\lambda - \bar{\lambda}) .$$

Then, for all $\lambda \in \mathbb{R}$ we have

$$\min \{\ell_1(\lambda),\ell_2(\lambda)\} \le v(\lambda) \le \max \{\ell_1(\lambda),\ell_2(\lambda)\}$$

(iii) v is strictly monotonous decreasing and Lipschitz continuous on $\mathbb{R}$

(iv) $v(\lambda) = 0$ if and only if $\lambda = \lambda^* = ||\varphi - r(z^*,\cdot)||_\infty$ with z^* the solution of (RAP) and (LRAP(λ^*)).

Remark: (ii) means $v(\lambda)$ remains in the double-cone defined by the lines ℓ_1, ℓ_2 and this is the case for every $\bar{\lambda} \in \mathbb{R}$. From this (iii) is obvious.

Theorem I.8 shows, that the rational approximation problem can be reduced to the problem of determining a zero of the value function $v(\lambda)$ of a linear parametric SIP of type

(LSIP(λ)) Maximize $c^T(\lambda)z$ subject to

$$a^T(\lambda,t)z \le b(\lambda,t) \quad , \quad t \in B$$

An efficient procedure to solve this problem is presented in the next section.

I.5.2 A generalized Newton-method for determining zeros of $v(\lambda)$

We start with the following problem:

Let a continuous function $f : \mathbb{R} \to \mathbb{R}$ and $\lambda^* \in \mathbb{R}$ be given such that the one sided derivatives $f'_+(\lambda^*)$, $f'_-(\lambda^*)$ exist and have nonzero values.

Consider the iteration

$$\lambda^{i+1} = \lambda^i - \frac{f(\lambda^i)}{s_i} \tag{1}$$

with numbers $s_i \neq 0$. (For differentiable f, $s_i = f'(\lambda^i)$ would yield Newton's method). The proof of the following theorem is almost trivial.

Theorem I.9 Suppose we have

(i) $\{\lambda^i\}$ generated by (1) converges to λ^*

(ii) For every subsequence $\{\lambda^{i_j}\}$ with $\lambda^{i_j} \leq \lambda^*$ $(\lambda^{i_j} \geq \lambda^*)$ the corresponding s_{i_j} converge to $f'_-(\lambda^*)$ $(f'_+(\lambda^*))$.

Then

$$\frac{\lambda^{i+1} - \lambda^*}{\lambda^i - \lambda^*} \to 0 ,$$

i.e. the sequence $\{\lambda^i\}$ converges superlinearly.

Proof. We have

$$\lambda^{i+1} - \lambda^* = \lambda^i - \lambda^* - \frac{f(\lambda^i)}{s_i}$$

$$= \frac{\lambda^i - \lambda^*}{s_i}\left[s_i - \frac{f(\lambda^i) - f(\lambda^*)}{\lambda^i - \lambda^*}\right]$$

which gives the rest of the proof immediately. ■

In order to apply the above iteration to the problem of finding a zero of the value function $v(\lambda)$ of a parametric linear SIP we must be able to construct s_i such that Assumption (ii) of the theorem is met. This raises the question of the existence and representation of one-sided derivatives of $v(\lambda)$.

To get a feeling of what can be expected, consider first the convex finite parametric problem (a linear SIP is a convex problem!)

$$K(\lambda) \qquad \min \{f(\lambda,z) \mid g^i(\lambda,z) \leq 0 \ , \ i \in I\}$$

With the usual Lagrangian

$$L(\lambda,z,u) = f(\lambda,z) + \sum_{i \in I_o} u_i g^i(\lambda,z)$$

corresponding to a solution z^o of $K(\lambda^o)$ with

$$I_o = \{i \mid g^i(\lambda^o,z^o) = 0\}$$

the Karush–Kuhn–Tucker system

$$L_z(\lambda,z,u) = 0 \tag{2}$$

$$g^i(\lambda,z) = 0 \ , \ i \in I_o \tag{3}$$

combines the Kuhn–Tucker condition with the requirement that constraints for $i \in I_o$ remain active.

Assuming that in a neighborhood of λ^o the solution $z(\lambda)$ of $K(\lambda)$ is differentiable and the set of active constraints is constant, we conclude

$$v'(\lambda) = \frac{d}{d\lambda} f(\lambda,z(\lambda)) = f_\lambda + f_z z_\lambda$$

$$\overset{(2)}{=} f_\lambda - \sum_{i \in I_o} u_i g_z^i z_\lambda \overset{(3)}{=} f_\lambda + \sum_{i \in I_o} u_i g_\lambda^i \ ,$$

i.e. we have

$$v'(\lambda) = L_\lambda(\lambda,z,u).$$

In general we cannot expect that the above assumptions always hold. So $v(\lambda)$ may not be differentiable everywhere. However, assuming Slaters condition and $S_p(\lambda) \neq \emptyset$ ($S_p(\lambda)$ the set of solutions of $K(\lambda)$) and uniformly bounded, it is a standard result in convex parametric programming that $v(\lambda)$ has one sided derivatives given by

$$v'_+(\lambda) = \inf_{z\in S_p(\lambda)} \sup_{u\in S_D(\lambda)} L_\lambda(\lambda,z,u)$$

$$v'_-(\lambda) = \sup_{z\in S_p(\lambda)} \inf_{u\in S_D(\lambda)} L_\lambda(\lambda,z,u), \tag{4}$$

where $S_D(\lambda)$ is the set of solutions of a properly defined dual, which is not relevant for our purposes. Of course, this theory can be applied to (SIP), with the dual now our (SIP_D), but the representation (4) is rather complicated in applications. In [15], the following could be shown.

Theorem I.10 Assume that $P(\lambda^o)$, $D(\lambda^o)$ are both superconsistent. Assume that $\lambda_i \nearrow \lambda^o$ and that $z(\lambda_i) \in S_p(\lambda_i)$, $z(\lambda_i) \to z^o$, and $\mu(\lambda_i) \in S_D(\lambda_i)$, $\mu(\lambda_i) \to \mu^o$ are given. Then

$$v'_-(\lambda^o) = L_\lambda(\lambda^o,z^o,\mu^o)$$

(and $v'_+(\lambda^o)$ is given correspondingly with a sequence $\lambda_i \searrow \lambda^o$). Moreover, $S_p(\lambda)$, $S_D(\lambda)$ are uniformly bounded (and compact) in a neighborhood of λ^o. (Remark: In the above theorem, μ is taken in the dual of C(B) and convergence $\mu(\lambda_i) \to \mu^o$ is in the weak* sense).

Theorem 2 suggests the choice $s_i = L_\lambda(\lambda^i,z^i,\mu^i)$ in the algorithm defined by (i). If we assume that the problem

LSIP(λ) $\quad \min c^T(\lambda)z$ subject to

$$a^T(\lambda,t)\, z \le b(\lambda,t) \ , \quad t \in B$$

and its dual are superconsistent in $\lambda = \lambda^*$ then Assumption (ii) in Theorem I.9 is true when we apply the algorithm to find the zero λ^* of the value function $v(\lambda)$ (assumed that $\lambda^i \to \lambda^*$). I.e. convergence of $\{\lambda^i\}$ implies superlinear convergence.

This result is applied in [11] to the rational approximation problem, yielding an algorithm which is always globally and superlinearly convergent under very weak assumptions (see [11] for details).

I.5.3. Defect minimization methods for eigenvalues of elliptic operators.

Let $G \subset \mathbb{R}^2$ be an open connected region with boundary B. Consider the eigenvalue problem

$$\Delta u + \lambda u = 0 \quad \text{in } G \tag{1}$$
$$u = 0 \quad \text{on } B, \tag{2}$$

which may be used to describe oscillations of a homogeneous elastic membrane stretched in a plane frame of shape B. (The following considerations apply also to more general elliptic problems, cf. [14]).

Now, for $\lambda \in \mathbb{R}_+$ we choose a space

$$V(\lambda) = \text{span } \{\varphi^1(\lambda,..),..,\varphi^N(\lambda,..)\}$$

such that

$$\Delta_x \varphi(\lambda,x) + \lambda\varphi(\lambda,x) = 0 \text{ in } G$$

for every $\varphi(\lambda,.) \in V(\lambda)$, i.e. (1) always holds. Now we look for an element $\varphi(\lambda,.) \in V(\lambda)$, which in additon satisfies the constraint $||\varphi(\lambda,.)|| = 1$, where

$$||\varphi||^2 = \int_G \varphi^2 \, dx \,/ \int_G dx, \tag{3}$$

such that the defect of $\varphi(\lambda^i,.)$ on B is minimized, (i.e. approximate the 0-function on B by $\varphi(\lambda,.) \in V(\lambda)$, $||\varphi(\lambda,.)|| = 1$ in the Chebyshev sense).

A simplified problem is

LSIP(λ) minimize z_n ($n=N+1$) subject to

$$\left.\begin{array}{l} \sum_{j=1}^{N} z_j \varphi^j(\lambda,t) - z_n \leq 0 \\ -\sum_{j=1}^{N} z_j \varphi^j(\lambda,t) - z_n \leq 0 \end{array}\right] \quad t \in B$$

$$-p_1 \leq -1$$

where a normalization $p_1 \geq 1$ is chosen instead of $||\varphi(\lambda,.)|| = 1$ to prevent the zero solution. Then the following error estimate can be proved. Assume that $z(\lambda)$ solves (LSIP(λ)). If we define

$$\varepsilon(\lambda) := \frac{z_n(\lambda)}{||\sum_{j=1}^{N} z_j(\lambda)\varphi^j(\lambda,\cdot)||} < 1,$$

then there exists an eigenvalue $\bar{\lambda}$ of (1) (2) such that

$$\frac{|\lambda - \bar{\lambda}|}{\bar{\lambda}} \leq \varepsilon(\lambda). \tag{4}$$

This follows from general theorems (cf. for instance [14]) which are proved by using maximum principles for solutions of elliptic equations.

This suggests that, in order to obtain approximations of eigenvalues of (1)(2) (together with error-bounds!) local minima of $\varepsilon(\lambda)$ should be computed.

For (1),(2) a possible choice of functions $\varphi^j(\lambda,.)$ are the functions (r,Θ polar coordinates)

$$J\nu(\sqrt{\lambda}\, r) \sin \nu\Theta$$
$$J\nu(\sqrt{\lambda}\, r) \cos \nu\Theta$$

with $J\nu$ Bessel functions of the first kind. For all $\nu \in \mathbb{R}_+$ these functions satisfy $\Delta u + \lambda u = 0$. The corresponding spaces $V(\lambda)$ can be adapted such that singularities in corners of G can be approximated with a given rate (cf. [14] for details) which makes the method especially interesting for regions with (re–entrant) corners and the computation of higher order eigenvalues.

In [5] elliptic membranes are considered (Fig. 1)

Using symmetries it is possible to keep the dimension of $V(\lambda)$ low and to reduce the approximation to a quarter of the ellipse.

Fig. 2. shows $\varepsilon(\lambda)$ on [0,200] for b = 0,7 and the four types of symmetry. Every local minimum corresponds to an eigenvalue of the membrane and all eigenvalues $\lesssim 200$ are identified with high accuracy. It should be mentioned that the corresponding solutions $\sum_{j=1}^{N} z_j \varphi^j(\lambda,X)$ give excellent approximations to the eigenfunctions (a computable error bound in the norm (4) is available).

In [5], b is regarded as an additional parameter allowing the study of the dependence of the frequencies on the excentricity of the ellipse. This requires the consideration of a nonlinear parametric (SIP).

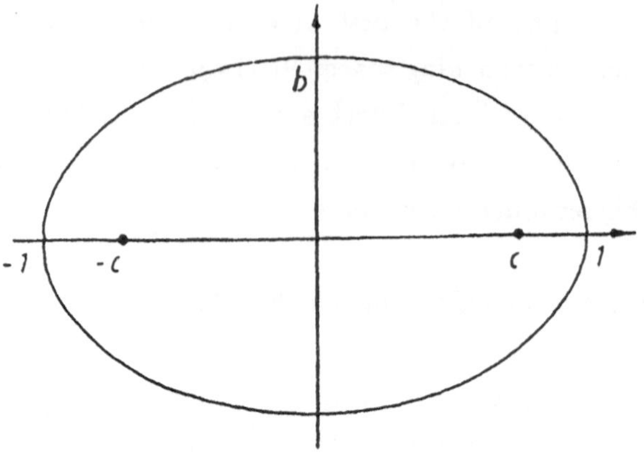

Figure 1

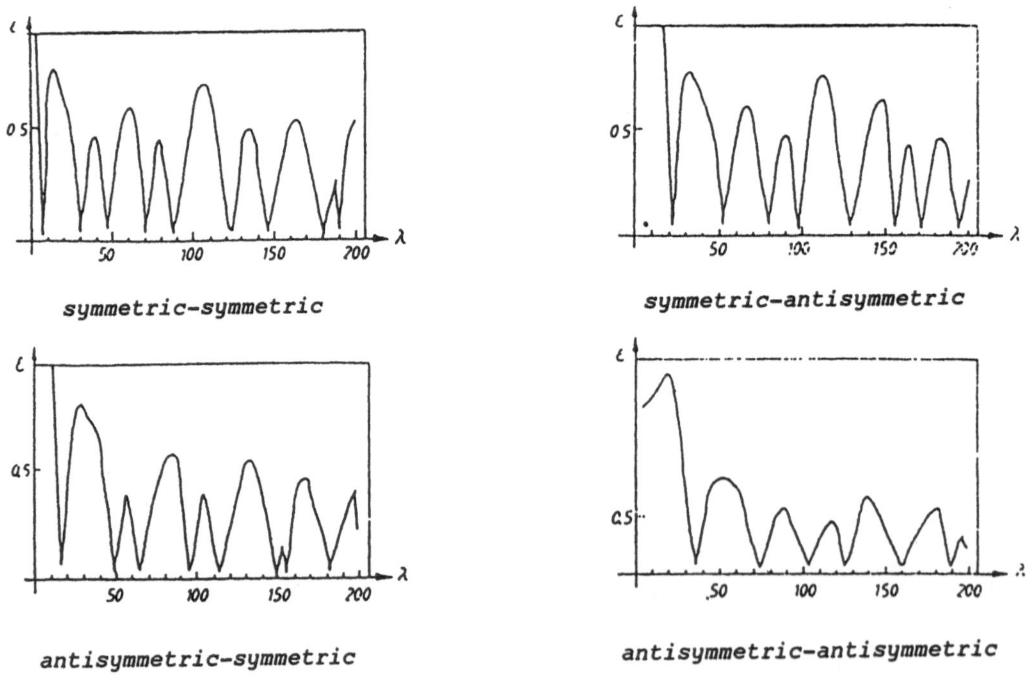

Figure 2

Part II Nonlinear SIP

II.1 Posing the Problem

Again our problem has the form

(SIP) $\max f(z)\ ,\ z \in Z = \{g(\cdot,t) \le 0, t \in B\} \subset \mathbb{R}^n$

with compact $B \subset \mathbb{R}^m$, for example given by

$$B = \{t \mid h^j(t) \le 0\ , j \in J\},\ |J| < \infty$$

and $f, g, h^j \in C^2$. We can express the constraint in a different way, as

$$z \in Z \Leftrightarrow v(z) \le 0$$

with $v(z)$ the value of the problem

$O(z)$ $\max g(z,t)$ subject to $t \in B$

where we consider z as a (fixed) parameter and maximize over the set parametrizing the constraints. This allows us to give (SIP) a simple appeareance:

(SIP) $\max f(z)$ under the constraint $v(z) \le 0$

However, the inconvenience is that the maximum–function v is not differentiable. In the sequel we will show, that this nondifferentiable problem often can be described alternatively by means of a finite number of smooth constraints.

II.2 The Kuhn–Tucker Theorem

We start our considerations with the basic first order optimality condition analoguous to I.3.3.

Given $\bar{z} \in Z$ let the set of active indices again be defined as

$$E(\bar{z}) = \{t \in B \mid g(\bar{z},t) = 0\}$$

The Mangasarian–Fromovitz constraint–qualification is now formulated in terms of the derivative.

(MFCQ) There exists $\xi \in R^n : \xi^T g_z(\bar{z},t) < 0$ for all $t \in E(\bar{z})$.

With that we can state a theorem analogous to the linear case:

Theorem II.1 Let (MFCQ) be true and $\bar{z}$ be an optimal point for (SIP). Then there exists

$$\mu : E(\bar{z}) \to \mathbb{R}_+ \text{ with } |\text{supp}(\mu)| \leq n$$

such that

$$f_z(\bar{z}) - \sum_{t \in E(\bar{z})} \mu(t) g_z(\bar{z},t) = 0$$

This is a system of n equations which is not sufficient to determine all variables $\bar{z}$, $\{t^1,..,t^p\} = \text{supp}(\mu)$ and $\mu(t_i)$. In the next section we will give such a system.

II.3. The Karush–Kuhn–Tucker system for (SIP)

The points $\bar{t}^1,..,\bar{t}^r$ in $E(\bar{z})$ are characterized as local solutions of the problem

$0(\bar{z})$ Maximize $g(\bar{z},t)$ subject to $t \in B$

$$B = \{t \mid h^j(t) \leq 0;\ j \in J\} \subset \mathbb{R}^m$$

Therefore, assuming (MFCQ) in each $\bar{t}^i$, the Kuhn–Tucker condition must hold for every i=1,..,r. Let

$$J_i = \{j \in J \mid h^j(\bar{t}^i) = 0\}.$$

Then there must be multipliers $\bar{\nu}^{i,j} \geq 0$, such that the following equations hold

$$(KKT_i) \qquad g_t(\bar{z},\bar{t}^i) - \sum_{j\in J_i} \bar{\nu}^{i,j} h_t^j(\bar{t}^i) = 0 \qquad (1)$$

$$h^j(\bar{t}^i) = 0, \; j \in J_i \qquad (2)$$

Note that, given $\bar{z}$, (1),(2) forms a system of $m + |J_i|$ equations for the same number of variables $\bar{t}^i \in \mathbb{R}^m$, $\bar{\nu}^{i,j} \in \mathbb{R}$, $j \in J_i$. (1), (2) is called the Karush–Kuhn–Tucker system of $O(\bar{z})$ in $\bar{t}^i$.

To come to a system which – under regularity assumptions – determines not only $\bar{t}^i$, $\bar{\nu}^{i,j}$ but also $\bar{z}$ and the multipliers $\mu(\bar{t}^i) = \bar{\mu}^i$ in the Kuhn–Tucker equations for (SIP) (Section II.2) we simply add the corresponding Karush–Kuhn–Tucker system

$$(KKT) \qquad f_z(\bar{z}) - \sum_{i=1}^{r} \bar{\mu}^i \, g_z(\bar{z},\bar{t}^i) = 0 \qquad (3)$$

$$g(\bar{z}, \bar{t}^i) = 0, \; i=1,..,r \qquad (4)$$

and call the system composed of (KKT), (KKT_i), i=1,..,r, the complete Karush–Kuhn–Tucker system (CKKT) of (SIP) in $\bar{z}$.

Sufficient conditions for (CKKT) to determine uniquely a solution of (SIP) (including $\bar{t}^i$ and multipliers $\bar{\mu}^i,\bar{\nu}^{i,j}$), consisting of the linear independence constraint qualification in $\bar{z}$ and all $\bar{t}^i$ together with strong second order sufficiency conditons, we omit here (cf. [10] for details).

II.4. O(z) as a parametric problem

We have constructed a function v(z), the value of

$$O(z) : \text{Max } \{g(z,t) \mid t \in B\}$$

such that z is feasible if and only if $v(z) \leq 0$. The problem $O(z)$ depends on the parameter z.

Consider again sets B of the form

$$B = \{t \mid h^j(t) \leq 0\,,\, j \in J\},\ |J| \leq \infty.$$

For a fixed point $\bar{z}$ let $t(\bar{z}) \in B$ be a solution of $O(\bar{z})$ and

$$\bar{J} = \{j \mid h^j(t(\bar{z})) \equiv =\}$$

the set of indices belonging to active constraints.

The (KKT) system to $t(\bar{z})$ is

$$g_t(\bar{z},t) - \sum_{j\in\bar{J}} \eta_j h_t^j(t) = 0 \quad \text{and } h^j(t) = 0,\ j \in \bar{J}$$

Varying z as a parameter, we would like to define functions $t(z)$, $\eta_j(z)$ in a neighbourhood of $\bar{z}$ giving a local solution and multipliers of $O(z)$. For this purpose, we use the implicit function theorem. The assumption about the functional matrix of (KKT) with respect to (t,η) has to be checked:

$$F = \begin{pmatrix} L_{tt} & H \\ H^T & 0 \end{pmatrix},\quad L_{tt} = g_{tt} - \sum_{j\in\bar{J}} \eta_j h_{tt}^j\,,\ H = (h_t^j)_{j\in\bar{J}}$$

To apply the theorem we need regularity of F. Sufficient conditions for that are:

(i) H has full rank, i.e. the linear independence constraint qualification (LICQ) holds

(ii) L_{tt} is positive definite on the orthogonal complement of span $\{h_t^j \mid j \in \bar{J}\}$. This is the strong second order sufficiency condition (SSOSC) for $t(\bar{z})$ as a solution of $O(\bar{z})$.

Thus if (LICQ) and (SSOSC) are true for a solution $t(\bar{z})$ of $O(\bar{z})$, then in a neighborhood of $\bar{z}$ there exist paths of solutions $t(z)$, $\eta(z)$ of the (KKT) system of $O(z)$. If in addition g and h^j are two times continuously differentiable, the functions $t(z), \eta(z)$ are in C^1.

A necessary condition for $t(z)$ to be a solution of the original problem $O(z)$, is that $\eta_j(z)$ are nonnegative, i.e. $\eta_j(z) \geq 0, \ j \in \bar{J}$.

If we require strict complementary slackness (SCS) $\eta_j = \eta_j(\bar{z}) > 0$, then this together with (SSOSC) is maintained for the functions $t(z)$, $\eta(z)$ solving the (KKT)-system . In some neighborhood of $\bar{z}$ thus we obtain solutions and Lagrange-multipliers for problem $O(z)$ in this neighborhood. In any case the uniqueness of $\bar{\eta}_j$ is guaranteed by (LICQ).

If we do not have (SCS), the actual solutions $t(z)$, $\eta_j(z)$ of $O(z)$ can still be shown to be Lipschitz-continuous (in contrast to the C^1-solutions of (KKT) and to possess directional derivatives.

We are able to calculate the derivatives of $t(z)$, $\eta_j(z)$ using the (KKT) system $\phi(z, t(z), \eta(z)) = 0$. We have

$$\phi_z + \underbrace{\phi_{t,\eta}}_{F} \begin{bmatrix} t_z \\ \eta_z \end{bmatrix} = 0$$

giving

$$\begin{bmatrix} L_{tt} & H \\ H^T & 0 \end{bmatrix} \begin{bmatrix} t_z \\ \eta_z \end{bmatrix} = -\phi_z = -\begin{bmatrix} g_{zt} \\ 0 \end{bmatrix}$$

and

$$t_z^T L_{tt} \, t_z = -t_z^T \, g_{zt}.$$

For the function $v(z) = g(z,t(z))$ we get the derivative

$$v_z = g_z + g_t\, t_z$$

and $g_t t_z$ vanishes of account of $H^T t_z = 0$, as is seen by means of the (KKT)–System.

For the second derivative we get

$$v_{zz} = g_{zz} + g_{tz} t_z = g_{zz} - t_z^T L_{tt} t_z$$

The last part of the right–hand–side is called shift–term, which is due to the continuously shifting active constraint. It is important for second–order optimality conditions and numerical methods converging superlinear. In addition it needs to be taken into consideration for descent–methods and penalty–functions. If we turn to a discrete problem, the shift term is lost.

II.5. Reduction of SIP to a finite problem

Looking at our restrictions $z \in Z$ we see that in the case considered in II.4 instead of taking all constraints corresponding to $t \in B$ we need only consider those belonging to the local extremes of $g(z,\cdot)$. We have

$$z \in Z \iff g(z,t^\ell(z)) \leq 0 \ , \quad \ell \in L$$

with $\{t^\ell(z)\}$, $\ell \in L$, the set of local solutions of $O(z)$.

If we can follow the paths $t^\ell(z)$ as in II.4 in a neighborhood $\bar{U}$ of $\bar{z}$, we can reduce (SIP) in $\bar{U}$ to the following finite problem.

(SIP_{red}) maximize $f(z)$ subject to

$$v^\ell(z) := g(z,t^\ell(z)) \leq 0 \ , \quad \ell \in L$$

with $v^\ell \in C^2$ under the conditions (LICQ), (SSOSC), (SCS).

III. Numerical methods

Basically there are three approaches to deal with semi–infinite problems numerically.

- Application of methods of nondifferentiable programming to the formulation (cf. II.1)

  $$\max \{f(z) \mid v(z) \leq 0\}$$

  and v(z) the value function of O(z). We will not pursue this approach here (cf. [13]).

- Discretization, i.e. B is replaced by (a sequence of) finite subsets $\bar{B} \subset B$. We will discuss some types of discretization methods in Section III.1.

- Reduction: Replace (SIP) successively by (locally equivalent) problems (SIP_{red}), which may depend on $z \in Z$, and apply methods of finite nonlinear programming (cf. Section III.2).

We close by giving an example in Section III.3 which illustrates the superiority of the latter approach over discretization methods.

III.1. Discretization methods

For $\bar{B} \subset B$, $|\bar{B}| < \infty$, we denote the discretized problem by

(SIP($\bar{B}$)) Maximize f(z) subject to $z \in Z(\bar{B})$,

$$Z(\bar{B}) = \{z \mid g(z,t) \leq 0,\ t \in \bar{B}\}.$$

According to the way $\bar{B}$ is constructed, we distinguish between grid methods (Subsection III.1.1) and exchange methods (Subsection III 1.2).

III.1.1. Grid methods

Let $h \in \mathbb{R}^m$ be a vector of mesh sizes. By G_h we denote a grid in $\mathbb{R}^m$ with mesh size h_j along the j-th coordinate, j=1,..,m, and by B_h

$$B_h = B \cap G_h$$

the part of G_h contained in B.

The most primitive is to choose h "sufficiently small" and to solve (SIP(B_h)). For practical problems this approach is not recommendable, as it requires to solve a (nonlinear) problem with a huge number of constraints which are considered uncorrelated instead of generated by a discretization of a continuous process. In III.3 we give an example where this approach is unacceptable.

More sophisticated (and recommendable) are methods which act on a number of successively refined grids which, however, are thinned out according to information gained from the solution on the preceding ones. The general step of such a method could proceed as follows:

Step ν. Given $\tilde{B}_{h_{\nu-1}}$ and the solution $z_{h_{\nu-1}}$ of (SIP($\tilde{B}_{h_{\nu-1}}$)).

(i) let $h_\nu = h_{\nu-1}/2$

(ii) choose $\tilde{B}_{h_{\nu-1}} \subset B_{h_{\nu-1}}$ such that at least all points $t \in B_{h_\nu}$ with

$$g(z_{h_{\nu-1}}, t) \geq -\eta_\nu$$

are in $\tilde{B}_{h_\nu}$ (η_ν given nonnegative levels).

(iii) compute a solution z_{h_ν} of (SIP($\tilde{B}_{h_\nu}$))

(if z_{h_ν} is not in Z ($\tilde{B}_{h_\nu}$), enlarge $\tilde{B}_{h_\nu}$ and repeat (ii),(iii))

For implementations of algorithms of this form cf [7,8]. In general, solutions on very fine grids can be computed with a moderate cost, as only small portion of the gridpoints has to be taken into acoount.

III.1.2. Exchange methods

Exchange methods are mainly used in the case of convex linearly constrained problems. A general (conceptional) algorithm proceeds as follows:

Step ν. Given $B_{\nu-1} \subset B$, $|B_{\nu-1}| < \infty$. Compute a solution z^ν of $(SIP(B_{\nu-1}))$ and local solutions $t^1,..,t^{r_\nu}$ of

$$O(z^\nu) \qquad \max \{g(z^\nu,t) \mid t \in B\}.$$

Stop, if $g(z^\nu,t^\ell) \leq \varepsilon$ (ε some chosen error bound), $\ell = 1,..,r_\nu$. Otherwise choose

$$B_\nu \subset B_{\nu-1} \cup \{t^1,..,t^{r_\nu}\}.$$

Examples

- Remes algorithms in linear Chebyshev approximation
- Cutting plane methods in convex programming
- Simplex-type methods in linear programming

Apart from special cases (with strongly unique solution) the convergence is rather slow. Proof of convergence is easy if $B_\nu = B_{\nu-1} \cup \{t^1,..,t^{r_\nu}\}$ is chosen.

III.2. Reduction methods

In Section II.5 we have seen that, assuming proper regularity conditions in a point $z \in \mathbb{R}^n$ in a neighborhood of $\bar{z}$ (SIP) can be replaced by a finite problem

(SIP_{red}) Maximize f(z) subject to

$$v^{\ell}(z) := g(z,t^{\ell}(z)) \leq 0, \ \ell \in L, \ |L| < \infty$$

with smooth functions v^{ℓ} and $t^{\ell}(z)$ local solutions of

$O(z)$ Maximize $g(z,t)$ subject to $t \in B$.

Assuming for simplicity that such a reduction is possible everywhere, conceptionally a reduction algorithm proceeds as follows:

Step ν. Given z^{ν}.

(i) Determine all local solutions $t^j(z^{\nu})$, $j = 1,..,r$ of $O(z^{\nu})$

(ii) Apply k_{ν} steps of a nonlinear programming algorithm to (SIP_{red}) ("follow" $t^{\ell}(z)$). Denote the iterates as $z^{\nu,1},..,z^{\nu,k_{\nu}}$

(iii) $z^{\nu+1} = z^{\nu,k_{\nu}}$.

The favourite methods to be used in (ii) are sequential quadratic programming (SQP) methods with quasi Newton approximation to the Hessian of the Lagrangian. If $t^j(z^{\nu})$ are computed sufficiently accurate, a superlinear rate of (local) convergence can be shown under standard assumptions. Global convergence proofs require assumptions. on the existence of the reduced problem along a piecewise linear path connecting all iterates (cf. [4] for details).

III.3. Example: Path tracking in Robotics

We close the paper with some numerical results on a problem from robotics which demonstrate the superiority of (SIP) methods over a sheer discretization approach.

Mathematically a robot can be described by a number of controllable robot–coordinates Θ_i, i =1,..,R, which may be angles in joints or lenghts of links. Suppose that a path $\Theta(\tau)$, $\Theta = (\Theta_1,..,\Theta_R)^T$, $\tau \in [0,1]$, is given. Then, a reparametrization $t = h(\tau)$ to time t is sought such that $\tilde{\Theta}(t) := \Theta(h^{-1}(\tau))$ becomes an executable movement. This means, that accelerations velocities etc. satisfy restrictions derived from the dynamic equations imposed by security requirements etc. In addition $h(\tau)$ is to be chosen such that the performance time is minimized.

Essentially, the above is an optimal control problem with rather complicated control and state restrictions. Dynamic programming methods have been suggested to solve this problem, which suffer however from a high computational cost even under low requirements on the accuracy.

Alternatively, in [12] a direct approach is proposed which restricts $h(\tau)$ to a (finite dimensional) cubic spline–space. In that way, the problem can be cast into the form of a nonlinear (SIP) with 6 R+1 functional constraints

$$g^i(z,\tau) \leq 0 \ , \ \tau \in [0,1] \ , \ j = 1,..,6R+1,$$

4 R of these nonlinear (see [5,12] for details). The unknowns z denote the n=N+4 spline coefficients where N is the number of interior knots in [0,1].

Discretization on for instance 4n points in [0,1] already for R=3, N=7 leads to a nonlinear programming problem with 4(n+4) (6 R+1) = 836 constraints and for N = 63 with 5=92 constraints. Common nonlinear programming algorithms (such as SQP etc.) perform very slowly on problems of this size.

Using the method given in III.2 a typical run for a robot with R=3 to obtain an accuracy of approximately 6 figures is accounted in Table 1 (the result for higher number of knots is obtained by taking the results for lower number of knots as starting points)

N	7	15	31	63
It	17	11	9	8
#CØ	87	57	75	135
#C max	196	92	120	245

Table 1

N: number of interior knots

It: number of SQP iterations

#CØ: average number of constraints in the successive reduced problems (SIP_{red})

#Cmax: the maximal number of constraints in one of the reduced problems (SIP_{red})

The improvement compared with discretization is obvious: the problems to be dealt with by the nonlinear programming method have much less constraints (for instance 5092 compared with a maximum of 245 for N = 63).

References

[1] E.J. Anderson and A.B. Philpott (eds), *Infinite programming,* Proc., Cambridge, 1984, Lecture Notes in Economics and Mathematical Systems 259, Springer-Verlag, Berlin–Heidelberg–New York–Tokyo, 1985

[2] A.V. Fiacco and K.O. Kortanek (eds), *Semi–infinite programming and applications,* Intern. Symp., Austin, TX, 1981, Lecture Notes in Economics and Mathematical Systems 215, Springer–Verlag, Berlin–Heidelberg–New York — Tokyo, 1983

[3] K.G. Glashoff and S.Å. Gustafson, *Linear optimization and approximation,* Springer–Verlag, Berlin–Heidelberg–New York, 1983

[4] G.M. Gramlich, *SQP–Methoden für semi–infinite Optimierungsprobleme,* Thesis (University of Trier), 1990

[5] E. Haaren–Retagne, K.O. Kortanek and M. Ries, *Robot trajectory planning with semi–infinite programming, to appear*

[6] R. Hettich (ed.), *Semi–infinite programming,* Proc. of a workshop, Bad Honnef, 1978, Lecture Notes in Control and Information Sciences 15, Springer— Verlag, Berlin–Heidelberg–New York–Tokyo, 1979

[7] ——, *An implementation of a discretization method for semi–infinite programming,* Mathematical Programming 34 (1986), pp 354–361

[8] —— and G. Gramlich, *A note on an implementation of a method for quadratic semi–infinite programming,* Mathematical Programming 46 (1990) pp 249–254

[9] ——, E. HAAREN, M. RIES AND G. STILL, *Accurate numerical approximations of eigenfrequencies and eigenfunctions of elliptic membranes,* ZAMM 67 (1987),pp 589–597

[10] —— AND P. ZENCKE, *Numerische Methoden der Approximation und semi-infiniten Optimierung,* Teubner Studienbücher Mathematik, Stuttgart, 1982

[11] ——, ——, *An algorithm for general rational Chebyshev approximation,* SIAM J. Numer. Anal. 27 (1990), pp 1024–1033

[12] S.P.MARIN, *Optimal parametrization of curves for robot trajectory design,* IEEE Trans. on Automatic Control, Vol. AC–33 (1988), pp 209–214

[13] E. POLAK, *On the mathematical foundations of nondifferentiable optimization in engineering design,* SIAM Review, 29 (1987), pp 21–89

[14] G. STILL, *Computable bounds for eigenvalues and eigenfunctions of elliptic differential operators,* Numer. Mathematik 54 (1988), pp 201–223

[15] P. ZENCKE and R. HETTICH, *Directional derivatives for the valuefunction in semi–infinite programming,* Math. Programming 38 (1987), pp 323–340

Vector Optimization: Theory, Methods, and Application to Design Problems in Engineering

Johannes Jahn
Institute of Applied Mathematics
of the University of Erlangen–Nürnberg
Martensstraße 3
8520 Erlangen, Germany

Abstract

This paper gives a short overview on important subjects of vector optimization and it examines present research directions in this area of optimization. In the theory we turn our attention to scalarization, optimality conditions and duality. Concerning the numerical methods we study only the class of interactive methods, a modified method of Polak and a method of reference point approximation. Finally, applications to problems of the design of a sandwich beam and a fluidized reactor-heater system are discussed.

1 Introduction

Vector optimization means minimization or maximization of a vector-valued mapping subject to constraints, or in a more general setting, optimization with ordering cones. In engineering or economics one speaks also of (Edgeworth-) Pareto optimization, multiobjective programming or multiple criteria decision making.

1.1 Problem Formulation

Throughout this paper we have the following standard assumption:

$$\left.\begin{array}{l}\text{Let } S \text{ be a nonempty subset of a real linear space } X;\\ \text{let } (Y,\leq) \text{ be a partially ordered real linear space (let } \leq \text{ denote a reflexive}\\ \text{transitive antisymmetric binary relation);}\\ \text{and let } f: S \to Y \text{ be a given objective mapping.}\end{array}\right\} \tag{1.1}$$

Although various results of this paper can be formulated in a more general setting, we restrict ourselves to the above assumption. The problem under investigation reads

$$\min_{x \in S} f(x). \tag{1.2}$$

Definition 1.1. Let the assumption (1.1) be satisfied. An element $\bar{x} \in S$ is called *minimal solution* of the problem (1.2) if there is no $x \in S$ with

$$f(x) \leq f(\bar{x})$$

and

$$f(x) \neq f(\bar{x}).$$

In this case $\bar{x}$ is also called *Edgeworth-Pareto optimal* or *efficient*. The image $f(\bar{x})$ is called a *minimal element* of the image set $f(S)$.

Maximal solutions are defined similarly. Figure 1 illustrates the set of minimal elements. In general, this set consists of infinitely many points.

Figure 1: Minimal elements of a set $f(S)$ for the natural partial order $\leq$ in $\mathbb{R}^2$.

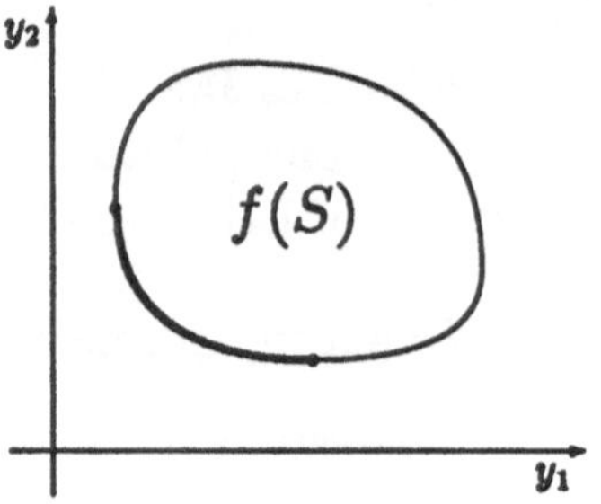

1.2 Historical Retrospect

The first papers in this research area were published by Edgeworth [5] (1881) and Pareto [18] (1896). The actual development of vector optimization begun with papers by Koopmans [15] (1951) and Kuhn-Tucker [16] (1951). An interesting and detailled article on the historical development of vector optimization is published by Stadler [21].

1.3 A Modification of the Minimality Notion

If $\leq$ denotes again the partial order in the linear space Y, then

$$C := \{y \in Y \mid 0 \leq y\}$$

describes a pointed convex cone (i. e., $\lambda C \subset C$ for all $\lambda \in \mathbb{R}_+$, $C + C \subset C$ and $C \cap (-C) = \{0\}$) and conversely, every pointed convex cone induces a partial order. An element $\bar{x} \in S$ is a minimal solution of the problem (1.2) if

$$f(S) \cap (\{f(\bar{x})\} - C) = \{f(\bar{x})\}$$

and

$$f(S) \cap (\{f(\bar{x})\} - (C \setminus \{0\})) = \emptyset,$$

respectively. If, in addition to the assumption (1.1), Y is a topological linear space and the interior $\text{int}(C)$ of C is nonempty, we can weaken the minimality notion if we replace the set $C \setminus \{0\}$ by the set $\text{int}(C)$. This leads to the so-called concept of weak minimality.

Definition 1.2. Let the assumption (1.1) be satisfied. In addition, let Y be a topological linear space, and let $\text{int}(C)$ be nonempty. An element $\bar{x} \in S$ is called a ***weakly minimal solution*** of the problem (1.2) if

$$f(S) \cap (\{f(\bar{x})\} - \text{int}(C)) = \emptyset.$$

In this case the image $f(\bar{x})$ is called a ***weakly minimal element*** of the image set $f(S)$.

Figure 2 illustrates weakly minimal elements of a special set. In applications one is not interested in these elements; they are only interesting from a mathematical point of view.

Remark 1.3. Under the assumptions formulated in Definition 1.2 every minimal solution is also a weakly minimal solution (the converse statement is not true in general, see Figure 2).

There are also other modifications of the minimality notion. For instance, the concept of ***proper minimality*** is used in special cases. At least 13 variants of this notion are known in the literature.

Figure 2: Weakly minimal elements of a set $f(S)$ for $C := \mathbb{R}^2_+$.

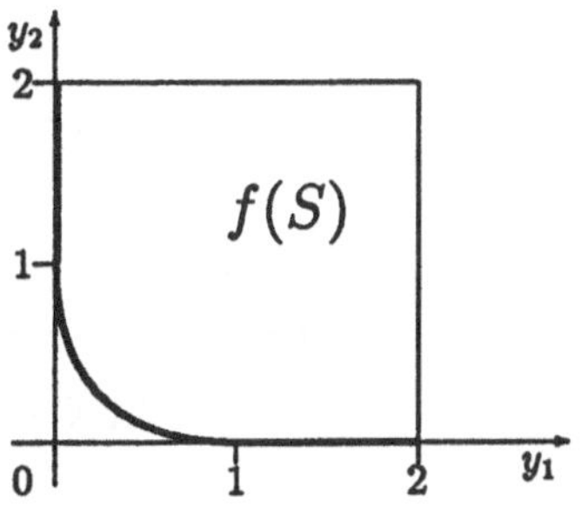

2 Theory

This section is devoted to a few fundamental results in vector optimization. It is the aim to present important principles in the area of scalarization, optimality conditions and duality.

2.1 Scalarization

In general, scalarization means the replacement of a vector optimization problem by a suitable scalar optimization problem which is an optimization problem with a real-valued objective functional. It is a basic principle in vector optimization that optimal elements of a subset of a partially ordered linear space can be characterized as optimal solutions of certain scalar optimization problems.

The standard assumption in this subsection reads as follows:

> Let S be a nonempty subset of a real linear space X;
> let Y be a real topological linear space partially ordered with the aid of a closed pointed convex cone C with nonempty interior $\text{int}(C)$;
> and let $f : S \to Y$ be a given objective mapping. (2.1)

Definition 2.1. Let the assumption (2.1) be satisfied.

(a) For every $y_1, y_2 \in Y$ with $y_1 \leq y_2$ the set

$$[y_1, y_2] := (\{y_1\} + C) \cap (\{y_2\} - C)$$

is called the *order interval* between y_1 and y_2.

(b) For every $a \in \text{int}(C)$ let $\|\cdot\|_a$ denote the Minkowski functional of the order interval $[-a, a]$, i. e.

$$\|y\|_a := \inf\{\lambda > 0 \mid \tfrac{1}{\lambda} y \in [-a, a]\} \text{ for all } y \in Y.$$

It can be seen easily that $\|\cdot\|_a$ is indeed a norm (see Figure 3), since the order interval $[-a, a]$ is absolutely convex, absorbing and algebraically bounded.

The following result gives a complete characterization of minimal and weakly minimal elements with the aid of the parametric norm $\|\cdot\|_a$. This theorem clarifies the relationship between vector optimization and approximation theory.

Figure 3: Illustration of the order interval $[-a, a]$.

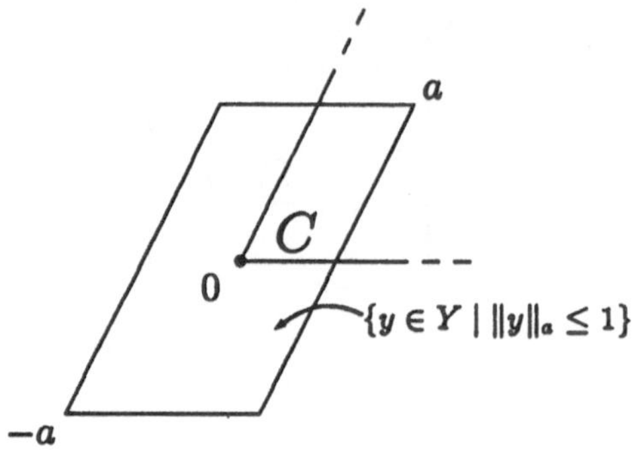

Theorem 2.2 *([9]). Let the assumption (2.1) be satisfied. Moreover, let an element $\hat{y} \in Y$ be given with the property*

$$f(S) \subset \{\hat{y}\} + \text{int}(C). \tag{2.2}$$

(a) An element $\bar{x} \in S$ is a minimal solution of the problem (1.2) if and only if there exists an element $a \in \text{int}(C)$ such that

$$\|f(\bar{x}) - \hat{y}\|_a < \|f(x) - \hat{y}\|_a \text{ for all } x \in S \text{ with } f(x) \neq f(\bar{x}). \tag{2.3}$$

(b) An element $\bar{x} \in S$ is a weakly minimal solution of the problem (1.2) if and only if there exists an element $a \in \text{int}(C)$ such that

$$\|f(\bar{x}) - \hat{y}\|_a \leq \|f(x) - \hat{y}\|_a \text{ for all } x \in S. \tag{2.4}$$

Proof. Let any $\hat{y} \in Y$ with the property (2.2) be chosen.
(a) If $\bar{x}$ is a minimal solution of the problem (1.2), then we have

$$(\{f(\bar{x})\} - C) \cap f(S) = \{f(\bar{x})\}$$

and

$$(\{f(\bar{x}) - \hat{y}\} - C) \cap (f(S) - \{\hat{y}\}) = \{f(\bar{x}) - \hat{y}\}, \tag{2.5}$$

respectively. With the inclusion (2.2) we obtain $\hat{y} - f(\bar{x}) \in -\text{int}(C)$, and we conclude that

$$f(S) - \{\hat{y}\} \subset \text{int}(C) \subset \{\hat{y} - f(\bar{x})\} + \text{int}(C) \subset \{\hat{y} - f(\bar{x})\} + C. \tag{2.6}$$

Consequently, the set equation (2.5) implies

$$(\{\hat{y} - f(\bar{x})\} + C) \cap (\{f(\bar{x}) - \hat{y}\} - C) \cap (f(S) - \{\hat{y}\}) = \{f(\bar{x}) - \hat{y}\}$$

and

$$[-(f(\bar{x}) - \hat{y}), f(\bar{x}) - \hat{y}] \cap (f(S) - \{\hat{y}\}) = \{f(\bar{x}) - \hat{y}\}. \tag{2.7}$$

If we notice the set equation

$$[-a, a] = \{y \in Y \mid \|y\|_a \leq 1\}$$

with $a := f(\bar{x}) - \hat{y}$, then (2.7) is equivalent to the inequality (2.3).

For the converse implication, let for any $a \in \text{int}(C)$ a solution $\bar{x} \in S$ of the inequality (2.3) be given, and assume that $\bar{x}$ is not a minimal solution of the problem (1.2). Then, there exists some $x \in S$ with $f(x) \neq f(\bar{x})$ and $f(x) \in (\{f(\bar{x})\} - C) \cap f(S)$. Consequently, we have

$$f(x) - \hat{y} \in \{f(\bar{x}) - \hat{y}\} - C$$

which implies

$$\|f(x) - \hat{y}\|_a \leq \|f(\bar{x}) - \hat{y}\|_a$$

by the definition of the parametric norm $\|\cdot\|_a$. But this is a contradiction to the inequality (2.3).
(b) Assume that $\bar{x}$ is a weakly minimal solution of problem (1.2). Then the set equation

$$(\{f(\bar{x})\} - \text{int}(C)) \cap f(S) = \emptyset$$

is satisfied, and with (2.6) we get

$$(\{\hat{y} - f(\bar{x})\} + \text{int}(C)) \cap (\{f(\bar{x}) - \hat{y}\} - \text{int}(C)) \cap (f(S) - \{\hat{y}\}) = \emptyset$$

and

$$\text{int}\,([-(f(\bar{x}) - \hat{y}), f(\bar{x}) - \hat{y}]) \cap (f(S) - \{\hat{y}\}) = \emptyset,$$

respectively. But this set equation implies that

$$\{y \in Y \mid \|y - \hat{y}\|_a < 1\} \cap (f(S) - \{\hat{y}\}) = \emptyset$$

for $a := f(\bar{x}) - \hat{y}$. So, the inequality (2.4) is satisfied.

Finally, we prove the converse statement. Let any $a \in \text{int}(C)$ be given, and assume that some $\bar{x} \in S$ solves the inequality (2.4) which is not a weakly minimal solution of problem (1.2). Then there is some $x \in S$ with

$$f(x) \in (\{f(\bar{x})\} - \text{int}(C)),$$

and we get

$$f(x) - \hat{y} \in \{f(\bar{x}) - \hat{y}\} - \text{int}(C).$$

By the definition of the parametric norm $\|\cdot\|_a$ for $a := f(\bar{x}) - \hat{y}$ this implies that

$$\|f(x) - \hat{y}\|_a < \|f(\bar{x}) - \hat{y}\|_a$$

which contradicts the inequality (2.4). □

Notice that, by Theorem 2.2, every minimal or weakly minimal element of a set can be characterized as a solution of a certain approximation problem with a parametric norm (see Figure 4). This result is even true for a nonconvex set. The only requirement formulated by the inclusion (2.2) says that the set $f(S)$ must have a strict lower bound $\hat{y}$.

If the preceding theorem is applied to a multiobjective optimization problem (with $Y = \mathbb{R}^m$ and $C = \mathbb{R}^m_+$) and a problem of the determination of minimal covariance matrices, we get special results summarized in the following two corollaries.

Corollary 2.3 *([4]). Let the assumption (2.1) be satisfied with $X := \mathbb{R}^n$, $Y := \mathbb{R}^m$, $C := \mathbb{R}^m_+$, and assume that there exists some $\hat{y} \in \mathbb{R}^m$ with*

$$\hat{y}_i < f_i(x) \text{ for all } x \in S \text{ and all } i \in \{1, \ldots, m\}.$$

(a) A vector $\bar{x} \in S$ is a minimal solution of the problem (1.2) if and only if there exist positive real numbers $a_1, \ldots, a_m$ such that

$$\max_{1 \leq i \leq m} \left\{ \frac{f_i(\bar{x}) - \hat{y}_i}{a_i} \right\} < \max_{1 \leq i \leq m} \left\{ \frac{f_i(x) - \hat{y}_i}{a_i} \right\} \text{ for all } x \in S \text{ with } f(x) \neq f(\bar{x}).$$

Figure 4: Illustration of the idea for proving Theorem 2.2.

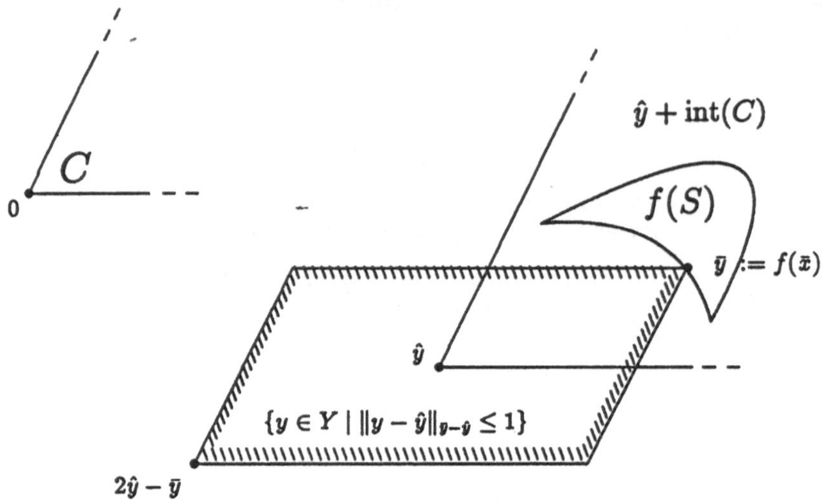

(b) *A vector $\bar{x} \in S$ is a weakly minimal solution of the problem (1.2) if and only if there exist positive real numbers $a_1, \ldots, a_m$ such that*

$$\max_{1 \leq i \leq m} \left\{ \frac{f_i(\bar{x}) - \hat{y}_i}{a_i} \right\} \leq \max_{1 \leq i \leq m} \left\{ \frac{f_i(x) - \hat{y}_i}{a_i} \right\} \text{ for all } x \in S.$$

Proof. This corollary is a direct consequence of Theorem 2.2, if we notice that for $a_1, \ldots, a_m > 0$

$$\|y\|_a = \max_{1 \leq i \leq m} \left\{ \frac{|y_i|}{a_i} \right\} \text{ for all } y \in \mathbb{R}^m.$$

□

It can be shown that the result of the preceding corollary remains also valid without any assumption on $\hat{y} \in \mathbb{R}^m$.

Corollary 2.4. *Let the assumption (2.1) be satisfied with $X := Y := \mathbb{R}^{(n,n)}$ (vector space of symmetric (n, n) matrices with real coefficients), $C := \{A \in Y \mid A \text{ positive semi definite}\}$ and $f := id$ (the identity). Moreover, let S be a set of covariance matrices.*

(a) *A covariance matrix $\bar{B} \in S$ is a minimal element of S if and only if there exists a positive definite matrix $A \in Y$ such that*

$$\sup_{x \neq 0} \left\{ \frac{x^{\mathsf{T}}(\bar{B} + I)x}{x^{\mathsf{T}}Ax} \right\} < \sup_{x \neq 0} \left\{ \frac{x^{\mathsf{T}}(B + I)x}{x^{\mathsf{T}}Ax} \right\} \text{ for all } B \in S \text{ with } B \neq \bar{B};$$

here, I denotes the identity matrix.

(b) *A covariance matrix $\bar{B} \in S$ is a weakly minimal element of S if and only if there exists a positive definite matrix $A \in Y$ such that*

$$\sup_{x \neq 0} \left\{ \frac{x^{\mathsf{T}}(\bar{B} + I)x}{x^{\mathsf{T}}Ax} \right\} \leq \sup_{x \neq 0} \left\{ \frac{x^{\mathsf{T}}(B + I)x}{x^{\mathsf{T}}Ax} \right\} \text{ for all } B \in S;$$

here, I denotes the identity matrix.

Proof. The cone C is pointed, convex and closed and it has a nonempty interior. Since every covariance matrix is positive semi definite we have $S \subset C$ and because of $I \in \text{int}(C)$ we conclude

$$S - \{-I\} = S + \{I\} \subset C + \text{int}(C) = \text{int}(C).$$

So, the inclusion (2.2) is fulfilled for $\hat{y} := -I$. Moreover, we have for every positive definite matrix $A \in Y$

$$\|B\|_A = \sup_{x \neq 0} \left\{ \frac{|x^T B x|}{x^T A x} \right\} \text{ for all } B \in Y.$$

Then we obtain the desired result with Theorem 2.2. □

With the following two theorems we present a special scalarization result which can be used for multiobjective optimization problems.

Theorem 2.5 ([3]). *Let the assumption (2.1) be satisfied with $X := \mathbb{R}^n$, $Y := \mathbb{R}^m$, $C := \mathbb{R}^m_+$, and let $\tilde{x} \in S$ be a given feasible element. If $\bar{x} \in S$ is a solution of the optimization problem*

$$\left.\begin{array}{l} \min \sum_{i=1}^m f_i(x) \\ \textit{subject to the constraints} \\ x \in S \\ f_i(x) \leq f_i(\tilde{x}) \textit{ for all } i \in \{1, \ldots, m\}, \end{array}\right\} \quad (2.8)$$

then $\bar{x}$ is a minimal solution of the problem (1.2). If $\tilde{x} \in S$ is already a minimal solution of the problem (1.2), then $\tilde{x}$ solves also the problem (2.8).

Proof. (a) Let $\bar{x} \in S$ be a solution of the problem (2.8), and assume that $\bar{x}$ is not a minimal solution of the vector optimization problem (1.2). Then there exists some $x \in S$ with $f(x) \neq f(\bar{x})$ and $f(x) \leq f(\bar{x})$. Consequently, we have

$$\sum_{i=1}^m f_i(x) < \sum_{i=1}^m f_i(\bar{x})$$

and

$$f_i(x) \leq f_i(\bar{x}) \leq f_i(\tilde{x}) \text{ for all } i \in \{1, \ldots, m\}$$

which contradicts the assumption that $\bar{x} \in S$ solves the problem (2.8).
(b) If $\tilde{x}$ is already a minimal solution of the vector optimization problem (1.2), then there does not exist some $x \in S$ with $f(x) \neq f(\tilde{x})$ and $f(x) \leq f(\tilde{x})$, i. e. there does not exist some $x \in S$ with

$$\sum_{i=1}^m f_i(x) < \sum_{i=1}^m f_i(\tilde{x})$$

and

$$f_i(x) \leq f_i(\tilde{x}) \text{ for all } i \in \{1, \ldots, m\}.$$

So, $\tilde{x}$ is a solution of the problem (2.8). □

The preceding theorem can be used, if, for instance, the image-uniqueness of a solution of the Chebyshev approximation problem formulated in Corollary 2.3 cannot be checked. But the application of Theorem 2.5 leads to certain difficulties. For instance, if $\tilde{x}$ is already a minimal solution of the vector optimization problem, then the inequality constraints become active which means that they are actually equality constraints. This leads to numerical difficulties in the case of nonlinear problems (the well-known Slater condition is not satisfied).

Theorem 2.6. *Let the assumption (2.1) be satisfied with $X := \mathbb{R}^n$, $Y := \mathbb{R}^m$ and $C := \mathbb{R}^m_+$. An element $\bar{x} \in S$ is a minimal solution of the problem (1.2) if and only if*

$$f_i(\bar{x}) = \min \{f_i(x) \mid x \in S, f_j(x) \leq f_j(\bar{x}) \text{ for all } j \in \{1,\ldots,m\} \setminus \{i\}\} \text{ for all } i \in \{1,\ldots,m\}.$$

Proof. The proof of this theorem goes very much in line with the proof of the preceding theorem. □

For the scalar optimization problem formulated in Theorem 2.6 difficulties may also arise, if $\bar{x}$ is already a minimal solution of the vector optimization problem (1.2) because the Slater condition is not fulfilled. Therefore this simple scalarization result is not suitable for the development of necessary optimality conditions or a duality theory.

Next we present a well-known scalarization result for the convex case. In the following let

$$C^* := \{\ell \in Y^* \mid \ell(c) \geq 0 \text{ for all } c \in C\}$$

denote the *dual ordering cone* (where Y^* denotes the topological dual space of Y), and we define the set

$$C^\# := \{\ell \in Y^* \mid \ell(c) > 0 \text{ for all } c \in C \setminus \{0\}\}.$$

By a theorem of Krein-Rutman (compare Thm. 3.38 in [8]) the set $C^\#$ is nonempty if, in addition, Y is a separable normed space and the ordering cone is closed.

Theorem 2.7. *Let the assumption (2.1) be satisfied, and let the set $f(S) + C$ be convex. Then $\bar{x} \in S$ is a weakly minimal solution of the vector optimization problem (1.2) if and only if there exists some continuous linear functional $\ell \in C^* \setminus \{0\}$ such that*

$$\ell(f(\bar{x})) \leq \ell(f(x)) \text{ for all } x \in S. \tag{2.9}$$

Proof. (a) Let $\bar{x} \in S$ be a weakly minimal solution of the vector optimization problem (1.2). It can be seen easily that $f(\bar{x})$ is then also a weakly minimal element of the set $f(S) + C$, i. e.

$$(\{f(\bar{x})\} - \text{int}(C)) \cap (f(S) + C) = \emptyset.$$

The set $f(S) + C$ is assumed to be convex. Then by a separation theorem there exist a continuous linear functional $\ell \in Y^* \setminus \{0\}$ and a real number $\alpha \in \mathbb{R}$ with

$$\ell(f(\bar{x}) - c_1) \leq \alpha \leq \ell(f(s) + c_2) \text{ for all } s \in S \text{ and } c_1, c_2 \in C.$$

Since C is a cone, we get $\ell \in C^* \setminus \{0\}$ and the first part of the assertion is obvious.

(b) Next we assume that for some $\bar{x} \in S$ and some $\ell \in C^* \setminus \{0\}$ the inequality (2.9) holds. Under the assumption that $\bar{x} \in S$ is not a weakly minimal solution there exists some $\tilde{x} \in S$ with $f(\tilde{x}) \in \{f(\bar{x})\} - \text{int}(C)$. But then we conclude $\ell(f(\tilde{x})) < \ell(f(\bar{x}))$ which is a contradiction to the inequality (2.9). So, $\bar{x} \in S$ is a weakly minimal solution of the problem (1.2). □

The actual importance of the weak minimality notion is based on the complete characterization given in the preceding theorem. For the minimality concept it is not possible to present a complete characterization in analogy to Theorem 2.7.

Figure 5: Illustration of the result of Theorem 2.7.

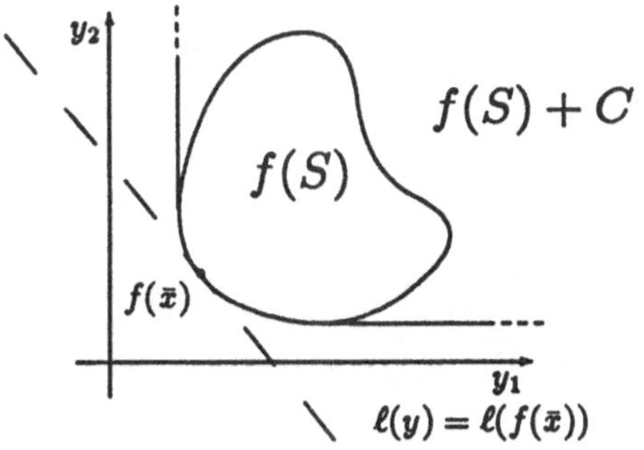

Theorem 2.8. *Let the assumptions of Theorem 2.7 be fulfilled.*

(a) If $\bar{x} \in S$ is a minimal solution of the vector optimization problem (1.2), then there exists a continuous linear functional $\ell \in C^* \setminus \{0\}$ such that

$$\ell(f(\bar{x})) \leq \ell(f(x)) \text{ for all } x \in S.$$

(b) If there exists some $\bar{x} \in S$ and some $\ell \in C^{\#}$ with

$$\ell(f(\bar{x})) \leq \ell(f(x)) \text{ for all } x \in S,$$

then $\bar{x}$ is a minimal solution of the vector optimization problem (1.2).

Proof. (a) If $\bar{x} \in S$ is a minimal solution of problem (1.2), it is also a weakly minimal solution. Then the assertion follows with Theorem 2.7, (a).
(b) Assume that $\bar{x} \in S$ is not a minimal solution of problem (1.2). Then there exists some $\tilde{x} \in S$ with $f(\tilde{x}) \in \{f(\bar{x})\} - (C \setminus \{0\})$. Consequently, we get for every $\ell \in C^{\#}$

$$\ell(f(\tilde{x})) < \ell(f(\bar{x})).$$

So, part (b) of the assertion is proved. □

There is also another possibility in order to formulate a sufficient condition for a minimal solution. For instance, if for some $\bar{x} \in S$ and some $\ell \in C^* \setminus \{0\}$

$$\ell(f(\bar{x})) \leq \ell(f(x)) \text{ for all } x \in S$$

and $f(\bar{x})$ is uniquely determined, the $\bar{x}$ is a minimal solution of the vector optimization problem (1.2) (see Figure 6).

In the last two theorems we used the convexity of the set $f(S) + C$ (in only one direction of the assertion). If for some mapping $f : S \to Y$ this set is convex, then f is also called *convex-like*. For instance, the vector function $f : [\pi, \infty) \to \mathbb{R}^2$ with

$$f(x) = \begin{pmatrix} x \\ \sin x \end{pmatrix} \text{ for all } x \geq \pi$$

is convex-like, but it is not convex.

Figure 6: Illustration of the result of Theorem 2.8.

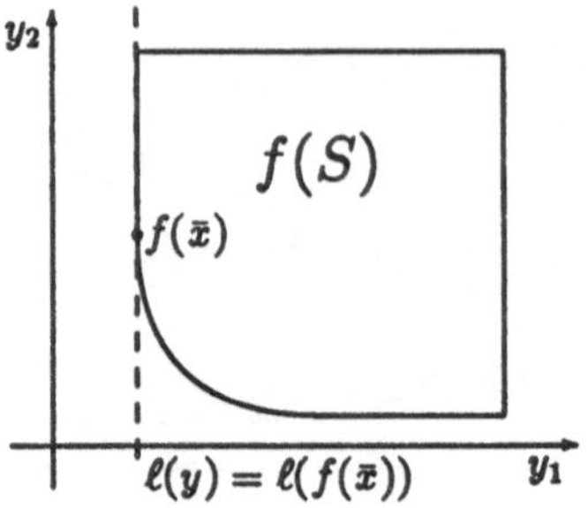

2.2 Optimality Conditions

In this subsection we consider a vector optimization problem with constraints in the form of inequalities and equalities. For such a problem we investigate the question how to formulate appropriate optimality conditions.

The standard assumption reads as follows:

$$\left.\begin{array}{l}\text{Let } X \text{ and } Z_2 \text{ be real Banach spaces;}\\ \text{let } Y \text{ and } Z_1 \text{ be real normed spaces partially ordered by the pointed convex}\\ \text{cones } C_Y \text{ and } C_{Z_1}\text{, respectively;}\\ \text{let } \operatorname{int}(C_Y) \text{ and } \operatorname{int}(C_{Z_1}) \text{ be nonempty;}\\ \text{let } \hat{S} \text{ be a convex subset of } X \text{ with } \operatorname{int}(\hat{S}) \neq \emptyset;\\ \text{let } f: X \to Y,\ g: X \to Z_1 \text{ and } h: X \to Z_2 \text{ be given mappings.}\end{array}\right\} \quad (2.10)$$

Then we consider the vector optimization problem

$$\min_{x \in S} f(x) \quad (2.11)$$

with

$$S := \{x \in \hat{S} \mid g(x) \in -C_{Z_1} \text{ and } h(x) = 0\}.$$

For a weakly minimal solution of this problem we present now a necessary optimality condition generalizing the multiplier rule of Lagrange.

Theorem 2.9. *Let the assumption (2.10) be satisfied, let $\bar{x} \in S$ be a weakly minimal solution of the vector optimization problem (2.11), let f and g be Fréchet differentiable at $\bar{x}$, let h be continuously Fréchet differentiable at $\bar{x}$, and let the image set $h'(\bar{x})(X)$ be closed.*

(a) Then there exist continuous linear functionals $t \in C_Y^*$, $u \in C_{Z_1}^*$ and $v \in Z_2^*$ with $(t, u, v) \neq 0$,

$$(t \circ f'(\bar{x}) + u \circ g'(\bar{x}) + v \circ h'(\bar{x}))(x - \bar{x}) \geq 0 \text{ for all } x \in \hat{S}$$

and

$$(u \circ g)(\bar{x}) = 0.$$

(b) If, in addition, there exists some $\hat{x} \in \operatorname{int}(\hat{S})$ with

$$g(\bar{x}) + g'(\bar{x})(\hat{x} - \bar{x}) \in -\operatorname{int}(C_{Z_1}),$$

$$h'(\bar{x})(\hat{x} - \bar{x}) = 0$$

and $h'(\bar{x})$ is surjective, then the assertion in part (a) holds with $t \neq 0$.

Proof. For the proof of this theorem we refer to the book [11]. □

Of course, it is also possible to formulate suitable sufficient optimality conditions (this will not be done in this paper). Next we investigate the important question whether it is better to use a scalar-valued or a vector-valued Lagrangian. A scalar-valued Lagrangian is defined as $l : X \to \mathbb{R}$ with

$$l := t \circ f + u \circ g + v \circ h$$

where t, u, v are given continuous linear functionals (the Lagrange multipliers), and a vector-valued Lagrangian is defined as $L : X \to Y$ with

$$L := f + U \circ g + V \circ h$$

where $U : Z_1 \to Y$ and $V : Z_2 \to Y$ are given continuous linear mappings. The preceding theorem presents a necessary optimality condition based on a scalar-valued Lagrangian. But the following lemma says in essential that there is an equivalence between the scalar and vector forms of the Lagrangian.

Lemma 2.10. *Let the assumption (2.10) be satisfied, and let f, g and h be Fréchet differentiable at some $\bar{x} \in S$. Then the following two assertions are equivalent:*

(a) There exist continuous linear functionals $t \in C_Y^*$, $u \in C_{Z_1}^*$ and $v \in Z_2^*$ with

$$(t \circ f'(\bar{x}) + u \circ g'(\bar{x}) + v \circ h'(\bar{x}))(x - \bar{x}) \geq 0 \text{ for all } x \in \hat{S} \tag{2.12}$$

and

$$(u \circ g)(\bar{x}) = 0.$$

(b) There exist continuous linear mappings $U : Z_1 \to Y$ and $V : Z_2 \to Y$ with $U(C_{Z_1}) \subset (\text{int}(C_Y) \cup \{0\})$,

$$(f'(\bar{x}) + U \circ g'(\bar{x}) + V \circ h'(\bar{x}))(x - \bar{x}) \notin -\text{int}(C_Y) \text{ for all } x \in \hat{S}$$

and

$$(U \circ g)(\bar{x}) = 0.$$

Proof. First, we assume that the statement (a) is true. By a property of the interior of C_Y (see Lemma 3.21, c) in [8]) there exists some $\tilde{y} \in \text{int}(C_Y)$ with $t(\tilde{y}) = 1$. Then we define the continuous linear mappings $U : Z_1 \to Y$ and $V : Z_2 \to Y$ by

$$U(z_1) = u(z_1)\tilde{y} \text{ for all } z_1 \in Z_1 \tag{2.13}$$

and

$$V(z_2) = v(z_2)\tilde{y} \text{ for all } z_2 \in Z_2.$$

It is obvious that

$$U(C_{Z_1}) \subset (\text{int}(C_Y) \cup \{0\}).$$

Furthermore, we obtain $t \circ U = u$ and $t \circ V = v$. Consequently, the inequality (2.12) can be written as

$$\big(t \circ (f'(\bar{x}) + U \circ g'(\bar{x}) + V \circ h'(\bar{x}))\big)(x - \bar{x}) \geq 0 \text{ for all } x \in \hat{S}.$$

Then we conclude with Theorem 2.7

$$(f'(\bar{x}) + U \circ g'(\bar{x}) + V \circ h'(\bar{x}))(x - \bar{x}) \notin -\text{int}(C_Y) \text{ for all } x \in \hat{S}.$$

Finally, with the equality (2.13) we get

$$(U \circ g)(\bar{x}) = (u \circ g)(\bar{x})\tilde{y} = 0.$$

So, the statement (b) is true.

For the second part of this proof we assume that the statement (b) is true. Then we have

$$(f'(\bar{x}) + U \circ g'(\bar{x}) + V \circ h'(\bar{x}))(x - \bar{x}) \notin -\mathrm{int}(C_Y) \text{ for all } x \in \hat{S}.$$

By Theorem 2.7 there exists a continuous linear functional $t \in C_Y^* \setminus \{0\}$ with the property

$$(t \circ f'(\bar{x}) + t \circ U \circ g'(\bar{x}) + t \circ V \circ h'(\bar{x}))(x - \bar{x}) \geq 0 \text{ for all } x \in \hat{S}.$$

If we define $u := t \circ U$ and $v := t \circ V$, we obtain

$$(t \circ f'(\bar{x}) + u \circ g'(\bar{x}) + v \circ h'(\bar{x}))(x - \bar{x}) \geq 0 \text{ for all } x \in \hat{S}$$

and

$$(u \circ g)(\bar{x}) = (t \circ U \circ g)(\bar{x}) = 0.$$

Furthermore, for each $z_1 \in C_{Z_1}$ it follows

$$u(z_1) = (t \circ U)(z_1) \geq 0$$

which implies $u \in C_{Z_1}^*$. This completes the proof. □

The use of a vector-valued Lagrangian seems to be formally more elegant. But from a practical point of view it is not more helpful because the image sets of the mappings U and V are one-dimensional subspaces of Y.

2.3 Duality

Until today there are many papers describing duality results in vector optimization. We give only a brief overview on a duality theory published in [8]. Since we aim only at the presentation of the main principles, we omit the proofs.

In this section we have the following standard assumption:

$$\left.\begin{array}{l} \text{Let } X \text{ be a real linear space;} \\ \text{let } Y \text{ and } Z \text{ be real topological linear spaces partially ordered by the pointed} \\ \text{convex cones } C_Y \text{ and } C_Z\text{, respectively;} \\ \text{let } \hat{S} \text{ be a nonempty convex subset of } X; \\ \text{let } f : \hat{S} \to Y \text{ and } g : \hat{S} \to Z \text{ be convex mappings;} \\ \text{and let the set } C_Y^{\#} := \{t \in Y^* \mid t(y) > 0 \text{ for all } y \in C_Y \setminus \{0\}\} \text{ be nonempty.} \end{array}\right\} \quad (2.14)$$

Under this assumption the so-called ***primal problem*** reads as follows:

$$\min_{x \in S} f(x)$$

with

$$S := \{x \in \hat{S} \mid g(x) \in -C_Z\}.$$

Next we assign the following ***dual problem*** to this primal problem:

$$\max_{y \in T} y$$

with

$$T := \{y \in Y \mid \exists\, t \in C_Y^{\#}, u \in C_Z^* \text{ with } (t \circ f + u \circ g)(x) \geq t(y)\ \forall x \in \hat{S}\}.$$

For the formulation of duality relations between these two problems we recall the concepts of normality and stability known from scalar optimization.

Definition 2.11. Let the assumption (2.14) be satisfied, and let $\varphi : \hat{S} \to \mathbb{R}$ be a convex functional.

(a) The scalar optimization problem

$$\inf_{x \in S} \varphi(x) \tag{2.15}$$

is called *normal* if

$$\inf_{x \in S} \varphi(x) = \sup_{u \in C_Z^*} \inf_{x \in \hat{S}} (\varphi + u \circ g)(x).$$

(b) The scalar optimization problem (2.15) is called *stable*, if it is normal and the problem

$$\sup_{u \in C_Z^*} \inf_{x \in \hat{S}} (\varphi + u \circ g)(x)$$

has at least one solution.

Normality ensures that there is not a duality gap between the primal and the dual problem. Next we present two duality theorems for the primal and dual problem (the proofs can be found in [8]).

Theorem 2.12 (Strong duality theorem). *Let the assumption (2.14) be satisfied, let $\bar{x} \in S$ be a minimal solution of the primal problem with an associated "supporting" functional $t \in C_Y^\#$ (i. e. $t(f(\bar{x})) \leq t(f(x))$ for all $x \in S$), and let the scalar optimization problem $\inf_{x \in S}(t \circ f)(x)$ be stable. Then $\bar{y} := f(\bar{x})$ is also a maximal solution of the dual problem.*

Theorem 2.13 (Converse duality theorem). *Let the assumption (2.14) be satisfied and, in addition, let Y be locally convex and let $f(S) + C_Y$ be closed. Let the dual problem have a feasible point, and let the scalar optimization problem*

$$\inf_{x \in S} (t \circ f)(x)$$

be normal for all $t \in C_Y^\#$. Then every maximal solution $\bar{y}$ of the dual problem is the image of a minimal solution $\bar{x}$ of the primal problem (i. e. $\bar{y} = f(\bar{x})$).

Notice that in the strong duality theorem special minimal elements are considered. These are minimal elements which can be characterized by a scalarization result based on a continuous linear functional $t \in C_Y^\#$. These elements are closely related to so-called properly minimal elements. In vector optimization it is not possible to give "symmetric" duality results for general convex problems because Theorem 2.12 works with special minimal solutions and Theorem 2.13 works with maximal solutions. The use of the weak minimality notion is also problematic (see [8]) because the primal problem has to be formulated as: Determine weakly minimal elements of the set $f(S) + C_Y$.

2.4 Recent Directions of Research

Further investigations are made on sensitivity and stability ([13], [1]). Optimality conditions for nonsmooth problems can be found in [22]. Zhuang [24] introduced a new concept called super efficiency which is related to the proper minimality notion. Complementary inequalities are obtained in [19] with the aid of special optimization results. A duality theory for set-valued mappings can be found in [17]. The habilitation thesis [23] presents a general theory for nonconvex separation.

3 Methods

In this section we do not give a survey of different classes of numerical methods but we aim only at one special method for the solution of bicriterial nonlinear optimization problems. After the introduction of interactive methods we discuss a modified method of Polak and a method of reference point approximation.

3.1 Interactive Methods

Very often a vector optimization problem is derived from a concrete decision problem, like a design problem in engineering or a problem of decision making in economics. The actual decision problem is certainly not solved with the determination of all minimal solutions of the vector optimization problem. Among all minimal solutions the decision maker has to select a solution which is the subjectively best one.

During the last 20 years so-called *interactive methods* have been developed which make it possible to combine the numerical iteration process with subjective considerations of the decision maker.

An optimal solution computed with the aid of an interactive method is, in a high degree, determined subjectively. Such a method is characterized by a periodic change between an objective computation phase and a subjective decision phase. Until today there are various methods for the interactive solution of vector optimization problems. In the following we restrict ourselves to the method of reference point approximation in the bicriterial case. For the description of this method we study first a modification of a method of Polak.

3.2 A Modified Method of Polak

The standard assumption of this section reads as follows:

$$\left.\begin{array}{l}\text{Let } X := \mathbb{R}^n,\ Y := \mathbb{R}^2,\ C := \mathbb{R}^2_+;\\ \text{let } S \text{ be a nonempty subset of } X;\\ \text{and let } f = (f_1, f_2) : S \to \mathbb{R}^2 \text{ be a given vector function.}\end{array}\right\} \qquad (3.1)$$

Then we investigate the following bicriterial problem:

$$\min_{x \in S} \begin{pmatrix} f_1(x) \\ f_2(x) \end{pmatrix}. \qquad (3.2)$$

For an appropriate approximation of the set M of all minimal solutions of the problem (3.2) (compare [20]) one determines the points

$$a := \min_{x \in S} f_1(x)$$

and

$$b := f_1(\tilde{x}) \text{ with } f_2(\tilde{x}) = \min_{x \in S} f_2(x)$$

on the y_1-axis (see Figure 7). Then one discretizes the interval $[a, b]$ by choosing points

$$y_1^{(k)} := a + k\frac{b-a}{m} \text{ for } k = 0, 1, 2, \ldots, m \text{ (with } m \in \mathbb{N} \text{ fixed).}$$

For every discretization point $y_1^{(k)}$ $(k = 0, 1, 2, \ldots, m)$ one solves the scalar optimization problem

$$\left.\begin{array}{l}\min f_2(x)\\ \text{subject to the constraints}\\ x \in S\\ f_1(x) = y_1^{(k)}.\end{array}\right\} \qquad (3.3)$$

Figure 7: Approximation of $f(M)$ with Polak's method.

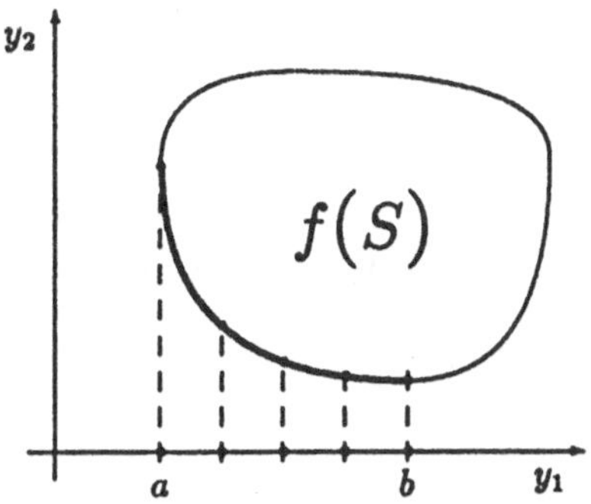

Then one gets points $y^{(k)} := f(x^{(k)})$ (where $x^{(k)}$ is a solution of the optimization problem (3.3)) which belong to the set $f(S)$. But in general it is not ensured that the preimages $x^{(k)}$ are in fact minimal solutions of the vector optimization problem (3.2).

Figure 8: Determination of minimal elements of $f(S)$.

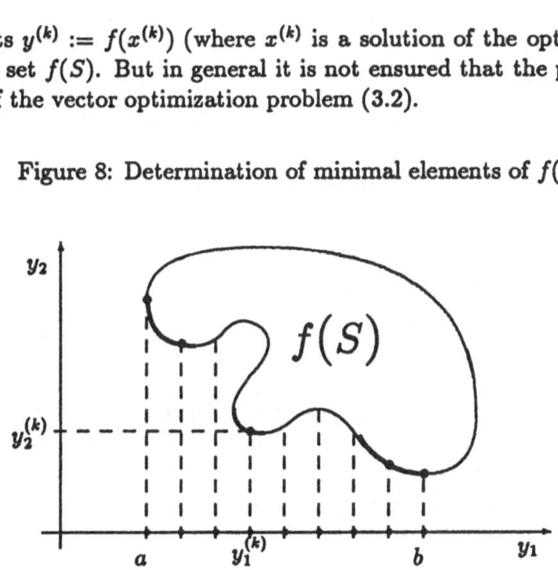

In order to obtain minimal solutions one has to select points such that

$$y_2^{(k_1)} > y_2^{(k_2)} > y_2^{(k_3)} > \cdots$$

describes a strictly monotonically decreasing sequence. So, the set $\{x^{(k_1)}, x^{(k_2)}, \ldots, x^{(k_{m'})}\}$ is an approximation of the set M of minimal solutions (see Figure 8).

For the solution of the scalar optimization problem (3.3) one needs a constrained optimization method, for instance a usual penalty method. Since most of these methods can be used for the calculation of only local minima and not for the determination of global minima, in general the method of Polak cannot be carried out satisfactorily. So, for the solution of practical problems one has to guarantee that the determined "solutions" of problem (3.3) are in fact also global solutions.

3.3 A Method of Reference Point Approximation

We consider again the bicriterial optimization problem (3.2) under the assumption (3.1) and we assume that the set M of all minimal solutions is already computed. Then we suppose that the decision maker can give a reference point $\hat{y} \in \mathbb{R}^2$ which should be approximated in a best way by the image of a minimal solution. In other words, we have to solve the approximation problem

$$\min_{x \in M} \|\hat{y} - f(x)\|$$

where $\|\cdot\|$ denotes a norm in $\mathbb{R}^2$ (e. g., the weighted Chebyshev norm). Notice that this problem is not always solvable because M does not need to be closed. But in the case of the existence of a solution we can present a minimal solution whose image is as close as possible to the subjective expectation of the decision maker. This process can be repeated several times by the decision maker.

The resulting algorithm reads as follows:

Algorithm 3.1 ([10]).

Part I. Computation Phase

Step 1: With the aid of the modified method of Polak compute, by an appropriate discretization, minimal solutions $x^{(0)}, x^{(1)}, \ldots, x^{(q)}$ (with $q \in \mathbb{N}$). Set $M := \{x^{(0)}, \ldots, x^{(q)}\}$.

Part II. Decision Phase

Step 2: The decision maker chooses the weights $t_1, t_2 > 0$ of the weighted Chebyshev norm $\|\cdot\|$ in $\mathbb{R}^2$.

Step 3: The decision maker chooses an arbitrary reference point $y^{(1)} \in \mathbb{R}^2$.

Part III. Computation Phase

Step 4: Set $i := 1$.

Step 5: Determine a point $\bar{x}^{(i)} \in M$ with the property

$$\|y^{(i)} - f(\bar{x}^{(i)})\| = \min\{\|y^{(i)} - f(x)\| \mid x \in M\}. \tag{3.4}$$

Part IV. Decision Phase

Step 6: The point $\bar{x}^{(i)} \in M$ is presented to the decision maker. If the decision maker accepts this point as the subjectively best one, then stop; otherwise proceed in the next step.

Step 7: Based on additional information about the original problem the decision maker proposes a new reference point $y^{(i+1)} \in \mathbb{R}^2$.

Part V. Computation Phase

Step 8: Set $i := i + 1$ and go to step 5.

Notice that the approximation problem (3.4) is always solvable because M is a discrete set. Part I of Algorithm 3.1 is the computer intensive part while the parts II–V can be carried out interactively in a very fast way. So, Algorithm 3.1 is normally applied in such a way that the part I is executed independently from the other parts. The actual interactive method starts then with the elements of the set M.

In step 7 it should be possible to provide the decision maker with all information obtained in the computation phases. For instance, the graphical presentation of the image set of minimal solutions supports the decision maker to select new reference points.

4 Application to Design Problems in Engineering

The method of reference point approximation presented in the previous section is now applied to a design problem in mechanical engineering and to a problem in chemical engineering. We study the optimal design of a sandwich beam and a fluidized reactor-heater system.

4.1 Design of a Sandwich Beam

We consider a sandwich beam consisting of a pitted aluminium core and covered by two aluminium coats. This beam is supported by five steel bars (see Figure 9). This problem is due to Eschenauer-Schäfer [7].

Figure 9: Sandwich beam.

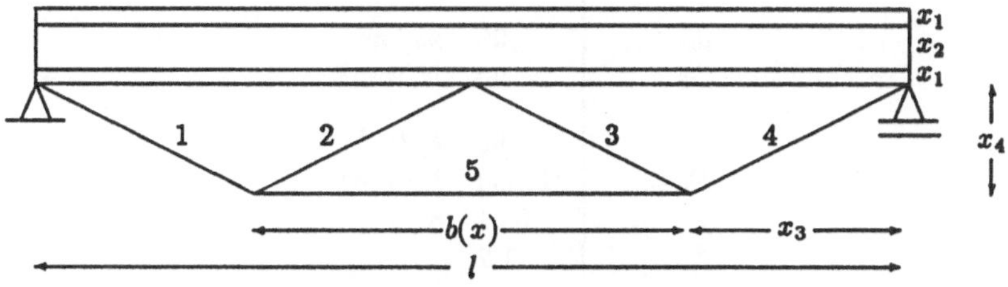

The design variables are

x_1 — thickness of the coat,
x_2 — height of the pitted core,
x_3, x_4— coordinates of the substructure,
x_5 — diameter of the bar no. 5,
x_6 — diameter of the bar no. 1 and no. 4,
x_7 — diameter of the bar no. 2 and no. 3.

It is the aim to minimize the weight f_1 of the whole structure and the deformation f_2 of the beam under its net weight. These two objectives are given as follows

$$\begin{aligned} f_1(x) &= \left[2x_6\sqrt{x_3^2+x_4^2} + 2x_7\sqrt{\frac{(l-2x_3)^2}{4}+x_4^2} + x_5(l-2x_3)\right]\rho_s g f_s \\ &\quad +(2x_1\rho_D + x_2\rho_k)\, b_s g f_s l \end{aligned}$$

and

$$f_2(x) = \sqrt{\frac{2}{3}\left[\max\left\{\omega(\xi,x)\mid 0\le\xi\le\frac{l}{2}\right\}\right]^2 + \frac{1}{3}\omega\left(\frac{l}{2},x\right)^2} \tag{4.1}$$

where

$$\begin{aligned} \omega(\xi,x) &= \frac{H_1(x)}{6B_s(x)}(\xi l^2-\xi^3) + \frac{q_0(x)}{24B_s(x)}(\xi^4-\xi l^3) \\ &\quad -\frac{H_2(x)}{48B_s(x)}\xi l^2 + \frac{q_0(x)}{2G_s(x)}(\xi l-\xi^2) + \frac{H_2(x)}{2G_s(x)}\xi \end{aligned}$$

with

$$\begin{aligned} H_1(x) &= \frac{F_2(x)}{2} + q_0(x)\frac{l}{2} + \frac{2x_4}{l}\left(F_4(x)\frac{x_3}{x_4} - S_5(x)\right), \\ H_2(x) &= F_2(x) + \frac{4x_4}{l}\left(F_4(x)\frac{x_3}{x_4} - S_5(x)\right), \end{aligned}$$

Table 1: Values of the used constants.

constant	value
l	2000 mm
b_s	200 mm
ρ_D	2700 kg/m^3
E_D	70000 N/mm^3
ρ_k	90 kg/m^3
G_k	30 N/mm^2
ρ_s	7860 kg/m^3
E_s	210000 N/mm^2
g	9.81 m/s^2
f_s	1.2
$\sigma_{k_{max}}$	300 N/mm^2
s	1.4

$$F_2(x) = x_7\sqrt{\frac{(l-2x_3)^2}{4} + x_4^2}\,\rho_s g f_s,$$

$$F_4(x) = \frac{1}{2}\left(x_6\sqrt{x_3^2+x_4^2} + x_7\sqrt{\frac{(l-2x_3)^2}{4}+x_4^2} + x_5(l-2x_3)\right)\rho_s g f_s,$$

$$q_0(x) = (2x_1\rho_D + x_2\rho_k)\, b_s g f_s,$$

$$S_5(x) = \frac{1}{e(x)}\left[F_2(x)\left(\frac{x_4 l}{12B_s(x)} + \frac{x_4}{G_s(x)l}\right) - F_4(x)\left(\frac{x_3}{x_4}r + \frac{4(x_3^2+x_4^2)^{3/2}}{Ex_6x_4l^2} + \frac{2x_3^2}{E_DA_Dx_4l}\right)\right.$$
$$\left. +q_0(x)\left(\frac{5x_4l^2}{96B_s(x)} + \frac{x_4}{2G_s(x)}\right)\right],$$

$$e(x) = \frac{8}{El^3}\left(\frac{(x_3^2+x_4^2)^{3/2}}{x_6} + \frac{\left(\frac{(l-2x_3)^2}{4}+x_4^2\right)^{3/2}}{x_7}\right)$$
$$+\frac{4x_3^2}{E_DA_Dl^2} + \frac{x_4^2}{3B_s(x)} + \frac{4x_4^2}{G_s(x)l^2} + \frac{l-2x_3}{Ex_5l},$$

$$B_s(x) = \frac{1}{2}E_Db_sx_1(x_1+x_2)^2,$$

$$G_s(x) = G_k\frac{b_s(x_1+x_2)^2}{x_2}.$$

The values of the constants are listed in Table 1.

The constraints of this problem are given as

$$0.1 \leq x_1 \leq 2$$

$$
\begin{aligned}
10 &\leq x_2 \leq 100 \\
0 &\leq x_3 \leq 990 \\
0 &(\leq) x_4 \leq 990 \\
10 &\leq x_5 \leq 100 \\
10 &\leq x_6 \leq 100 \\
10 &\leq x_7 \leq 100
\end{aligned}
$$

and

$$\frac{|H_1(x)\frac{l}{2} - q_0(x)\frac{l^2}{8}|}{(x_1 + x_2)b_s x_1} + \frac{|-F_4(x)\frac{x_3}{x_4} + \frac{2x_3}{l}\left(F_4(x)\frac{x_3}{x_4} - S_5(x)\right)|}{b_s x_1} \leq \frac{\sigma_{k_{max}}}{s}. \tag{4.2}$$

So, we have to minimize the two functions f_1 and f_2 subject to these constraints. The application of part I of Algorithm 3.1 leads to estimates of minimal elements of the image set of (f_1, f_2) listed in Table 2 (see also [10]).

With the aid of the iterated minimal elements we obtain an approximation of the set of minimal elements illustrated in Figure 10. If one chooses the weights $t_1 = t_2 = 1$ in part II of Algorithm 3.1, one gets estimates as best approximations from the set of minimal elements given in Table 3.

The nonlinear optimization problems arising as subproblems in Algorithm 3.1 were solved by a penalty method where the BFGS method was used for the solution of the unconstrained problems.

Figure 10: Approximation of the set of minimal elements.

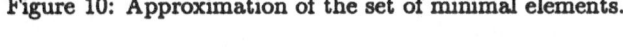

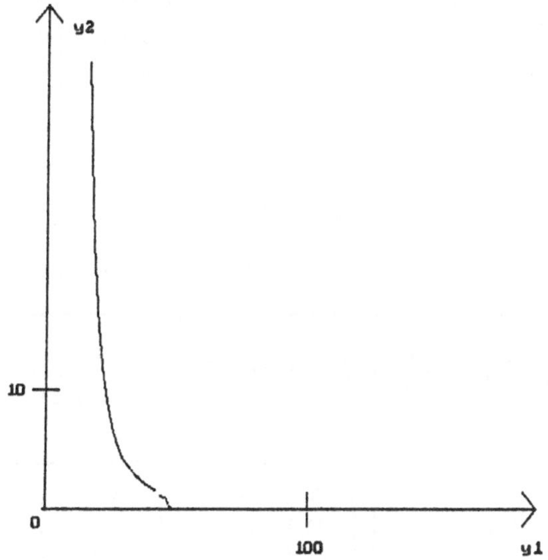

We should notice that this structural optimization problem is in fact a nonsmooth problem (see (4.1) and (4.2)). But nevertheless it is possible to work with the BFGS method. It is well-known that this method is not very sensitive for problems with mainly smooth functions. Of course this statement is based on numerical experience and cannot be generalized.

Table 2: Iterated minimal elements (where the $y_1^{(k)}$'s are chosen non-equidistantly).

$f_1(x^{(k)})$	$f_2(x^{(k)})$	$f_1(x^{(k)})$	$f_2(x^{(k)})$
8.170743	1613.325880	65.485503	0.032772
10.170946	335.297128	68.483001	0.031190
12.173212	129.391623	71.483001	0.029269
14.169546	63.646828	74.483001	0.027133
16.168230	37.121296	77.483001	0.024913
18.170628	22.107509	80.483001	0.022716
20.170631	15.165513	83.482979	0.020618
22.170864	11.005089	86.483001	0.018664
24.168286	8.316745	89.482979	0.016877
26.170787	6.480834	92.485484	0.015261
28.170658	5.187719	92.903848	0.015049
30.170622	4.241161	94.903869	0.014126
32.168918	3.534406	115.455054	0.011496
34.170188	2.988244	118.452682	0.009315
36.169843	2.558842	121.455039	0.008211
38.168907	2.215119	124.455107	0.007615
40.168608	1.935677	126.454850	0.007393
42.168488	1.705571	128.456033	0.007245
44.170746	1.160908	131.455268	0.007108
46.170746	1.006852	134.455584	0.007021
47.485498	0.271472	137.456063	0.006949
49.485507	0.109936	140.456706	0.006872
52.482961	0.054953	143.457513	0.006780
55.480575	0.041851	146.457594	0.006669
58.482930	0.037930	149.394510	0.006540
61.483003	0.034422	151.392035	0.006455
63.485505	0.033628		

Table 3: Compromise solutions.

reference point	(50,10)	(10,10)	(30, 5)	(50, 2)
estimate as best approximation from the set of minimal elements	$\begin{pmatrix} 42.168488 \\ 1.705571 \end{pmatrix}$	$\begin{pmatrix} 18.170628 \\ 22.107509 \end{pmatrix}$	$\begin{pmatrix} 30.170622 \\ 4.241161 \end{pmatrix}$	$\begin{pmatrix} 49.485507 \\ 0.109936 \end{pmatrix}$
preimage of this estimate	$\begin{pmatrix} 1.408978 \\ 11.848378 \\ 989.930847 \\ 0.248698 \\ 10.000272 \\ 10.049928 \\ 10.000246 \end{pmatrix}$	$\begin{pmatrix} 0.466426 \\ 11.755359 \\ 989.930632 \\ 0.248085 \\ 10.000682 \\ 10.071396 \\ 10.000504 \end{pmatrix}$	$\begin{pmatrix} 0.937273 \\ 11.840433 \\ 989.930879 \\ 0.248696 \\ 10.000241 \\ 10.046709 \\ 10.000213 \end{pmatrix}$	$\begin{pmatrix} 0.257025 \\ 97.324846 \\ 0.022438 \\ 989.977789 \\ 10.001214 \\ 10.001056 \\ 10.001630 \end{pmatrix}$

4.2 Design of a Fluidized Reactor-Heater-System

We consider a fluidized reactor-heater-system in the exothermic case consisting of a reactor, a heat exchanger and a cooler (see [12]).

The design variables are

$$x_1 \text{ — extent of reaction,}$$
$$x_2 \text{ — temperature.}$$

It is the aim to minimize the total investment costs f_1 and the net operating costs f_2 of the system. The objective functions read as follows

$$f_1(x_1,x_2) = \begin{cases} p_{01}V(x_1,x_2)^\alpha x_2^\beta + p_{02}x_1^{-\gamma}, & \text{if } x_1 = \varphi(x_2) \\ p_{01}V(x_1,x_2)^\alpha x_2^\beta + p_{02}x_1^{-\gamma} + p_{04} + p_{06}(x_1 - \varphi(x_2))^\delta, & \text{if } x_1 > \varphi(x_2) \end{cases}$$

$$f_2(x_1,x_2) = p_{03}\frac{1}{x_1} - p_{05}(x_1 - \varphi(x_2))$$

where

$$\varphi(x_2) = \frac{p_{52}x_2 - 1}{p_{53} - p_{51}x_2},$$

and

$$V(x_1,x_2) = \frac{p_{61}}{x_1}\int_0^{x_1}\left[55\left(\frac{1-s}{2-s}\right)^2 e^{-\frac{4770}{x_2}} - 0.000014\,\frac{s}{2-s}e^{-\frac{19270}{x_2}}\right]^{-1} ds.$$

The following constants are used:

$$\begin{array}{llll} p_{01}= 1750, & p_{02}= 15000, & p_{03}= 6550, & p_{04}= 3000, \\ p_{06}= 33000, & p_{11}= 63.8, & p_{21}= 0.00036, & p_{31}= 0.85, \\ p_{41}= 0.00121, & p_{51}= 0, & p_{52}= 0.00206, & p_{53}= 2.76, \\ p_{61}= 0.0362, & & & \\ \alpha = \beta = \gamma = \delta = 0.6\,. & & & \end{array}$$

The constraints of this problem can be written in the following way:

$$\begin{aligned} (P_{31}V(x_1,x_2))^{\frac{2}{3}} &\leq \frac{x_2^2}{p_{11}x_1} \\ p_{11}p_{21}(1+x_1) &\leq x_2 \\ x_2 &\leq \frac{1}{p_{41}} \\ (p_{51}x_1 + p_{52})x_2 &\leq 1 + p_{53}x_1 \\ 0 \leq x_1 &\leq 0.99999\,. \end{aligned}$$

An approximation of the set of images of minimal solutions is illustrated in Figure 11 (see [14]), and Figure 12 presents an approximation of the image set $f(S)$ obtained by the images of an appropriate grid of the constraint set. Further numerical results concerning the approximation of suitable reference points can be found in [14].

4.3 Further Investigations of Nonlinear Problems

A complicated vector optimization problem concerning the design of plates was proposed and investigated by Brosowski-Conci [2]. A comprehensive collection of real-world vector optimization problems is published in the book [6] of Eschenauer-Koski-Osyczka.

Figure 11: Approximation of the set of minimal elements ([14]).

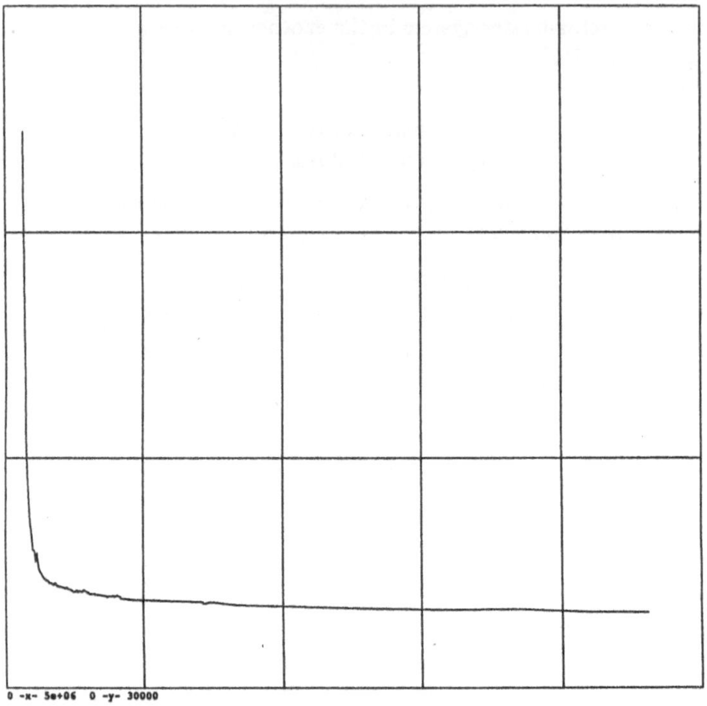

Figure 12: Approximation of the image set $f(S)$.

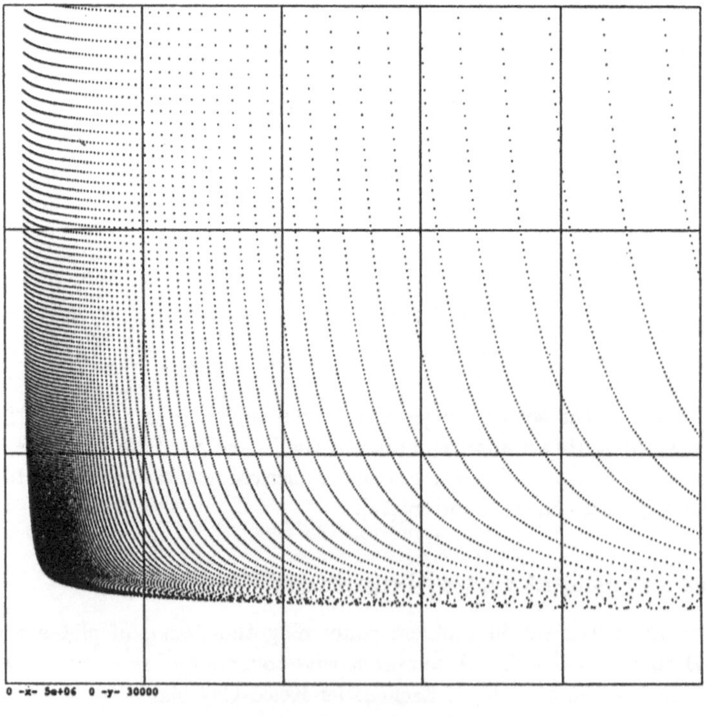

References

[1] M. Bleß, *Sensitivität in der Vektoroptimierung* (dissertation, University of Göttingen, 1990).

[2] B. Brosowski and A. Conci, "On the Optimal Design of Stiffened Plates", *Proceedings COBEM 83* (1983), p. 169–178.

[3] A. Charnes and W. Cooper, *Management models and industrial applications of linear programming, Vol. 1* (Wiley, New York, 1961).

[4] W. Dinkelbach, "Über einen Lösungsansatz zum Vektormaximumproblem", in: M. Beckmann (ed.), *Unternehmensforschung Heute* (Springer, Berlin 1971), p. 1–13.

[5] F. Y. Edgeworth, *Mathematical Psychics* (Kegan Paul, London, 1881).

[6] H. Eschenauer, J. Koski and A. Osyczka (eds.) *Multicriteria Design Optimization* (Springer, Berlin, 1990).

[7] H. Eschenauer and E. Schäfer, "Sandwichbalken mit Fachwerkunterbau" (manuscript, University of Siegen, 1989).

[8] J. Jahn, *Mathematical Vector Optimization in Partially Ordered Linear Spaces* (P. Lang, Frankfurt, 1986).

[9] J. Jahn, "Parametric Approximation Problems Arising in Vector Optimization", *JOTA 54* (1987) 503–516.

[10] J. Jahn and A. Merkel, "A Method of Reference Point Approximation for the Solution of Bicriterial Nonlinear Optimization Problems", *JOTA* (to appear).

[11] A. Kirsch, W. Warth and J. Werner, *Notwendige Optimalitätsbedingungen und ihre Anwendung* (Springer, Berlin, 1978).

[12] H. Kitagawa, N. Watanabe, Y. Nishimura and M. Matsubara, "Some Pathological Configurations of Noninferior Set Appearing in Multicriteria Optimization Problems of Chemical Processes", *JOTA 38* (1982) 541–563.

[13] J. Klose, "Sensitivity Analysis Using the Tangent Derivative" (manuscript, University of Erlangen-Nürnberg, 1990).

[14] J. Klose, "On the Numerical Solution of a Bicriterial Optimization Problem from Chemical Engineering" (manuscript, University of Erlangen-Nürnberg, 1990).

[15] T. C. Koopmans, "Analysis of Production as an Efficient Combination of Activities", in: T. C. Koopmans (ed.), *Activity Analysis of Production and Allocation* (Wiley, New York, 1951), p. 33–97.

[16] H. W. Kuhn and A. W. Tucker, "Nonlinear Programming", in: J. Neyman, *Proceedings of the Second Berkeley Symposium on Mathematical Statistics and Probability* (Berkeley, 1951), p. 481–492.

[17] D. T. Luc and J. Jahn, "Axiomatic Approach to Duality in Optimization" (manuscript, University of Erlangen-Nürnberg, 1990).

[18] V. Pareto, *Cours d'Economie Politique* (F. Rouge, Lausanne, 1896).

[19] M. Petschke, *Extremalstrahlen konvexer Kegel und komplementäre Ungleichungen* (dissertation, Technical University of Darmstadt, 1989).

[20] E. Polak, "On the Approximation of Solutions to Multiple Criteria Decision Making Problems", in M. Zeleny (ed.), *Multiple Criteria Decision Making, Kyoto 1975* (Springer, Berlin, 1976), p. 271–281.

[21] W. Stadler, "Initiators of Multicriteria Optimization", in: J. Jahn und W. Krabs (eds.), *Recent Advances and Historical Development of Vector Optimization* (Springer, Berlin, 1987), p. 3–47.

[22] T. Staib, *Notwendige Optimalitätsbedingungen in der mehrkriteriellen Optimierung mit Anwendung auf Steuerungsprobleme* (dissertation, University of Erlangen–Nürnberg, 1989).

[23] P. Weidner, *Ein Trennungskonzept und seine Anwendung auf Vektoroptimierungsverfahren* (habilitation thesis, University of Halle, 1990).

[24] D. Zhuang, *Regularity and maximality properties of set-valued structures in optimization* (dissertation, Dalhousie University, Halifax, 1989).

NONCONVEX OPTIMIZATION AND ITS STRUCTURAL FRONTIERS

Hubertus Th. Jongen [1] & Gerhard-W. Weber

RWTH Aachen
Lehrstuhl C für Mathematik

Abstract: This is a tutorial paper on the static and dynamical aspect of finite dimensional (constrained) optimization. The static feature refers to critical point theory, whereas the dynamical aspect aims at transitions of one optimization problem into another one.

1.INTRODUCTION.

In this tutorial paper we will mainly consider optimization problems of the following type:

$$(\mathcal{P}) \qquad \text{Minimize } f \text{ on the feasible set } M[H,G] \tag{1}$$

where

$$M[H,G] = \{x \in \mathbb{R}^n \mid h_i(x) = 0,\ i \in I\ , g_j(x) \geq 0,\ j \in J\}, \tag{2}$$

and where $f, h_i, g_j : \mathbb{R}^n \to \mathbb{R}$ are continuous functions, $|I| < n$, $|J| < \infty$, and $H = (h_i, i \in I)$, $G = (g_j, j \in J)$.

When dealing with structural frontiers of nonconvex optimization problems, two aspects will play an essential role. On one hand we have the relation of critical points (local minima, local maxima, saddlepoints) with the topological structure of the feasible set. Viewing at the latter as a static description of optimization features, a dynamical ingredient is discovered when looking at the transition from one optimization problem $(\mathcal{P}1)$ into another one, say

[1]The contribution of H. Th. Jongen was made when visiting the International Computer Science Institute at Berkeley, Spring 1991.

($\mathcal{P}2$). The transition will be governed by means of a real parameter t on which the defining functions f, H, G will depend additionally. This is the second aspect to which we will pay attention. Along the presentation we will stick to main ideas and connections and we try to avoid unnecessary technicalities which, however, can be found in the accompaning references. The paper is organized as follows. In Section 2 we consider critical point theory without constraints : Morse Lemma, the deformation principle, topological changes of lower level sets when passing a critical value, Morse relations. In Section 3 we take constraints into account; now Kuhn-Tucker points play the role of critical points in the unconstrained case. Section 4 consists of some remarks concerning the possibility, resp. difficulties when trying to make a critical point theory available for nondifferentiable problems. In particular, functions of maximum-type are treated. In Section 5 we consider the generic one parametric transitions from one optimization problem into another one. Special attention is paid to the evolution of the set of (generalized) critical points. Based on that information, Section 6 deals with the fundamental problem of tracing branches of local minima. At the end of each section we make a compilation of some relevant references and connections with other areas. Altogether we try to give an overall static and dynamic view on nonconvex (finite dimensional) optimization, thereby approaching its structural frontiers.

2. UNCONSTRAINED OPTIMIZATION, CRITICAL POINTS, MORSE RELATIONS.

We start with building up a function f of n variables by means of structural components. To this aim let f belong to $C^2(\mathbb{R}^n, \mathbb{R})$, the space of twice continuously differentiable functions from the Euclidean $\mathbb{R}^n$ to $\mathbb{R}$. By Df and $D^2 f$ we denote the row vector of first partial derivatives and the matrix of second derivatives (Hessian), resp.
The first structural component describes the behaviour of f in the neighbourhood of a point $\bar{x} \in \mathbb{R}^n$ at which Df does not vanish. Then, the function f is locally linearizable by means of a local coordinate transformation of class C^2 . In fact, we foliate a neighbourhood of $\bar{x}$ by means of level surfaces of the function f (cf. Fig.1). To this aim we may assume that the partial derivative $\partial f/\partial x_1$ does not vanish at $\bar{x}$, and we put

$$y = \Phi(x), \text{ where } \begin{cases} y_1 = f(x) - c & , \text{with } c = f(\bar{x}) \\ y_i = x_i & , i = 2, \cdots, n. \end{cases} \tag{3}$$

Then, Φ is of class C^2 with nonsingular Jacobian matrix at $\bar{x}$, and hence, a local coordinate transformation. From (3) it follows $\Phi(\bar{x}) = 0$, and

$$f \circ \Phi^{-1}(y_1, \cdots, y_n) = y_1 + c. \tag{4}$$

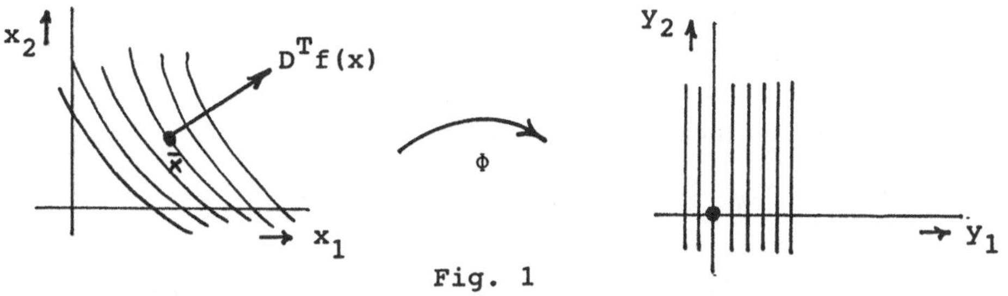

Fig. 1

A point $\overline{x} \in \mathbb{R}^n$ with $Df(\overline{x}) = 0$ is called a <u>critical point</u>. If in addition, the Hessian $D^2f(\overline{x})$ is nonsingular, the critical point is called <u>nondegenerate</u>. The second structural component describes the behaviour of f in the neighbourhood of a nondegenerate critical point $\overline{x} \in \mathbb{R}^n$. This is the content of the famous <u>Morse Lemma</u>. The latter states that there exists a local C^1- coordinate transformation Φ, sending $\overline{x}$ to the origin, such that in these coordinates the function f takes the following simple (quadratic) form, with $c = f(\overline{x})$:

$$f \circ \Phi^{-1}(y_1, \cdots, y_n) \;=\; -\sum_{i=1}^{k} y_i{}^2 + \sum_{j=k+1}^{n} y_j{}^2 + c\,. \tag{5}$$

The number k in (5) equals the number of negative eigenvalues of the Hessian $D^2f(\overline{x})$, also called the <u>(quadratic) index</u> of the critical point. The proof of the Morse Lemma can be given with the aid of a Homotopy-Ansatz which is fundamental in the theory of singularities (cf. [36], [73], [318]). We shortly explain the idea (cf. [158] for details).
Without loss of generality we may assume that $\overline{x} = 0, f(0) = 0, Df(0) = 0,\ D^2f(0)$ nonsingular. Let g denote the second order Taylor expansion of f at the origin; hence we have $g(x) = \frac{1}{2}x^T D^2f(0)x$. By means of a linear coordinate transformation we can transform g into the sum of squares as in (5). So it remains to transform f into its truncated Taylor expansion g. We connect f and g by means of a (linear) one parameter family $H(\cdot, u)$; this is the mentioned Homotopy-Ansatz :

$$H(x, u) \;=\; (1-u)f(x) \;+ ug(x). \tag{6}$$

We obviously have $H(\cdot,0) = f$, $H(\cdot,1) = g$. In the space $\mathbb{R}^n \times \mathbb{R}$, the last component corresponding to the parameter u , the following vector field is introduced :

$$F(x,u) = \begin{cases} (-D_u H \frac{D_x H}{\|D_x H\|^2}, 1)^T & , x \neq 0 \\ (0,1)^T & , x = 0. \end{cases} \tag{7}$$

In time one integration of F we come from the level $u = 0$ to the level $u = 1$. Important to note is the fact that the function H is <u>constant</u> on the trajectories of the vector field F , since

$$DH \cdot F = -D_u H \cdot \frac{D_x H}{\|D_x H\|^2} \cdot D_x^T H \ + 1 \cdot D_u H \equiv 0. \tag{8}$$

Therefore, by integrating the vector field F in time one, each level of the function f is automatically shifted to the corresponding level of g, thereby keeping the origin fixed (cf. Fig.2) .

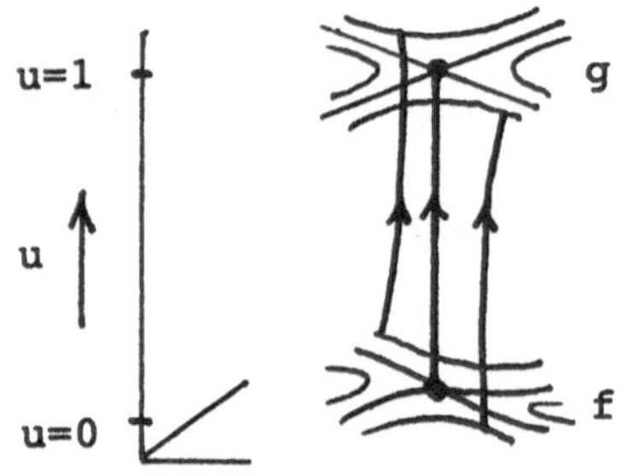

Fig. 2

The crucial point which remains to be addressed, is the smoothness of the vector field F. In fact, note that $D_x H(0,u) \equiv 0$. So in (7) there appears a singularity factor of order $\|D_x H\|^{-1}$ as we approach points $(0,u)$. However, the partial derivative $D_u H$ equals $g - f$ which vanishes faster than $\|x\|^2$ vanishes (Taylor expansion !). From the nondegeneracy of the critical point $\overline{x} = 0$ it follows that $\|D_x H\|^{-1}$ is of order $\|x\|^{-1}$ as x tends to zero. Altogether this shows that the vector field F is differentiable, and from an analogous estimate on the partial derivatives of F it follows that F is even of class C^1. But then its (local) flow is of class C^1 as well, which finally establishes the desired C^1-coordinate transformation. For a generalization on degenerate critical points we refer to [318]; under appropriate assumptions the

Homotopy-Ansatz also works in infinite dimensions; cf. [44], [113].
Now, suppose that all critical points of f are nondegenerate. Then f can be seen as having been built up from two types of structural components, corresponding to the so-called normal form (4) and (5), resp. The question now arises which combinations can occur. A key tool for this is the study of lower level sets M^a for varying level a, where $M^a = \{x \in I\!R^n \mid f(x) \leq a\}$. Consider Fig.3. We see that, up to continuous deformation, the structure of the lower level set M^a only changes at the values $a \in \{a_2, a_5, a_7\}$. These are precisely the values of f corresponding to a critical point (i.e. the critical values of f).

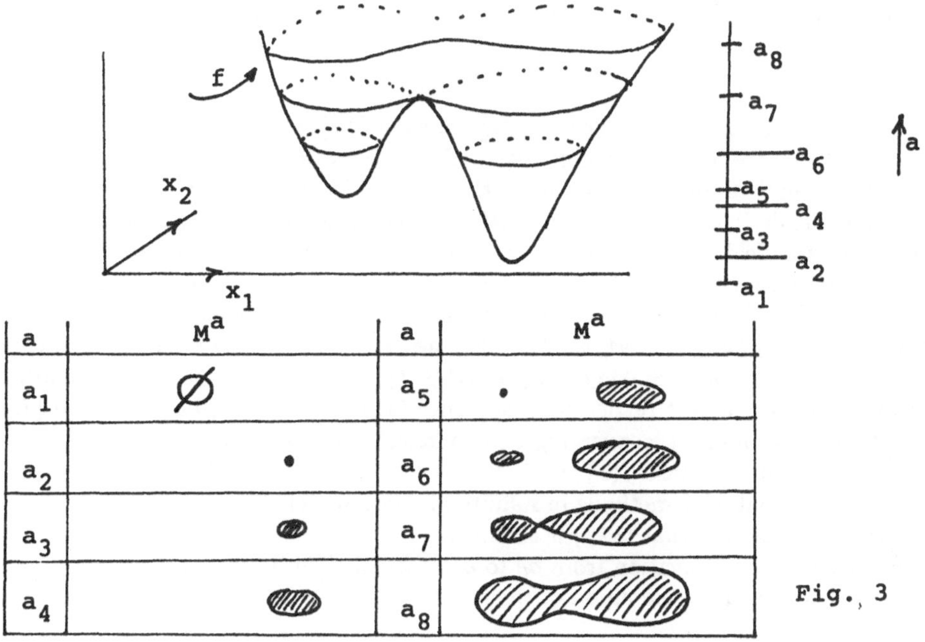

Fig. 3

Let us discuss the transitions from level a_3 to a_4, and from a_6 to a_8, separately. Put $M_a^b = \{x \in I\!R^n \mid a \leq f(x) \leq b\}$. Now, if M_a^b is compact and contains no critical points, then M^b can be deformed continuously to the set M^a. This can be achieved along trajectories of the vector field of steepest descent (cf. Fig.4):

$$\dot{x} = -D^T f(x), \tag{9}$$

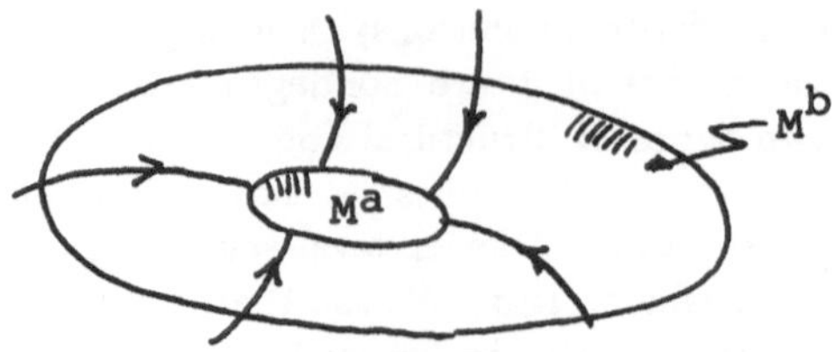

Fig. 4

In fact, since M_a^b is assumed to be compact and $Df(x) \neq 0$ for all $x \in M_a^b$, we see that $\inf\{\|Df(x)\| \mid x \in M_a^b\} > 0$. Therefore, in a neighbourhood of M_a^b, we can consider the scaled vector field $\mathcal{F}(x)$,

$$\mathcal{F}(x) = -D^T f(x)/\|Df(x)\|^2, \tag{10}$$

which is <u>bounded</u> (around M_a^b). For $x \in M_a^b$ we have $Df(x) \cdot \mathcal{F}(x) = -1$; hence, if we start to integrate $\mathcal{F}$ at $x \in M_a^b$, we are precisely after integration time $t := f(x) - f(a)$ in the set M^a. The boundedness of $\mathcal{F}$ guarantees that we can integrate $\mathcal{F}$ over longer times, which finally yields the desired deformation (cf. [158], [217] for details). If M_a^b is not compact, then the scaled vector field $\mathcal{F}$ needs not to be bounded, and the foregoing deformation breaks down. For example consider $f(x) = e^x$. Then f has no critical points at all; but for all $\epsilon > 0$ we have $M_{-\epsilon}^{\epsilon}$ is not compact and, moreover, $M^{-\epsilon} = \emptyset$, whereas $M^{\epsilon} \neq \emptyset$. So, in order to deal with <u>noncompact</u> sets M_a^b as well, it is necessary to impose an additional condition on f. The so-called <u>Palais-Smale condition</u> (P-S) ([236],[241],[302]; see also Remark 3.1) is sufficient to this aim and it reads: "any sequence (x_k) with $(f(x_k))$ bounded and $(Df(x_k))$ tending to zero, contains a converging subsequence". The sequence $-1, -2, -3, \cdots$ shows that (P-S) is violated in case $f(x) = e^x$. It should be noted that (P-S) plays a crucial role when studying variational methods in infinite dimensions (cf. [71], [252], [332]).
From now on we assume that (P-S) is satisfied.
Next we focus on the transition from a_6 to a_8 in Fig.3. The lower level set M^{a_8} can continuously be deformed into the union of the set M^{a_6} with an interval, cf. Fig.5.

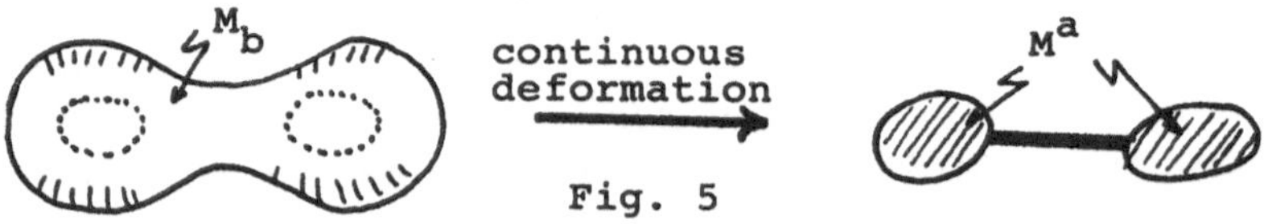

Fig. 5

Let $D^k := \{(z_1, \cdots, z_k) \in I\!R^k \mid \|z\| \leq 1\}$ denote the k-dimensional unit ball in $I\!R^k$ with boundary $S^{k-1} := \{z \in I\!R^k \mid \|z\| = 1\}$; we put $S^{-1} := \emptyset$. A homeomorphic copy of D^k is called a <u>k-cell</u>.

So, we see that M^{a_8} can be continuously deformed to M^{a_6} with a 1-cell attached along its boundary. The 1-cell represents the eigenspace of the Hessian $D^2 f$ at the critical point corresponding to the negative eigenvalue. The above situation can be generalized as follows. Suppose that M_a^b contains exactly one critical point $\overline{x}$. Moreover, suppose that $a < f(\overline{x}) < b$ and that the index of the critical point $\overline{x}$ equals k. Then, M^b can be continuously deformed into M^a with a k-cell attached along its boundary. The latter can roughly be seen as follows (cf. [158], [217] for details). Let $\mathcal{O}$ be a neighbourhood of $\overline{x}$ in which f can be represented by means of the normal form (5). Outside $\mathcal{O}$ we perform a deformation from M^b to M^a as in the foregoing case without critical points. Within $\mathcal{O}$ we take the splitting $I\!R^n = I\!R^k \times I\!R^{n-k}$ into account, and deform M^b – parallel w.r.t. the last coordinates – into M^a with an attached k-cell; the latter k-cell lies in $I\!R^k \times \{0_{n-k}\}$ (see Fig.6).

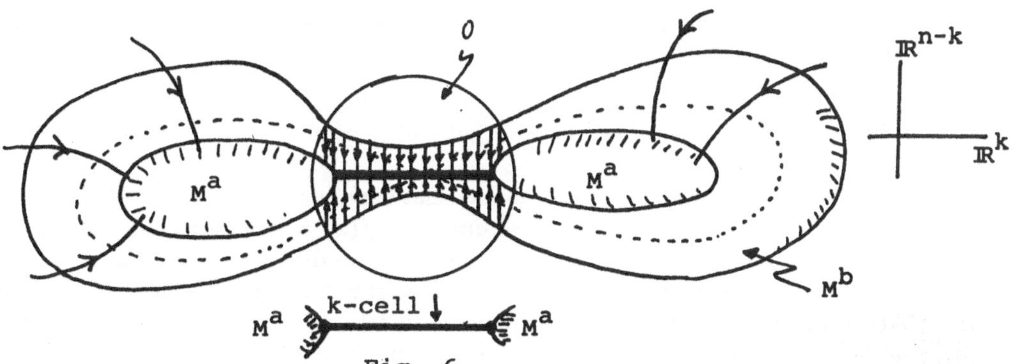

Fig. 6

Now we have to describe the topological effect of attaching a k-cell. To this aim consider the attachment of a 2-cell as depicted in Fig.7. The space to which the cell is attached, is a rectangle with a hole in it. We see two alternatives : either a 2-dim. hole is closed, or a 3-dim. hole is created.

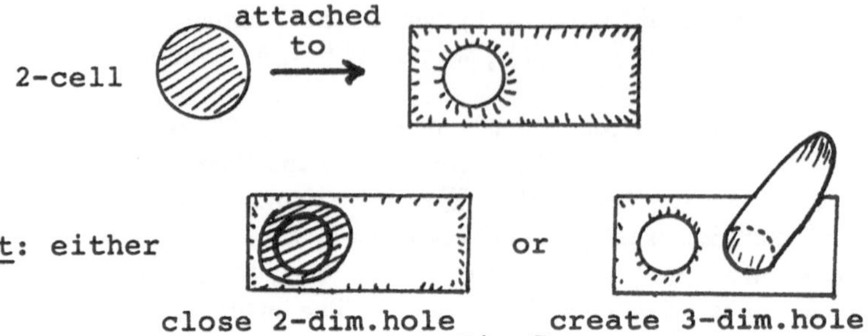

Fig.7

To be precise : by a k-dim. hole of a topological space X we mean a basis element of $H_{k-1}(X)$, the $(k-1)$ singular homology space of X over the real number field; cf. [158], [314]. (In particular, $H_0(X)$ counts the number of path-connected components of X; if one would like a 1-dimensional hole being measured by means of two path-components, then working with the reduced homology space would be preferable.) In the latter sense the alternative as suggested in Fig.7 is valid in general. So, attaching a k-cell to a topological space X results in the following alternative :

$$\begin{cases} \text{-Either the number of } k\text{-dim. holes of X goes down by one.} & \text{(I)} \\ \text{-Or the number of } (k+1)\text{-dim. holes of X goes up by one.} & \text{(II)} \end{cases} \tag{11}$$

Now we can deduce necessary relations (Morse relations) which have to be fulfilled by the critical points of f. For simplicity, let us assume that $f(x) = \|x\|^2$ for x sufficiently large. Moreover, we assume that distinct critical points of f have distinct critical values. Since all critical points of f are nondegenerate, there is a finite number of them. As suggested in Fig.3, consider the topological behaviour of the lower level sets M^a as a increases. The number of holes in M^a only changes if we pass a critical value. In the latter case we have precisely one of the alternatives (I), (II) in (11) with $X = M^a$. Let c_k denote the number of critical points of index k. Moreover, let c_k^- and c_k^+ be the number of critical points of index k at which the alternative (I) and (II), resp., hold, when passing the corresponding critical level. For a exceeding the highest critical level, the set M^a has the (deformation) type of a convex set (cf. assumed asymptotic behaviour of f). A convex set is continuously deformable into one point, and continuous deformations do not affect the number of k-dim. holes. A convex set has no k-dim. holes for $k \geq 2$, and it has one path-connected component. A simple counting argument now shows that the following equalities (Morse equalities) should hold:

$$\begin{cases} c_0^+ - c_1^- = 1 \\ c_i^+ - c_{i+1}^- = 0 \quad , \text{ for } i \neq 0. \end{cases} \tag{12}$$

We note that $c_0^+ = c_0$ and $c_{i+1} = 0$ for $i \geq n$. From (12) we learn that for the minimization problem the points of index 1 are highly important. In fact, having found a critical point of index 1 there are two directions of quadratic descent (cf. (5)); in case that alternative (I) in (11) holds, one gets into two different connected components in each of which the search for minima can proceed. However, it might occur that alternative (II) holds; see Fig.8 for the significant difference (the integers in Fig.8 denote the index of the corresponding critical point).

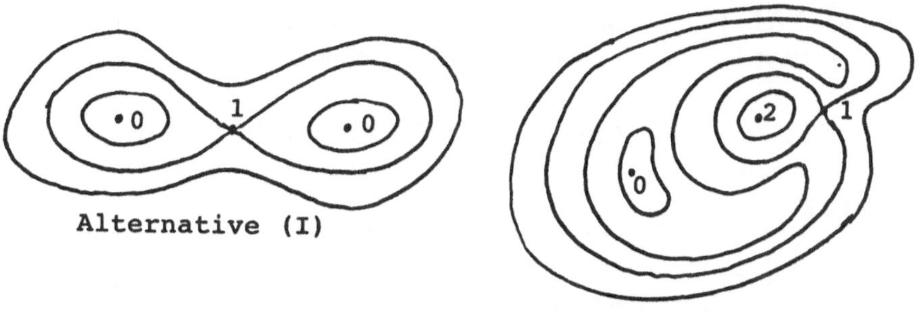

Fig. 8

To our knowledge there is no other local search concept for critical points of index 1 than the stabilizing differential equation, as introduced in [168], based on partial reflection of the gradient of f. In case that there exist many local minima, say due to stochastic influences, it might nevertheless be true that there exists a certain "averaged" function $\tilde{f}$ with only a few number of critical points describing the essential global behaviour of f (cf. also [117] within this context).
Throughout this section we assumed that all critical points of f are nondegenerate. This, however, is not a severe restriction, since the family of all those functions is C_S^2-open and dense in $C^2(\mathbb{R}^n, \mathbb{R})$ (cf. [159]). The main idea is based on Sard's Theorem, which implies that almost all (in the sense of the Lebesgue measure) linear perturbations turn f into the nondegenerate family (see [219]; cf. also [43], [146], [342]). It remains to define the C_S^2-topology. In fact, a typical neighbourhood $\mathcal{N}_\epsilon$ of the zero-function is generated by means of

a continuous positive function $\epsilon : \mathbb{R}^n \to \mathbb{R}$ (cf. [146], [159]) :

$$\begin{aligned} \mathcal{N}_\epsilon \quad := \quad & \{\phi \in C^2(\mathbb{R}^n, \mathbb{R}) \mid |D^\alpha \phi(x)| < \epsilon(x) \\ & \forall x \in \mathbb{R}^n, \forall \alpha = (\alpha_1, \cdots, \alpha_n) \in (\mathbb{N} \cup \{0\})^n \text{ with} \sum_{i=1}^{n} \alpha_i \leq 2\} \end{aligned} \tag{13}$$

In (13) the symbol $D^\alpha \phi$ stands for $\frac{\partial^{\alpha_1 + \cdots + \alpha_n} \phi}{\partial x_1^{\alpha_1} \cdots \partial x_n^{\alpha_n}}$, α as multi-index.
A final remark should be made concerning infinite dimensional problems. In that case one might think about the occurence of critical points of infinite index. However, such points cannot be identified by counting the number of holes. More precisely, attaching a cell of infinite dimension does not change the number of holes. This is due to the fact that, e.g., the unit ball D in the Hilbert space ℓ_2 can be deformed continuously into its boundary ∂D, keeping ∂D fixed (cf. [158]). The latter is not possible with a k-cell. It reflects also the fact that the famous fixed point theorem of Brouwer on convex sets is only true in finite dimensions.

Remark 2.1 The dimensions of the spaces $H_k(X)$, called Betti numbers (see [229], [314]), do have an interesting (integral-) geometrical meaning (cf. [46], [286]). They occur in homology theory (see [229], [292], [314], cf. also [300]), here with respect to the function f. This theory is in (so-called Poincaré) duality (cf. [86], [229]) with cohomology theory (cf. [314], and also [275], [294]) with respect to the function $-f$. Both theories are fundamental in many fields of mathematics like algebraic geometry and graph theory (cf. also [20], [94], and [326], resp.) and they will guide us in the subsequent sections.
Searching (finite) bounds for these geometrical indices (see [344]; cf. also [93], [131], [218]) gives some insight into the (quantitative) complexity of problems in nonconvex optimization (see [343], [345]; cf. also [278]) tracing it back to the convex case (cf. [27], [46], [144]).
As indicated above, algebraic topology together with convex analysis (see [27]) is an important tool in discrete optimization (see [29], [233]). In nonlinear optimization, convexity ([90], [269], [316]) has more and more been weakened (see e.g. [19], [80], [185], [186], [188], [207], [209], [270], [342]), and we are going to treat much more broader classes of functions (cf. Remarks 3.1, 4.1). Moreover, for two other interesting researches related to Morse theory as treated here, namely in convex analysis and in data fitting, see [246] and [235], resp., and for a Morse theory in the context of several objective functions see [304] (cf. also [188]). We underline the difficulty of carrying over facts from finite dimensional to infinite dimensional nonlinear analysis by pointing to another basic topological (homotopy- and perturbation-) invariant, i.e. index, namely the topological degree of a mapping. Indeed, going from dim.

$< \infty$ (Brouwer degree) to dim. $= \infty$ (Leray-Schauder degree) one focuses on compact (nonlinear) operators ([133] and [78], [100], [203], [232], [290]). For some applications of these degrees in optimization (dim. $< \infty$) and control theory (dim. $= \infty$) see [6], [104], [250] and [234]. In the next sections we will come back to the degree. In parallel to our concept we will pay attention to the development of the index theory due to M.F. Atiyah and I.M. Singer. In this way we falthogether hope to give a good account on stability phenomena on one hand, and one the theory of invariants on the other hand. The latter theory was developped for elliptic operators (cf. [3], [14], and in a variational setting [200], [226]) firstly w.r.t. finite dimensional compact manifolds without boundary. It was also studied studied from the viewpoint of K-theory (cf. [13]) and from a cohomological viewpoint (see, e.g., [14] and [255]).

3.CONSTRAINED OPTIMIZATION, KUHN-TUCKER POINTS, MORSE RELATIONS.

In this section we take constraint sets $M := M[H, G]$ as defined in (2) into account, and we assume throughout that the mappings H, G , as well as the function f are of class C^2. We discuss two constraint qualifications, LICQ and MFCQ.
The linear independence constraint qualification (LICQ) is said to hold at $\overline{x} \in M[H, G]$ if the set of vectors $\{Dh_i(\overline{x}), Dg_j(\overline{x}), i \in I, j \in J_0(\overline{x})\}$ is linearly independent. Here, $J_0(\overline{x})$ stands for the index set of active (binding) inequality constraints,

$$J_0(\overline{x}) = \{j \in J \mid g_j(\overline{x}) = 0\}. \tag{14}$$

Under LICQ the set M takes, in new C^2-coordinates around $\overline{x}$, the form $\mathbb{H}^p \times \mathbb{R}^q$ around the origin, where $p = |J_0(\overline{x})|$, $q = n - m - p$, and $\mathbb{H}^p := \{z \in \mathbb{R}^p \mid z_i \geq 0 \ , \ i = 1, \cdots, p\}$ (the nonnegative orthant in $\mathbb{R}^p$); cf. Fig.9. In fact, assuming without loss of generality that $J_0(\overline{x}) = \{1, \cdots, p\}$, put

$$\begin{cases} y_i = h_i(x) \ , & i = 1, \cdots, m \\ y_j = g_{j-m}(x) \ , & j = m+1, \cdots, m+p \\ y_k = \xi^T(x - \overline{x}) \ , & k = m+p+1, \cdots, n \end{cases} \quad , \text{ shortly } y = \Phi(x) \ , \tag{15}$$

where the vectors $\xi_k \in \mathbb{R}^n$ are chosen such that $\{D^T h_i(\overline{x}), D^T g_j(\overline{x}), \xi_k, \ i \in I, \ j \in J_0(\overline{x}), k \in \{m+p+1, \cdots, n\}\}$ is a basis for $\mathbb{R}^n$.

Let us call a coordinate transformation of the type (15) canonical.

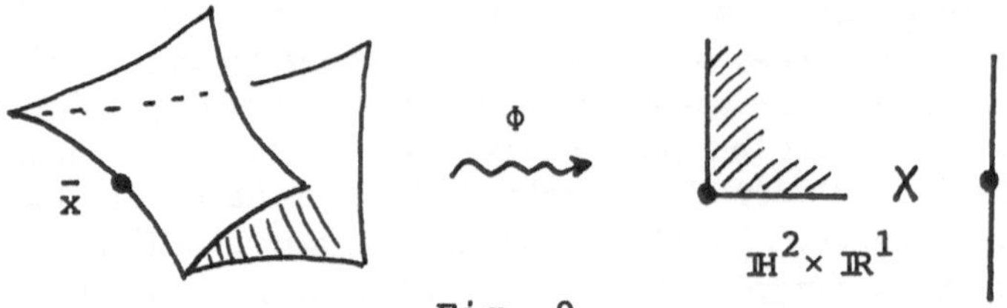

Fig. 9

Now, assume that LICQ is satisfied. A point $\bar{x} \in M$ is called a critical point for $f|_M$ if the following relation holds :

$$Df = \sum_{i\in I} \bar{\lambda}_i Dh_i + \sum_{j\in J_0(\bar{x})} \bar{\mu}_j Dg_j \,|_{\bar{x}} \tag{16}$$

The (unique) reals $\bar{\lambda}_i, \bar{\mu}_j$ are called Lagrange multipliers, and the corresponding function $L := f - \sum_{i\in I} \bar{\lambda}_i h_i - \sum_{j\in J_0(\bar{x})} \bar{\mu}_j g_j$ the Lagrange function. If all $\bar{\mu}_j$ in (16) are nonnegative, then $\bar{x}$ is called a Kuhn-Tucker point (KT-point). Under LICQ local minima are necessarily KT-points. A critical point is called nondegenerate if the following two conditions hold :

$$\left\{ \begin{array}{ll} \text{ND1 (linear)} & \mu_j \neq 0,\ j \in J_0(\bar{x}) \\ \text{ND2 (quadratic)} & D^2L(\bar{x})|_{T_{\bar{x}}M} \text{ nonsingular ,} \end{array} \right. \tag{17}$$

where the nonsingularity of $D^2L(\bar{x})|_{T_{\bar{x}}M}$ in (17) means that the matrix $V^T \cdot D^2L(\bar{x}) \cdot V$ is nonsingular, the columns of V forming a basis for the tangent space $T_{\bar{x}}M$,

$$T_{\bar{x}}M = \bigcap_{i\in I} Ker Dh_i(\bar{x}) \cap \bigcap_{j\in J_0(\bar{x})} Ker Dg_j(\bar{x}). \tag{18}$$

The condition ND1 is often called the strict complementarity condition. In a neighbourhood of a nondegenerate critical point an analogous normal form as in (5) can be achieved by means of a canonical C^1-transformation Φ , and we get (cf. [158]) :

$$\left\{ \begin{array}{lllll} f \circ \Phi^{-1}(y_1, \cdots, y_n) = & -\sum y_{i_1} & + \sum y_{i_2} & - \sum y_{i_3}^2 + \sum y_{i_4}^2 + c \\ & y_{i_1} \geq 0 & y_{i_2} \geq 0 \ , & \end{array} \right. \tag{19}$$

where the coordinates $y_{i_1}(y_{i_2})$ and $y_{i_3}(y_{i_4})$ correspond to the negative (positive) Lagrange multipliers and negative (positive) eigenvalues of $D^2L(\bar{x})|_{T_{\bar{x}}M}$, resp. The quadratic index at $\bar{x}$ is defined as the number of negative eigenvalues of $D^2L(\bar{x})|_{T_{\bar{x}}M}$. Note that $\bar{x}$ becomes a local minimum iff all $\bar{\mu}_j$ are positive and $D^2L(\bar{x})$ positive definite on $T_{\bar{x}}M$ (i.e., the quadratic index vanishes).
For constrained minimization problems we can also derive (Morse) relations. Now, the Kuhn-Tucker points play the role of critical points in the unconstrained case. In fact, assume that all critical points of $f|_M$ are nondegenerate and, for simplicity, that M is compact. (For noncompact M an appropriate variant of the Palais-Smale condition can be introduced, cf. [125], [166], but also [332]). Moreover, put $M^a = \{x \in M \mid f(x) \leq a\}$, $M_a^b = \{x \in M \mid a \leq f(x) \leq b\}$, and we will use the precise concept of homotopy-equivalence (cf. [314]) instead of "continuous deformation". As a preparation for the Morse relations we have to consider two cases (cf. [158] for details).
Case 1: M_a^b contains no Kuhn-Tucker points. Then, M^b and M^a are homotopy-equivalent. The proof is based on the fact that now, for $\bar{x} \in M_a^b$ the following system of inequalities is solvable :

$$\begin{cases} Df(\bar{x})\xi < 0 \\ Dh_i(\bar{x})\xi = 0\,, & i \in I \\ Dg_j(\bar{x}) > 0\,, & j \in J_0(\bar{x})\,. \end{cases} \tag{20}$$

Any solution vector ξ of (20) is a direction of linear descent for f , as well as a tangent vector for a differentiable curve $\{x(t)\ , t \geq 0\} \subset M$. In this way a vector field can be constructed which yields the desired continuous deformation.
Case 2: M_a^b contains exactly one Kuhn-Tucker point $\bar{x}$, and $a < f(\bar{x}) < b$ and the quadratic index of $\bar{x}$ equals k. Then, M^b is homotopy-equivalent to M^a with a k-cell attached.
For the proof, let us shortly explain the idea of D. Braess (cf. [32]). In view of (19) we may assume that M has (locally) the form $\mathbb{H}^p \times \mathbb{R}^q$ and that f takes the form :

$$\sum_{i=1}^{p} x_i \; - \sum_{j=p+1}^{p+k} x_j^2 \; + \sum_{\ell=p+k+1}^{p+q} x_\ell^2 \; + \; c. \tag{21}$$

The idea of Braess consists in shifting the Kuhn-Tucker point into the interior of $\mathbb{H}^p \times \mathbb{R}^q$ by pushing the graph of the linear function $\sum_{i=1}^{p} x_i$ down. In this way the positive linear terms transform into a nondegenerate local minimum in the interior of $\mathbb{H}^p$. The latter, however, does not contribute to the total index, since a nondegenerate local minimum has index zero.

Moreover, note that by pushing the linear graph downwards, at the boundary of $\mathbb{H}^p \times \mathbb{R}^q$ will occur additional critical points, but they are not Kuhn-Tucker points (cf. Fig.10).

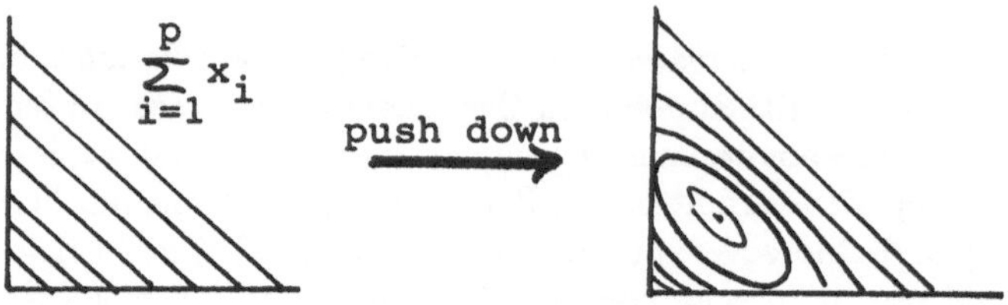

Fig. 10

Now, the Morse relations are easily established. In fact, suppose that all KT-points have different functional values. Let c_k be the number of KT-points with quadratic index k, let $c_k^-(c_k^+)$ denote the number of them for which alternative (I) ((II)) in (11) w.r.t. the lower level set holds when exceeding the corresponding (Kuhn-Tucker) level. Moreover, let r_k denote the number of $(k+1)$-dimensional holes of M. Then, similarly as (12) was derived, we obtain the following Morse relations :

$$c_i^+ \; - \; c_{i+1}^- \; = \; r_i \, , \; i = 0, 1, \cdots , \tag{22}$$

where $c_0^+ = c_0$ and $c_s = 0$ for $s > n - m$ (the dimension of M). We can get rid of the (+) and (−) signs in (22) by adding all equations in (22) with alternating sign, noting that $c_k = c_k^+ + c_k^-$:

$$(c_0^+ - c_1^-) - (c_1^+ - c_2^-) + (c_2^+ - c_3^-) - \cdots = r_0 - r_1 + r_2 \cdots$$

and we obtain

$$c_0 - c_1 + c_2 \cdots + (-1)^s c_s = r_0 - r_1 + r_2 \cdots + (-1)^s r_s \; , \quad s = n - m. \tag{23}$$

Note that the righthandside in (23) only depends on the topology of the feasible set M; in fact that number is called the Euler characteristic of M. On one hand, the topology of M gives relations on the number of KT-points of various indices. On the other hand, with a given function $f|_M$ information about the topological structure of M can be obtained. In fact, formula (22) represents both aspects. As an example, consider the 2-sphere $S^2 := \{x \in \mathbb{R}^3 \mid x_1^2 + x_2^2 + x_3^2 = 1\}$. The usual height function $g(x_1, x_2, x_3) := x_3$ has two nondegenerate KT-points on S^2, namely the global minimum S and the global maximum N (cf. Fig.11).

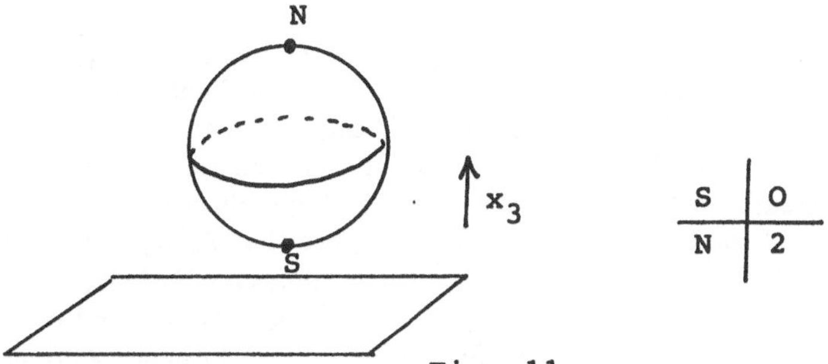

Fig. 11

There are no KT-points of quadratic index 1 ; hence, we obtain $r_0 = c_0^+ = 1$, $r_1 = -c_2^- = 0$ and $r_2 = c_2^+ = 1$. Now, let $f \in C^2(\mathbb{R}^3, \mathbb{R})$ be any function, such that $f|_{S^2}$ has only nondegenerate critical points. It then follows, together with (23), that $c_0 - c_1 + c_2 = r_0 - r_1 + r_2 = 2$, i.e. #(local minima) − #(saddle points) + # (local maxima) = 2, # standing for the cardinality.

In case that we only would add the first ℓ equations of (22) with alternating sign, we obtain the so-called Morse inequalities, for example:

first equation ($c_0^+ = c_0$) yields : $c_0 \geq r_0$
first, second equation : $r_1 - r_0 = (c_1^+ - c_2^-) - (c_0^+ - c_1^-) \leq c_1 - c_0$, etc.

The above theory on KT-points can be generalized assuming a weaker constraint qualification on one hand and a weaker nondegeneracy concept for KT-points on the other hand : The Mangasarian-Fromovitz constraint qualification (MFCQ) is said to hold at $\bar{x} \in M[H, G]$ if MF 1 and MF 2 are satisfied :

MF 1 . The vectors $Dh_i(\bar{x}), i \in I$, are linearly independent.
MF 2 . There exists a vector $\xi \in \mathbb{R}^n$ such that

$$\begin{cases} Dh_i(\bar{x})\xi = 0, & i \in I \\ Dg_j(\bar{x})\xi > 0, & j \in J_0(\bar{x}). \end{cases} \tag{24}$$

(cf. [205]). The constraint qualification MFCQ can be interpreted as a "positive" linear independence condition w.r.t. the inequality constraints. In fact, in virtue of linear duality we have : MFCQ is satisfied at $\bar{x} \in M$ if and only if the following implication is valid :

$$\left\{ \begin{array}{c} \sum_{i \in I} \lambda_i Dh_i(\bar{x}) + \sum_{j \in J_0(\bar{x})} \mu_j Dg_j(\bar{x}) = 0 \\ \mu_j \geq 0, j \in J_0(\bar{x}) \end{array} \right\} \Longrightarrow \text{ all } \lambda_i, \mu_j \text{ vanish.} \tag{25}$$

A point $\bar{x} \in M$ is called a Kuhn-Tucker point if (16) holds for some reals $\bar{\lambda}_i, i \in I$, $\bar{\mu}_j, j \in J_0(\bar{x})$ with all μ_j nonnegative. Now the numbers $\bar{\lambda}_i, \bar{\mu}_j$ need not be unique anymore. However, the set of all $\bar{\lambda}_i, i \in I$ and all nonnegative $\bar{\mu}_j$ is compact if and only if MFCQ is satisfied (cf. [109]; see also [195], [317]). Moreover, if $\bar{x} \in M$ is a local minimum and MFCQ is satisfied, then $\bar{x}$ is necessarily a KT-point. The constraint qualification MFCQ is basic in view of global stability of the feasible set. In fact, suppose that $M[H,G]$ is compact. Then there exists a C^1_S-neighbourhood $\mathcal{O}$ of (H,G) with the property that $M[\tilde{H},\tilde{G}]$ is homeomorphic with $M[H,G]$ for all $(\tilde{H},\tilde{G}) \in \mathcal{O}$ if and only if MFCQ is valid at all points of $M[H,G]$ (cf. [125]). Moreover, in the latter case, $M[H,G]$ is a Lipschitzian manifold (with boundary) of dimension $n-m$. M. Kojima introduced in [180] the concept of strongly stable KT-points. A KT-point $\bar{x}$ is called strongly stable if it persists and depends continuously on the data (f,H,G) under local C^2-perturbations. Under MFCQ, Kojima characterized strong stability in terms of first and second derivatives of (f,H,G). In particular, the vanishing of Lagrange multipliers $\bar{\mu}_j$ (linear part) is stabilized by means of nonsingularity of certain quadratic forms with constant index (cf. [180]), called the stationary index. In [125] it is shown that the Morse relations again are valid where the quadratic index is to be replaced by the mentioned stationary index. Hence, with respect to the minimization problem, KT-points with stationary index equal to one play again a basic role. It should be mentioned that the idea of attaching cells is further generalized in [115] where some complicated structures as entities are considered. (For further reading see also [61], [130], [175], [327] and [245].)

Remark 3.1 There are a lot of investigations on the stability (regularity) of the feasible set in nonconvex optimization actually being based on parametric and convex processes (see [121], [258], [260], [263], [265], [266]; cf. also [74], [148], [152], [153], [154], [170], [243], [331]). For an adjacent phenomenon we mention the structural stability of a (constrained) optimization problem $(\mathcal{P})$ ([121]). This means that all lower level sets are related to corresponding lower level sets of any C^2_S-slightly perturbed problem by means of a continuous family of homeomorphisms. In case that $M[H,G]$ is compact and different KT-points possess different critical values, structural stability is equivalent to MFCQ holding at each $x \in M[H,G]$ together with the strong stability of all KT-points of $(\mathcal{P})$ (see [171]). A concept similar to

the strong stability of a KT-point (cf. also [59], [170], [181], [194], [272]) is the concept of strong regularity (see [261], [262], [264], and cf. [178]).
Morse theory does moreover reflect geometry in its different branches ([51]); see [30], [217]. The articles [236] and [302] give an extension of it to Hilbert manifolds (perhaps with dim. $= \infty$; cf. [238]). For a study in the case of a Riemannian manifold with measurement of nondegeneracy and degeneracy of a differentiable germ f (singularity theory again) see [118]. Now, we immediately arrive at the "normal forms" given by Mather's result on (universal) unfoldings ([11], [36], [170], [210], [329]) and - additionally - at parametrizations of the Morse lemma (cf. [10], [12], [41], [170]; for a related question of representation see [184]).
We note that Morse theory is intimately connected with Ljusternik-Schnirelman theory (cf. [100], [201], [207], [289], [290]), which also takes account of the condition (P-S). The latter represents critical values by minimax expressions and also estimates the number of critical points; it can be given for Banach manifolds, too (cf. [239]). For condition (P-S), arising in both theories, and Ekeland's variational principle ([87]) see [222], [224], [225], [226] (cf. also [15], [17]), and for more see [84], [150], [240], [277], [301]. The two theories may also be regarded as parts of the rapidly developping index theory (cf. [21], [30], [50], [60], [77], [83], [92], [127], [132], [179], [183], [197], [226], [237], [249], [293], [299]) that proves to be a useful tool for the study of ordinary and partial differential equations. So, let us at first note works on the existence (and the number) of periodic solutions of Hamiltonian systems: [8], [22], [88], [89], [221], [251], [336]; cf. also [24], [25], [58], [64].
In the last years the Conley index as a generalization of the Morse index got a growing acceptance for the investigation of (semi-) flows around (isolated) invariant sets due to compact metric spaces (cf. [65], [66], [67], [68], [232], [309]; see also [220]). For the relation of the latter theory with the index winding number (cf. [33], [330]) see [28], [55], [291]. Moreover, for a model discretization (cf. [103]) see [228], and for a discussion in the case of noncompactness (e.g. under (P-S)) cf. [23] and [282]. In [333] the index of [23] has been used in bifurcation theory (see, e.g., [114], [134], [208], [284], [309]), instead of pushing the case dim. $= \infty$ back to the case dim. $< \infty$ by the Liapunov-Schmidt method or by using the center manifold theorem (see [47], [53], [91], [283]; cf. also [69]). For (level-) portraits similar to such arising in the study of structural stability see [37] and [38].
Turning again to the index Leray-Schauder degree (cf. Remark 2.1) we note as a version the coincidence degree (cf. [105], [213], [319]) and its extension to the case of a compact multivalued mapping, the so-called Webb degree (cf. [334], [335]). This degree is a strong tool, even under nondifferentiability, in the (existence-) theory of optimal control and for the research on periodic solutions of Hamiltonian inclusions ([106], [107], [244], see also [70]), both in a convex model (Rockafellar formulation; cf., e.g., [271]) and (by relaxation) in a nonconvex model (Clarke formulation; cf., e.g., [56]).

Finally we remark that the Atiyah-Singer index theory has been generalized to the case of manifolds with boundary (cf., e.g., [237] and [255]), too.

4. A REMARK ON NONDIFFERENTIABLE PROBLEMS.

In the foregoing section the functions under consideration were assumed to be continuously differentiable. In case that some of them are nondifferentiable, carrying over the previous ideas strongly depends on the "structure" of nondifferentiability. Let us restrict to the unconstrained case.
As a first example consider a function ρ of maximum-type :

$$\rho(x) := \max_{k \in K} f_k(x) \ , \tag{26}$$

where K is a finite index set and $f_k \in C^2(\mathbb{R}^n, \mathbb{R})$ for all k. The function ρ in (26) is continuous, but in general nondifferentiable. In particular, ρ needs not to be differentiable at its (local) minima, and that situation might be stable under small perturbations of the defining functions f_k (cf. Fig.12).

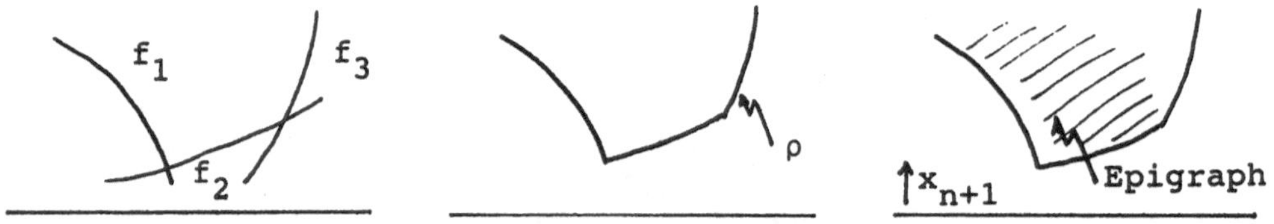

Fig. 12

Now, the minimization problem of ρ can be transformed into a differentiable one by means of minimizing the height-function $(x_1, \cdots, x_n, x_{n+1}) \mapsto x_{n+1}$ on the epigraph of ρ, where $Epigraph(\rho) := \{(x, x_{n+1}) \in \mathbb{R}^{n+1} \mid x_{n+1} \geq \rho(x)\}$. Taking the special structure of ρ into account, the transformed problem in $\mathbb{R}^{n+1}$ reads :

$$\left.\begin{array}{l} \text{minimize } x_{n+1} \\ \text{subject to } x_{n+1} - f_k(x) \geq 0 \ , \ k \in K \end{array}\right\} \tag{27}$$

(cf. [102]). The transformed problem (27) is a differentiable minimization problem of the type as we discussed in Section 3; Kuhn-Tucker points and quadratic index for ρ can be defined via (27). However, the topological connection between the two problems has to be established. This will be settled by means of the following simple observation. Let X be a topological space and $F : X \to \mathbb{R}$ a continuous function. Then the lower level sets X^a and $\tilde{X}^a$ are homotopy equivalent for all $a \in \mathbb{R}$, where
$X^a = \{x \in X \mid F(x) \leq a\}$ and $\tilde{X}^a = \{(x, z) \in Epigraph(F) \mid z \leq a\}$
(cf. [35], [158]). The latter homotopy-equivalence is performed by means of a deformation as sketched in Fig.13.

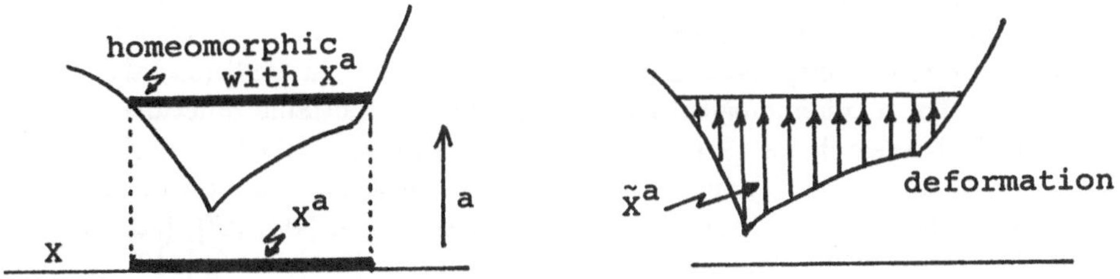

Fig. 13

Note that the function ρ in (26) is a continuous selection of the functions $f_k, k \in K$. In general, one might consider a function f which is a continuous selection of $f_k, k \in K$, shortly written $f = CS(f_k, k \in K)$. Such a function is locally Lipschitz, so we have the subdifferential ∂f at hand (cf. [57]). A Kuhn-Tucker point (or critical point) is now defined to be a point at which ∂f contains the zero vector. In [165] it is shown that for a generic choice of $f_k, k \in K$, in suitable local Lipschitz-coordinates Φ around its KT-points $f := CS(f_k, k \in K)$ takes the following form :

$$f \circ \Phi^{-1}(x_1, \cdots, x_n) = CS(x_1, \cdots, x_r, -\sum_{i=1}^{r} x_i) - \sum_{j=r+1}^{r+k} x_j^2 + \sum_{\ell=r+k+1}^{n} x_\ell^2 + c. \qquad (28)$$

From (28) it becomes transparent that, apart from the quadratic index k, also the linear part $CS(x_1, \cdots, x_r, -\sum_{i=1}^{r} x_i)$ gives a certain contribution to the attaching of structures at lower level sets when passing the level c. However, if the linear part equals $max(x_1, \cdots, x_r, -\sum_{i=1}^{r} x_i)$, then the contribution of the linear part does not count. In that case, only a

k-cell is attached, as we saw in the former discussion of functions of maximum-type. For general local Lipschitz functions a complete critical point theory cannot be expected. In fact, the most striking point is the behaviour of such a function in a neighbourhood of its KT-points. Here, the "structure" of nondifferentiability has to be cleared up, the structure of a "continuous selection" (cf. also [129]) being only one of the many diverse possibilities. Outside the Kuhn-Tucker set a deformation along descent flows is easily established, even in the infinite dimensional case (cf. [50], [332]; see also [21], [77], [83]). The distance of the subdifferential to the zero-vector now gives rise to an appropriate variant of the Palais-Smale condition.

Remark 4.1 In Remark 3.1 we mentioned the nondifferentiable (but locally Lipschitzian), or even discontinuous (but lower semicontinuous) Clarke formulation (cf. also [80]). Nondifferentiability and discontinuity have already been "measured" in [342] (see also [144]) and [103], [231], resp. For a cohomological research on nondifferentiable functions see [82], for a Morse theoretical research see [31], and for a more Fourier analytic one cf. [253]. The following works (among others) admit a further insight into both the arising and the treatment of nondifferentiability in nonconvex optimization: [48], [49], [57], [63], [145], [170], [177], [192], [194], [204], [216], [230], [242], [273], [298], [322], [341], [347], [348].
The (non-)differentiability and representability of special sets should be discussed, too. In the research of S.M. Robinson certain piecewise linear manifolds associated with polyhedral convex sets arise, so-called normal manifolds. They are related with (global or local) Lipschitzian homeomorphisms that arise quite naturally in nonconvex optimization (see [268], cf. Remark 3.1). For more work in this field see [101], [190], [256], [285]; now we are soon within the context of the various versions of the implicit function theorem (cf. [1], [40], [57], [164], [193], [196], [202]). As another class of interesting manifolds, we have in this context the invariant manifolds of dynamical systems (see, e.g., [2], [34], [52], [54], [98], [142], [214], [227], [303], [309], and Remark 3.1). For their representability as graphs of Lipschitzian functions see [97], [147]. Here, the existence of the Lipschitz mapping is implicitely verified by means of fixed point theory. However, due to the more special inertial manifolds (cf. [69], [79], [320]) given for certain partial differential equations (becoming ordinary differential equations after restricting to those manifolds) a (constructive) formula for the Lipschitzian mapping can be given in terms of spectral theory (cf. [69]).
With (26) one is again led to a minimax-problem (see [174] for a generic study). These problems turn out to be special cases of semi-infinite optimization (see [176]) where J is replaced by an index set, say having the form $Y = M[U, V]$ for which LICQ holds at each of its points. Hence, $M[H, G]$ is still a Lipschitzian manifold. Moreover, in the compact case it remains globally stable (cf. [169]), and under a strengthened (namely excisional)

stability concept (cf. [278]) this fact is true even in the noncompact case, too. For a generic study of the feasible set - without the so-called Reduction-Ansatz ([169]) - see [172], and for Morse theory in semi-infinite optimization see [173]. It is quite technical to go towards a characterization of structural stability under noncompactness of $M[H, G]$ without sticking to the Reduction-Ansatz. Within the semi-infinite model, optimality conditions can be found in [26], [138], [176], [212], [337] (see [39], [111], [112], [126], [139], [170], [187], [189], [295] additionally); some open problems are stated in [96].
For the Atiyah-Singer index theory there is a version in the noncompact case, too. It is given by Roe ([274], [275]) who takes up the idea of regular exhaustion. For more reading, e.g. an application in Schoen-Yau theory (minimal surfaces, curvature; cf. [287], [288]), see the cohomological work [276]. Roe announced further research in connection with Witten's contribution to Morse theory (cf. [288]). By consideration of a certain supersymmetric quantum mechanics Hamiltonian E. Witten introduced an index and Betti numbers, and he proved Morse inequalities (cf. [339]). For other work in mathematical physics that is related to our structural and stability analysis see [85], [255], [338], [340], and in particular with emphasis on the different types of manifolds in dim. = 4 see [99].

5.ONE-PARAMETRIC TRANSITIONS.

In this section we consider the transition of one optimization problem into another one by means of one real parameter t. As explained in the foregoing sections, the critical points (resp. Kuhn-Tucker points) play a key role in the problem structure. So, now we will discuss the evaluation of those points as the parameter t increases. In general, one-dimensional curves in (x, t)-space can be expected. We will look at the following one-parametric problem $\mathcal{P}(t)$:

$$\mathcal{P}(t): \text{ Minimize } f(\cdot, t) \text{ on the feasible set } M(t), \tag{29}$$

where

$$M(t) = \{x \in I\!R^n \mid h_i(x,t) = 0, i \in I,\ g_j(x,t) \geq 0, j \in J\}, \tag{30}$$

and where $f, h_i, g_j : I\!R^n \times I\!R \to I\!R$ are continuously differentiable functions, $|I| < n, |J| < \infty$.

In order to study the evaluation of critical points we have to introduce the concept of a generalized critical point. Let $z = (x,t)$ denote a point in $\mathbb{R}^n \times \mathbb{R}$. A point $(\bar{x},\bar{t})$ with $\bar{x} \in M(\bar{t})$ is called a generalized critical point (g.c. point) if the set of vectors $\{D_x f, D_x h_i, D_x g_j, i \in I, j \in J_0(\bar{z})\}|_{\bar{z}}$ is linearly dependent. Here, $D_x f$ stands for the partial derivative with respect to the variable x. If $\bar{z}$ is a g.c. point we have the following linear relation (not all $\lambda, \lambda_i, \mu_j$ vanish):

$$\lambda D_x f = \sum_{i\in I} \lambda_i D_x h_i + \sum_{j \in J_0(\bar{z})} \mu_j D_x g_j|_{\bar{z}=(\bar{x},\bar{t})} \cdot \tag{31}$$

In case that λ and all μ_j can be chosen to be nonnegative the point $\bar{z}$ is called a Fritz John point, and, if in addition $\lambda > 0$, the point is called a Kuhn-Tucker point. For given functions f, h_i, g_j, let $\Sigma \subset \mathbb{R}^n \times \mathbb{R}$ denote the set of generalized critical points.
The basic theorem about Σ is the following (cf. [161]; let f, h_i, g_j belong to $C^3(\mathbb{R}^n \times \mathbb{R}, \mathbb{R})$): There exists a C^3_S- open and -dense (in particular generic) subset $\mathcal{F} \subset C^3(\mathbb{R}^n \times \mathbb{R}, \mathbb{R})^{1+|I|+|J|}$ such that the following holds : if $(f,\cdots,h_i,\cdots,g_j,\cdots) \in \mathcal{F}$, then each point of Σ is exactly one of five types.
We will proceed with a brief discussion of the above announced 5 types. For details and precise characterizations see [120], [155], [159], [160], [161].
Type 1. A point of Type 1 stands for a nondegenerate critical point. In this case the the set Σ of g.c. points can be (locally) parametrized in virtue of the implicit function theorem. Let $\bar{z} = (\bar{x},\bar{t})$, with $\bar{x} \in M(\bar{t})$, be the point of consideration, and suppose that $\bar{x}$ is a critical point for $f(\cdot,\bar{t})|_{M(\bar{t})}$, i.e. (31) is satisfied with $\lambda = 1$. Moreover, the constraint qualification LICQ is assumed to be satisfied. The fact that $\bar{x}$ is a critical point, say with Lagrange multipliers $\bar{\lambda}_i, \bar{\mu}_j$, can be restated by the fact that the following map $\mathcal{T}$ vanishes at $(\bar{x},\bar{\lambda},\bar{\mu})$, where $\mathcal{T}$ maps $\mathbb{R}^{n+r+1}$ to $\mathbb{R}^{n+r}$, $r = |I| + |J_0(\bar{z})|$:

$$\mathcal{T} : \begin{pmatrix} x \\ \lambda \\ \mu \\ t \end{pmatrix} \mapsto \begin{pmatrix} D_x^T f(x,t) - \sum_{i\in I} \lambda_i D_x^T h_i(x,t) - \sum_{j\in J_0(\bar{x},\bar{t})} \mu_j D_x^T g(x,t) \\ -h_i(x,t),\ i \in I \\ -g_j(x,t),\ j \in J_0(\bar{x},\bar{t}) \end{pmatrix} \tag{32}$$

Since the defining functions f, h_i, g_j are of class C^3, the mapping $\mathcal{T}$ is of class C^2 with partial Jacobian matrix $D_{(x,\lambda,\mu)}\mathcal{T}$ at $(\bar{x},\bar{\lambda},\bar{\mu},\bar{t})$:

$$D_{(x,\lambda,\mu)}\mathcal{T}|_{\begin{pmatrix}\overline{x}\\ \overline{\lambda}\\ \overline{\mu}\\ \overline{t}\end{pmatrix}} = \left(\begin{array}{c|c} A & B \\ \hline B^T & 0 \end{array}\right), \tag{33}$$

where $A = D_x^2 L(\overline{x},\overline{t})$ (the Hessian of the Lagrangian w.r.t. the variable x), and $B = (\cdots, -D_x^T h_i, \cdots, -D_x^T g_j, \cdots)|_{(\overline{x},\overline{\lambda},\overline{\mu},\overline{t})}$. Since LICQ is satisfied, the columns of the matrix B are linearly independent. Note that the linear space orthogonal to the columns of B is precisely the tangent space $T_{\overline{x}}M(\overline{t})$ (cf. (18)). Now, the partial Jacobian matrix in (33) is nonsingular if and only if $D_x^2 L|_{T_{\overline{x}}M(\overline{t})}$ is nonsingular. In the latter case, we can apply the implicit function theorem, and we obtain locally defined C^2- functions $x(t), \lambda(t), \mu(t)$, such that $\mathcal{T}(x(t), \lambda(t), \mu(t), t) \equiv 0$. Hence, for t near $\overline{t}$, the point $x(t)$ is a critical point of $f(\cdot,t)|_{M(t)}$, and the set Σ is locally represented by means of the map $t \mapsto (x(t), t)$, see Fig.14. In the special case of local minima cf. also [95].

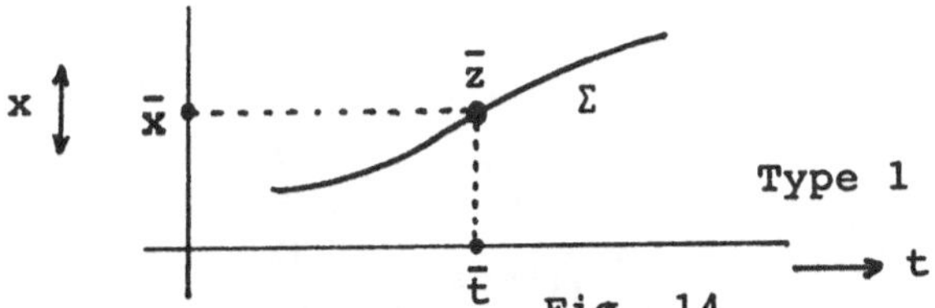

Fig. 14

The characteristics of a point $\overline{z} = (\overline{x},\overline{t})$ of Type 1 are the following: LICQ is satisfied at $\overline{x} \in M(\overline{t})$ and $\overline{x}$ is a nondegenerate critical point for $f(\cdot,\overline{t})|_{M(\overline{t})}$. Now, the nondegeneracy condition (ND2) in (17) implies that the partial Jacobian matrix in (33) is nonsingular which establishes the existence of the mappings $x(t), \lambda(t), \mu(t)$ as above. By continuity, (ND2), but also (ND1), remains valid at the points $x(t)$ for t near $\overline{t}$. Hence, in a neighbourhood of $\overline{z}$ the set Σ consists entirely of nondegenerate critical points with constant local structure according to (29), up to the critical value $c = c(t)$, which now depends on t.

The points of Type 1 form an open and dense set in Σ . The points of Type $2-5$ are isolated points in $\mathbb{R}^n \times \mathbb{R}$. In contrast to nondegenerate critical points, they correspond to the following possible degeneracies (cf. also [247], [248])

$$\left\{\begin{array}{lll} \text{D1. Violation of LICQ :} & & \text{Type 4 , Type 5} \\ \text{D2. Violation of strict complementarity} & \text{(ND1 in (17)) :} & \text{Type 2} \\ \text{D3. Singularity of } D_x^2 L|_{T_{\bar{z}}M(\bar{t})} & \text{(ND2 in (17)) :} & \text{Type 3.} \end{array}\right. \quad (34)$$

Type 2. At a point $\bar{z} = (\bar{x}, \bar{t})$ of Type 2 LICQ is satisfied and we have a critical point $\bar{x} \in M(\bar{t})$ for problem $\mathcal{P}(\bar{t})$

$$D_x f = \sum_{i \in I} \bar{\lambda}_i D_x h_i + \sum_{j \in J_0(\bar{z})} \bar{\mu}_j D_x g_j \quad |_{\bar{z}}. \quad (35)$$

The only degeneracy that occurs is the vanishing of exactly one Lagrange multiplier corresponding to an active inequality constraint, say $\bar{\mu}_p = 0$, where $p \in J_0(\bar{z})$. Then, $\bar{x}$ is a nondegenerate critical point for the problems $\mathcal{P}_1(\bar{t})$ and $\mathcal{P}_2(\bar{t})$: problem $\mathcal{P}_1(\bar{t})$ differs from $\mathcal{P}(\bar{t})$ by omitting the inequality constraint g_p , whereas in $\mathcal{P}_2(\bar{t})$ the constraint g_p is regarded as an equality constraint. Hence, the set Σ around the point $\bar{z}$ now consists of the feasible part of two curves, cf. Fig.15.

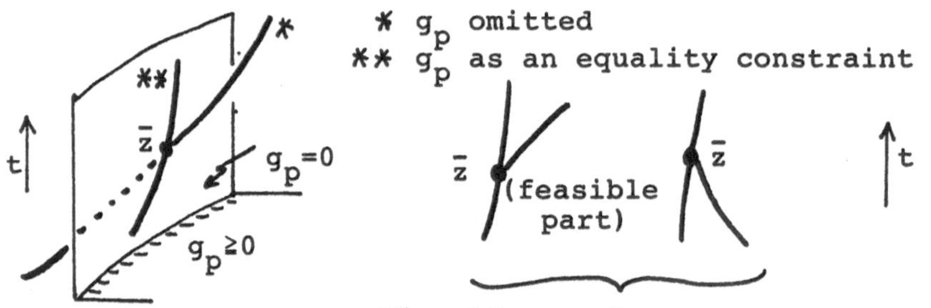

Fig. 15

We note that along the critical curve where g_p is regarded as equality constraint, the Lagrange multiplier corresponding to g_p changes sign when passing the point $\bar{z}$. This implies that only one half of that curve can consist of Kuhn-Tucker points.

Type 3. At a point $\bar{z} = (\bar{x}, \bar{t})$ of Type 3 again LICQ is satisfied and we have a critical point $\bar{x} \in M(\bar{t})$, cf. (35). The only degeneracy occuring consists in the vanishing of exactly one eigenvalue of $D_x^2 L(\bar{z})|_{T_{\bar{z}}M(\bar{t})}$. At the point $\bar{z}$ the set Σ has a quadratic turning point; more precisely, around $\bar{z}$ the set Σ is a one-dimensional C^2-manifold and at $\bar{z}$ the projection $(x,t) \mapsto t$ has a nondegenerate (local) minimum or maximum (Fig.16). In particular, if the functions f, h_i, g_j are smooth (i.e. of class C^∞), then there exists a local smooth coordinate transformation Φ of the type

$$\Phi : (x,t) \mapsto (\varphi_1(x,t), \varphi_2(t)), \quad \varphi_2'(t) > 0 \tag{36}$$

such $f(\cdot,t)$ and $M(t)$ in the new coordinates take the form :

$$\begin{aligned} & x_1^3 \pm tx_1 + \sum_{i=2}^{p} \pm x_i^2 + \sum_{j=p+1}^{p+q} \pm x_j + \delta(t) \\ & x_j \geq 0,\ j = p+1, \cdots, p+q \quad (q = |J_0(\bar{z})|,\ p+q = n - |I|). \end{aligned} \tag{37}$$

The coordinate transformation Φ in (36) preserves the special role of the parameter t. Moreover, it is canonical for each fixed value of t; we will call Φ again <u>canonical</u>.

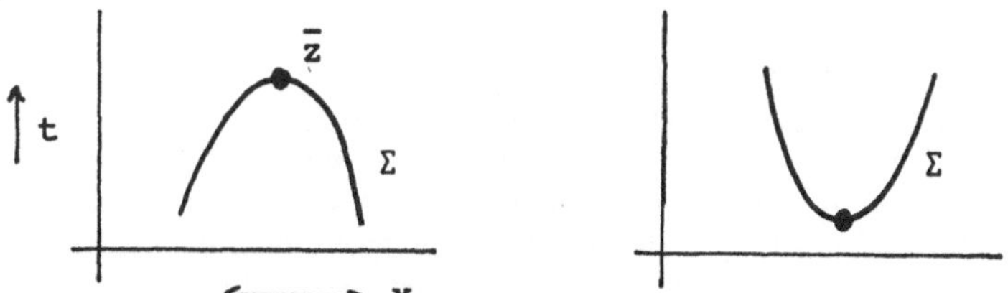

Fig. 16

<u>Type 4</u>. At a point $\bar{z} = (\bar{x}, \bar{t})$ of 4 the linear independence constraint qualification is violated, and moreover the number of active constraints ($= |I| + |J_0(\bar{z})|$) is less than $n+1$ ($=$ dimension(x) + dimension(t)). By means of a canonical coordinate transformation this case is essentially reducable to the case of exactly one (in)equality constraint. Let us consider only the case of one equality constraint h. Then, the point of Type 4 is characterized by means of the following conditions at $\bar{z}$:

$$\left.\begin{array}{ll} C1. & h = 0,\ D_x h = 0,\ D_t h \cdot D_x^2 h \text{ nonsingular} \\ C2. & D_x f \cdot [D_t h \cdot D_x^2 h]^{-1} \cdot D_x^T f \neq 0 \end{array}\right\} \tag{38}$$

Note that $D_x f(\bar{z}) \neq 0$, but $D_x h(\bar{z}) = 0$. From this it follows that the multiplier λ in (31) necessarily vanishes! Condition $C1$ implies that the feasible sets $M(t) = \{x \in \mathbb{R}^n \mid h(x,t) = 0\}$ around $\bar{x}$ behave like level sets of a function around a nondegenerate critical point. For either $t > 0$ or $t < 0$ each level produces a nondegenerate critical point for $f(\cdot,t)|_{M(t)}$ (cf. Fig.17). In (x,t)-space this results in a quadratic turning point for the set Σ; however, its topological nature is essentially different from the situation in Type 3.

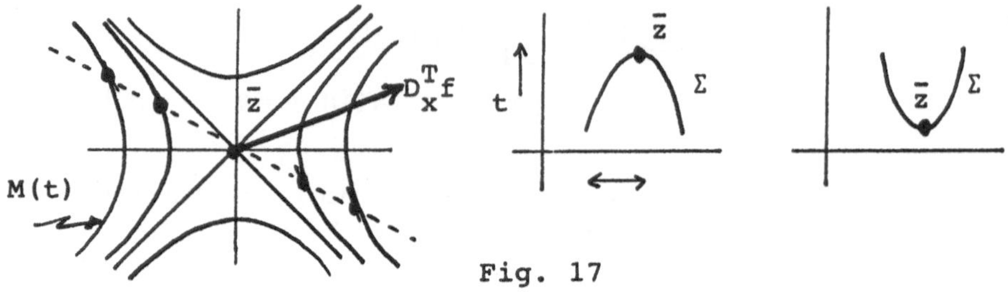

Fig. 17

Apart from the point $\bar{z}$, the points of Σ are nondegenerate critical points for $f(\cdot,t)|_{M(t)}$. So, we can represent these points in the form (35) and we obtain Lagrange multiplier vectors λ, μ at each point. When approaching the point $\bar{z}$ along Σ the lenght $\|\lambda\| + \|\mu\|$ tends to infinity. In particular, all components of μ become unbounded, and they switch sign when passing $\bar{z}$. So, if $J_0(\bar{z}) \neq \emptyset$, then only half of the curve Σ around $\bar{z}$ can consist of Kuhn-Tucker points.

Type 5. At a point $\bar{z} = (\bar{x}, \bar{t})$ of Type 5 the linear independence constraint qualification is violated, but in contrast to Type 4, the number of active constraints equals $n+1$. Let $p = |J_0(\bar{z})|$ denote the number of active inequality constraints. Then, around $\bar{z}$, w.r.t. the orientation of t, the set Σ consists of exactly p (half) curves, each of them either emanating

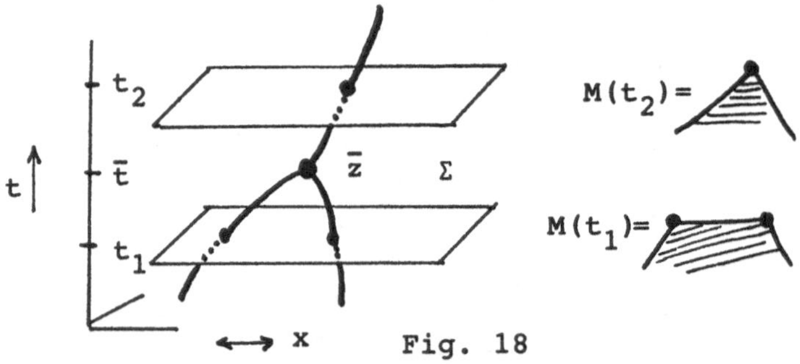

Fig. 18

from $\bar{z}$ or ending at $\bar{z}$ (cf. Fig.18). Moreover, only two (half) curves can be Kuhn-Tucker curves, not both of them ending at $\bar{z}$, resp. emanating from $\bar{z}$.

In Fig.19 we have depicted all possibilities of the structure of the set Σ of generalized critical points (we took only one orientation of the parabolas, etc.). Moreover, in the same figure the closure of the Kuhn-Tucker set, shortly $\overline{KT}$ is sketched. The Mangasarian-Fromovitz constraint qualification plays an interesting role here. In fact, a point $\bar{z} \in \Sigma$ is a boundary point of the set $\overline{KT}$ if and only if both MFCQ is violated and $J_0(\bar{z}) \neq \emptyset$. The set $\overline{KT}$ is a piecewise continuously differentiable one-dimensional manifold.

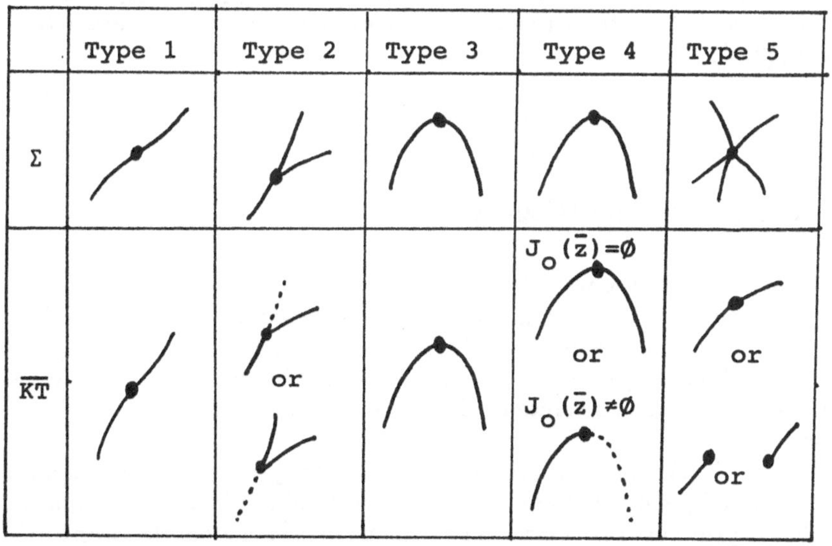

Fig. 19

<u>**Remark 5.1**</u> In the theory of differential equations one arrives at portraits similar to the ones above, too; cf. [67], [116], [325]. Especially <u>bifurcation theory</u> (see Remark 3.1; cf. also [72], [75], [76], [200]) is intimately related to our parametric research. We refer to [281], and to the classification (types) of Poore and Tiahrt ([248], [323]) who also turn to the multi-parametric case (cf. [247]). For an investigation of the multi-parametric Kuhn-Tucker set (and the set of generalized critical points, under generic assumptions) see [162] (for more reading cf. [81], [124], [170]), and as one generalization of the fundamental work [181] on one-parametric programs and strong stability of KT-points see [211] (under LICQ), dealing

with the local change of the stationary index. The ideas of [181] are generalized in [296] by means of a two-parametric study.
For a number of different approaches to parametric optimization see [18], [39], [42], [108], [122], [123], [170], [195], [308], [315] and [62]. Turning to a semi-infinite optimization in one parameter, there exists a modification of our five types' scheme ([280]). Moreover, in one-parametric finite optimization we can give a variant of MFCQ called the Mangasarian Fromovitz constraint qualification at infinity (MFCQI) as a sufficient condition for the feasible set to remain of the same homeomorphy type after a parameter variation over a given interval. For a generalization of this fact to the semi-infinite case cf. [167].
Finally we remark that for the Atiyah-Singer index theory there is a parametrized setting, too (see, e.g., [255]).

6.TRACING OF LOCAL MINIMA.

In this section we discuss the way of tracing local minima along the set of generalized critical points; basic references are [120], [156]. It will be based on the results of Section 5. So, throughout this section we assume that the defining functions f, h_i, g_j belong to the class $\mathcal{F}$ for which the five types of g.c. points are defined. Moreover, we assume that the feasible set $M(t)$ remains compact for all values of the parameter t. Again, we will treat the different types separately.
Type 1. Let $\bar{z} = (\bar{x}, \bar{t})$ represent a local minimum $\bar{x}$ for $f(\cdot, \bar{t})|_{M(\bar{t})}$, and suppose that $\bar{z}$ is of Type 1. Then, all points of Σ around $\bar{z}$ represent local minima as well, and we can trace them by following the curve Σ. This can be done by means of a pathfollowing procedure (cf. [6]), since Σ is locally defined by means of a system of equations.

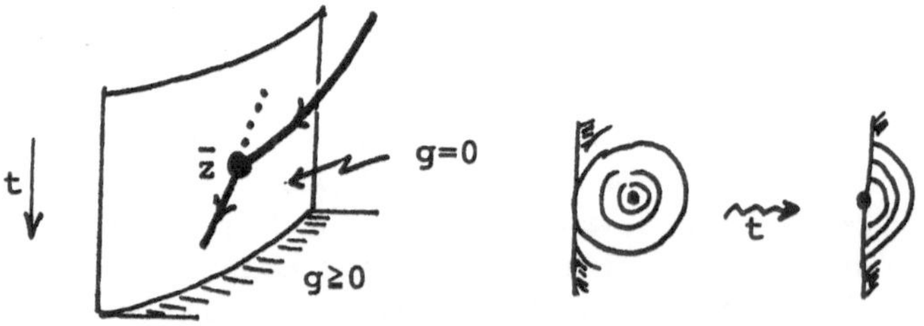

Fig. 20

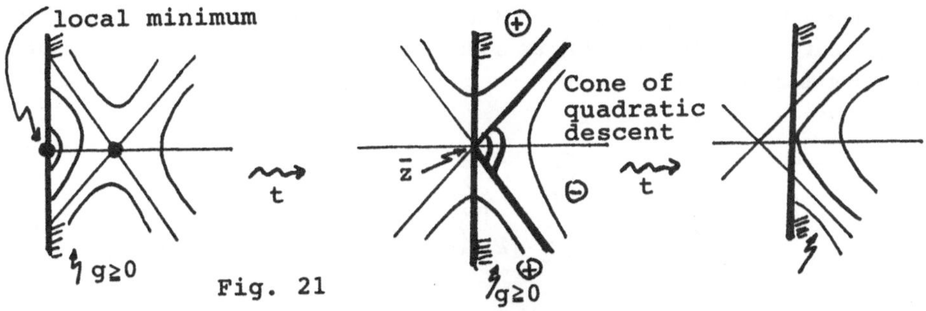

Fig. 21

Type 2. Let us restrict to the case of one inequality constraint g. There are two cases possible.

Case 2.a. As the parameter t increases, a local minimum proceeds from the interior, hits the boundary, and remains on it (or vice versa). See Fig.20.

Case 2.b. As t increases, a local minimum on the boundary disappears when reaching the point $\bar{z}$ (Fig.21). Then, at $\bar{x} \in M(\bar{t})$ we can start minimizing $f(\cdot,\bar{t})|_{M(\bar{t})}$ at $\bar{x}$ by means of a descent procedure. Together with the compactness of $M(\bar{t})$, this guarantees that we reach another branch of Σ, since the disappearing local minimum cannot be a global one (Fig.22). The descent at the point $\bar{x}$ is possible since a cone of directions of quadratic descent is available. Altogether a jump from one branch of local minima to another one is performed.

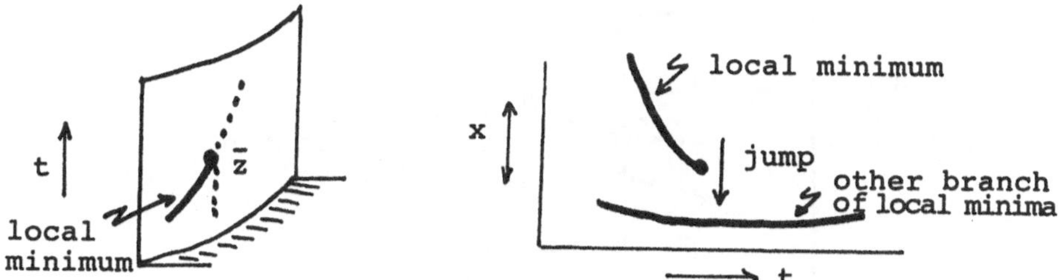

Fig. 22

In [156] a normal form for the situation of Case 2.b is derived, when all functions are smooth. As pointed out by Alina Ruiz ([279]), an additional coordinate transformation brings it into the following elegant form :

$$-(x_1+t)^2+\sum_{i=2}^{p} x_i + \sum_{j=p+1}^{p+q} x_j^2 + \delta(t) \tag{39}$$
$$x_i \geq 0, \quad i=1,\cdots,p \quad (p=|J_0(\bar{z})|,\ p+q=n-|I|).$$

It should be noted that it is easily decidable within the original coordinates which situation (Case 2.a or 2.b) is at hand.

Type 3. Let $\bar{z}=(\bar{x},\bar{t})$ be a point of Type 3 . When passing the point $\bar{z}$ along the path of local minima, the local minimum $x_m(t)$ switches into a Kuhn-Tucker point $x_s(t)$ of quadratic index one. Hence the path of local minima stops at $\bar{z}$, and for t near $\bar{t}$ the local minimum $x_m(t)$ for $f(\cdot,t)|_{M(t)}$ cannot be a global one. Now, the vector $v(t)/\|v(t)\|$, where $v(t)=x_s(t)-x_m(t)$, converges, for t tending to $\bar{t}$, to a direction v which is a direction of cubic descent. This provides the possibility to jump to another branch of local minima, by means of a descent procedure for $t=\bar{t}$, starting at the point $\bar{x}$ (cf. Fig.23).

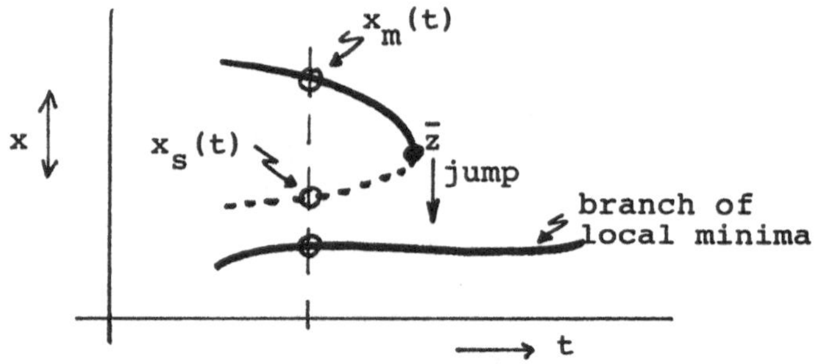

Fig. 23

Type 4. For simplicity, we will explain the situation with the aid of one inequality constraint g , which is supposed to be smooth. However, we note that the subsequent decision whether a jump is possible, remains valid in the general case. In appropriate canonical coordinates around the point $\bar{z} = (\bar{x}, \bar{t})$ of Type 4 we may assume that the function g takes the following form :

$$g(x,t) = \sum_{i=1}^{k} x_i^2 - \sum_{j=k+1}^{n} x_j^2 + \alpha \cdot t \quad (\geq 0), \tag{40}$$

where $\alpha \in \{-1, +1\}$. Now, we assume that we are approaching the point $\bar{z} = (0,0)$ along a branch of local minima (with <u>increasing</u> parameter t).

<u>Case 4.a : $\alpha = +1$</u>. Then it can be shown that necessarily $k = 1$. Moreover, the partial derivatives $f_i := \partial f / \partial x_i(\bar{z})$ satisfy the cone inequality $-f_1^2 + \sum_{j=2}^{n} f_j^2 < 0$. When passing the point $\bar{z}$, the branch of local minima switches into a branch of local maxima. From the geometrical interpretation in Fig.24 we see that for $t < 0, t \approx 0$, the value of f at the corresponding local maximum $x_{max}(t)$ is less than the value at the corresponding local minimum. In fact, the value of f decreases when passing $\bar{z}$ along Σ. Hence, starting for $t < 0, t \approx 0$, at the point $x_{max}(t)$ with a descent procedure for $f(\cdot,t)|_{M(t)}$, we will end up in a local minimum which is different from $x_{min}(t)$; so, by means of this jump we have reached another branch of local minima.

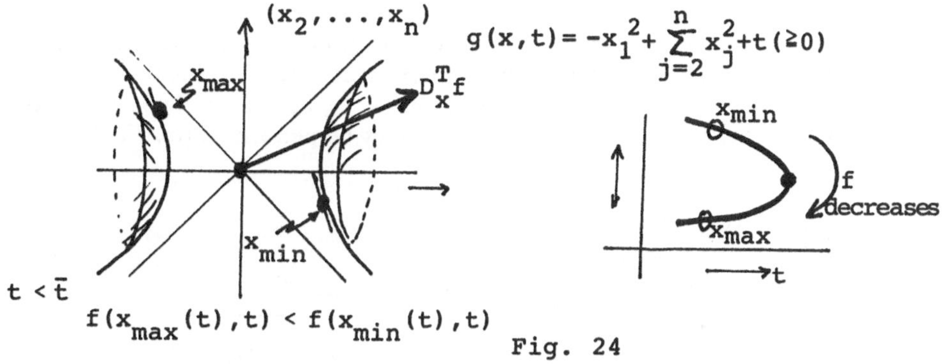

Fig. 24

<u>Case 4.b : $\alpha = -1$</u>. Then, it can be shown that $k = 0$. The inequality $g(x,t) \geq 0$ now takes the form : $\|x\|^2 + t \leq 0$. This means that the feasible set becomes empty as t increases to zero ($= \bar{t}$); in general, the actual connected <u>component</u> of the feasible set becomes empty. In contrast with the situation in Fig.24, for $t < 0, t \approx 0$, the value of f at the corresponding local maximum $x_{max}(t)$ is greater than the value at the corresponding local minimum $x_{min}(t)$ (cf. Fig.25). Moreover, f increases when passing along Σ from the branch of local minima into the branch of local maxima. In this case, we cannot expect to perform a jump to another

branch of local minima. In fact, if we would start at $x_{max}(t)$, $t < 0$, and $t \approx 0$, with a descent procedure, we will necessarily end up in the same local minimum $x_{min}(t)$.

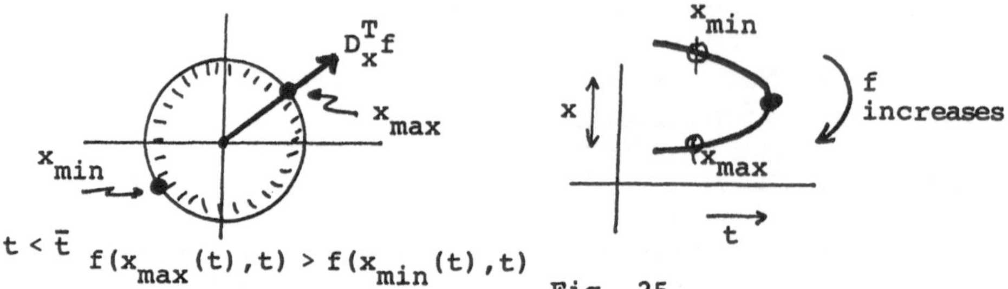

Fig. 25

Type 5. Let $\bar{z} = (\bar{x}, \bar{t})$ be a point of Type 5 . Recall that the set Σ consists of exactly p (half) curves, where $p = |J_0(\bar{z})|$, each of them either emanating from $\bar{z}$ or ending at $\bar{z}$ for increasing t . If we arrive at $\bar{z}$ along a (half) curve of local minima then two situations may occur : In case that MFCQ is satisfied at $\bar{x} \in M(\bar{t})$, then precisely one of the emanating (half) curves consists of local minima. Then we can proceed with tracing local minima. However, if MFCQ is violated, then the actual component of the feasible set becomes empty. In the latter case, the branch of local minima stops at $\bar{z}$, and, as in Type 4 , Case 4.b , a simple jump to another branch of local minima cannot be performed (cf. Fig.26).

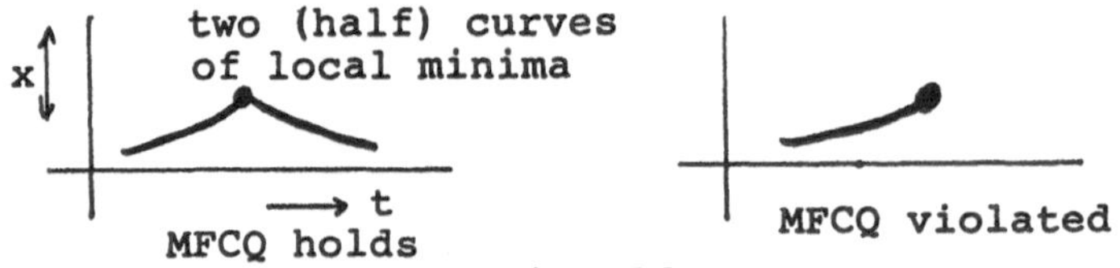

Fig. 26

Remark 6.1 For pathfollowing techniques the Newton method play an important role (cf. [5]). As a survey for the latters we mention [7], [45], [63], [157], [159], [163], [191], [215], [254], [257], [305], [306], [307], [310]. More information on continuation(-like) or embedding methods and on related researches concerning algorithms see [4], [6], [16], [76], [110], [141], [182], [249], [311], [312], [313], [328] and [128], [223], [267] as well as [170], [199], [297], [324], resp.

For embeddings in the context of the Atiyah-Singer index see [274], and for numerical methods in semi-infinite optimization see [9], [135], [136], [137], [139], [149], [151].
As indicated in Remark 4.1, discontinuity, e.g. jumps, is an important phenomenon in the whole area of nonconvex analysis which is associated with (un-)parametrized optimization. For example, we refer to the bang-bang principle in control theory (cf. e.g. [198], [206]), to catastrophes (cf. [140], [141], [143], [170], [308], [321], [346]), and to shocks and jumps in the analysis of partial differential equations (cf. [119], [309], and [134], resp.).

Acknowledgement. The first author would like to thank Lenore Blum, Mike Shub and Steve Smale for their kind hospitality during his stay at ICSI, Berkeley. Moreover, we would like to thank all members of Lehrstuhl C für Mathematik (RWTH Aachen) for technical support during the preparation of this tutorial paper.

REFERENCES.

[1] R. Abraham, J.E. Marsden and T. Ratiu, *Manifolds, Tensor Analysis, and Applications* (Springer-Verlag, 1988).

[2] V.S. Afraimovich, On smooth changes of variables, Selecta Math. Sov. 9 (1990) 205-213.

[3] S. Agmon, *Lectures on Boundary-Value Problems* (D. van Nostrand Company, Princeton, Toronto - New York - London, 1965).

[4] J.C. Alexander, The topological theory of an embedding method, in: *Continuation Methods*, Hj. Wacker, ed. (Academic Press, 1978).

[5] E.L. Allgower and K. Georg, Predictor corrector and simplicial methods for approximating fixed points and zero points of nonlinear mappings, in: *Mathematical Programming, the State of the Art*, A. Bachem et al., eds. (Springer-Verlag, 1983) 15-56.

[6] E.L. Allgower and K. Georg, *Numerical Continuation Methods* (Springer-Verlag, 1990).

[7] W. Alt, The Lagrange-Newton method for infinite-dimensional optimization problems, Num. Funct. Anal. Opt. 11 (1990) 201-224.

[8] A. Ambrosetti, Recent developments in the theory of the existence of periodic orbits of Hamiltonian systems, in: *Advances in Hamiltonian Systems*, J.P. Aubin, ed. (Birkhäuser-Verlag, Boston-Basel-Stuttgart, 1983) 1-22.

[9] E.J. Anderson, A new primal algorithm for semi-infinite linear programming, in: *Infinite Optimization*, E.J. Anderson and A.B. Philpott, eds., Lect. N. Econ. Math. Syst. 259 (Springer-Verlag, 1985) 108-122.

[10] V.I. Arnold, Wave front evolution and equivariant Morse lemma, Comm. Pure Appl. Math. XXIX (1976) 557-582.

[11] V.I. Arnold, S.M. Gusein-Zade and A.N. Varchenko, *Singularities of Differentiable Maps* I (Birkhäuser-Verlag, Boston-Basel-Stuttgart 1985).

[12] V.I. Arnold, S.M. Gusein-Zade, A.N. Varchenko, *Singularities of Differentiable Maps* II (Birkhäuser-Verlag, Boston-Basel-Berlin 1988).

[13] M.F. Atiyah, *K-Theory* (W.A. Benjamin, New York 1967).

[14] M.F. Atiyah, *Elliptic Operators and Compact Groups*, Lect. N. Math. 401 (Springer-Verlag, 1974).

[15] H. Attouch, *Variational Convergence for Functions of Operators* (Pitman Adv. Publ. Progr., Boston-London-Melbourne 1984).

[16] H. Attouch and H. Riahi, The epi-continuation method for minimazation problems. Relation with the degree theory of F. Browder for maximal monotone operator, in : *Partial Differential Equations and the Calculus of Variations* I, F. Colombini, et al., eds. (Birkhäuser-Verlag, Boston-Basel-Berlin 1989) 29-58.

[17] G. Auchmuty, Variational principle for operator equations and initial value problems, Nonl. Anal., M.T.A. 12 (1988) 531-564.

[18] B. Bank, J. Guddat, D. Klatte, B. Kummer and K. Tammer, *Non-Linear Parametric Optimization* (Akademie-Verlag, Berlin 1982).

[19] B. Bank and R. Mandel, *Parametric Integer Programming* (Akademie-Verlag, Berlin 1988).

[20] W. Barth, C. Peters and A. Van de Ven, *Compact Complex Surfaces* (Springer-Verlag, 1984).

[21] P. Bartolo, V. Benci and D. Fortunato, Abstract critical point theorems and applications to some nonlinear problems with "strong" resonance at infinity, Nonl. Anal., T.M.A. 7 (1983) 981-1012.

[22] V. Benci, The direct method in the study of periodic solutions of Hamiltonian systems with prescribed period, in: *Advances in Hamiltonian Systems*, Nonl. Anal., T.M.A. 7 (1983) 929-931.

[23] V. Benci, A new approach to the Morse-Conley theory, in: *Recent Advances in Hamiltonian Systems*, G.F. Dell'Antonio and B. D'Onofrio, eds. (World Scientific, Singapore 1986) 1-52.

[24] V. Benci, A. Capozzi and D. Fortunato, On asymptotically quadratic Hamiltonian systems, Nonl. Anal., T.M.A. 7 (1983) 929-931.

[25] V. Benci and P.H. Rabinowitz, Critical point theorems for indefinite functionals, Invent. math. 52 (1979) 241-273.

[26] A. Ben-Tal, M. Teboulle and J. Zowe, Second order necessary optimality conditions for semi-infinite programming problems, in: *Semi-Infinite Programming*, R. Hettich, ed., Lect. Notes Contr. Inf. Sc. (Springer-Verlag, 1979) 17-30.

[27] M. Berger, Convexity, Am. Math. Month. 97 (1990) 650-678.

[28] M.L. Bertotti, A note on a theorem of Conley and Zehnder, in: *Recent Advances in Hamiltonian Systems*, G.F. Dell'Antonio and B. D'Onofrio, eds. (World Scientific, Singapore 1986) 135-145.

[29] A. Björner, Topological methods (a chapter for "Handbook of Combinatorics", R. Graham et al., eds.), Preprint, Dep. Math., Roy. Inst. Tech., Stockholm (1990).

[30] R. Bott, Lectures on Morse theory, old and new, Bull. (New Series) Amer. Math. Soc. 7 (1982) 331-358.

[31] M.L. Bougeard, Morse theory for some lower-C^2 functions in finite dimensions, Math. Progr. 41 (1988) 141-159.

[32] D. Braess, Morse-Theorie für berandete Mannigfaltigkeiten, Math. Ann. 208 (1974) 133-148.

[33] H. Brauner, *Differentialgeometrie* (Vieweg-Verlag, Braunschweig-Wiesbaden 1981).

[34] H. Brézis, On a characterization of flow-invariant sets, Comm. Pure Appl. Math. XXIII (1970) 621-263.

[35] J. Brink-Spalink and H.Th. Jongen, Morse theory for optimization problems with functions of maximum type, Meth. Oper. Res. 31 (1979) 121-134.

[36] Th. Bröcker and L. Lander, *Differentiable Germs and Catastrohes*, London Math. Soc. Lect. N. Series 17 (Cambridge University Press, 1975).

[37] H.W. Broer, Bifurcations of parametrically forced oscillators, Preprint, University of Groningen, Dept. Math. (1990).

[38] H.W. Broer and G. Vegter, Bifurcational aspects of parametric resonance, Preprint, University of Groningen, Dept. Math. (1990).

[39] B. Brosowski, *Parametric Semi-Infinite Optimization* (Peter Lang Verlag, Frankfurt a.M.-Bern-New York 1982).

[40] A.B. Brown, Functional dependence, Trans. Am. Math. Soc. 38 (1935) 379-394.

[41] J.W. Bruce and R.M. Roberts, Critical points of functions on analytic varieties, Topology 27 (1988) 57-90.

[42] N. Bryson, Applications on the parametric programming procedure, Preprint (1989), Working paper, Inf. Syst. & Anal. Dept. Howard Univ., Washington, DC, submitted for publication (1989).

[43] K. Buchner, Sard's theorem on Banach manifolds, in: *Topics in Differential Geometry* I, J.Szenthe and L. Tamássy, eds., Colloq. Math. Soc. János Bolyai 46 (North-Holland, 1988) 241-244.

[44] M. Buchner, J. Marsden and S. Schecter, Examples for the infinite dimensional Morse lemma, SIAM J. Math. Anal. 14 (1983) 1045-1055.

[45] J.C.P. Bus, Newton-like methods for solving nonlinear simultaneous equations, Meth. Oper. Res. 31 (1976) 143-152.

[46] H. Busemann, *Convex Surfaces* (Interscience Publishers, Inc., New York 1958).

[47] J. Carr, *Applications of Centre Manifold Theory*, Appl. Math. Sc. 35 (Springer-Verlag, 1981).

[48] R.W. Chaney, Optimality conditions for piecewise C^2 nonlinear programming, J. Opt. Th. Appl. 61 (1989) 179-202.

[49] R.W. Chaney, Piecewise C^k functions in nonsmooth analysis, Nonl. Anal., T.M.A. 15 (1990) 649-660.

[50] K.-C. Chang, Variational methods for non-differentiable functionals and their applications to partial differential equations, J. Math. Anal. Appl. 80 (1981) 102-129.

[51] S.-S. Chern, What is geometry? , Am. Math. Month. 97 (1990) 679-686.

[52] S.-N. Chow and J.K. Hale, *Methods of Bifurcation Theory* (Springer-Verlag, 1982).

[53] S.-N. Chow and R. Lauterbach, A bifurcation theorem for critical points of variational problems, Nonl. Anal., T.M.A. 12 (1988) 51-61.

[54] K.N. Chueh, C.C. Conley and J.A. Smoller, Positively invariant regions for systems of nonlinear diffusion equations, Indiana Univ. Math. J. 26 (1977) 373-392.

[55] R.C. Churchill, Invariant sets which carry cohomology, J. Diff. Eq. 13 (1973) 523-550.

[56] F.H. Clarke, Admissible relaxation in variational and control problems, J. Math. Anal. Appl. 51 (1975) 557-576.

[57] F.H. Clarke, *Optimization and Nonsmooth Analysis* (J. Wiley, 1983).

[58] F.H. Clarke and I. Ekeland, Hamiltonian trajectories having prescribed minimal period, Comm. Pure Appl. Math. XXXIII (1982) 101-116.

[59] F.H. Clarke and P.D. Loewen, The value function in optimal control: sensitivity, controllability, and time-optimality, SIAM J. Contr. Opt. 24 (1986) 243-263.

[60] C.V. Coffman, Lyusternik-Schnirelman theory: complementarity principles and the Morse index, Nonl. Anal., T.M.A. 12 (1988) 507-529.

[61] L. Collatz and W. Wetterling, *Optimierungsaufgaben* (Springer-Verlag, 1971).

[62] F. Colonius and K. Kunisch, Sensitivity analysis for optimization problems in Hilbert spaces with bilateral constraints, Schwerpunktprogramm der Deutschen Forschungsgem. "Anwendungsbezogene Opt. u. Steuerung", Preprint 267 (1991).

[63] R. Cominetti and R. Correa, A generalized second-order derivative in nonsmooth optimization, SIAM J. Contr. Opt. 28 (1990) 789-809.

[64] C.C. Conley, A disk mapping associated with the satellite problem, Comm. Pure Appl. Math. XVII (1964) 237-243.

[65] C.C. Conley, On a generalization of the Morse index, in: *Ordinary Differential Equations*, L. Weiss, ed. (Academic Press, 1971) 27-33.

[66] C. Conley, *Isolated Invariant Sets and the Morse Index*, Conf. Board Math. Sc., Regional Conference Series in Mathematics 38 (Amer. Math. Soc., Providence 1978).

[67] C.C. Conley and S. Smoller, Topological techniques in reaction-diffusion equations, in : Biological Growth and Spread, Lect. Notes Biomath. 38 (1980) 473-483.

[68] C.C. Conley and E. Zehnder, Morse-type index theory for flows and periodic solutions for Hamiltonian equations, Comm. Pure Appl. Math. XXXVII (1984) 207-253.

[69] P. Constantin, C. Foias, B. Nicolaenko and R. Temam, *Integral Manifolds and Inertial Manifolds for Dissipative Partial Differential Equations* (Springer-Verlag, 1989).

[70] R. Conti, I. Massabò and P. Nistri, Set-valued perturbations of differential equations at resonance, Nonl. Anal., T.M.A. 4 (1983) 1031-1041.

[71] J.-N. Corvellec, Sur une propriété de deformation en théorie der points critiques, C. R. Acad. Sci. Paris, t.310 Série I (1990) 61-64.

[72] M.G. Crandall and P.H. Rabinowitz, Bifurcation from simple eigenvalues, J. Funct. Anal. 8 (1971) 321-340.

[73] J.N. Damon, Topological triviality and versality for subgroups of A and K, Memoirs of the Amer. Math. Soc. 75 , No. 389 (1988).

[74] J.W. Daniel, Remarks on perturbations in nonlinear inequalities, SIAM J. Num. Anal. 12 (1975) 770-772.

[75] D.W. Decker and H.B. Keller, Multiple limit point bifurcation, J. Math. Anal. Appl. 75 (1980) 417-430.

[76] D.W. Decker and H.B. Keller, Path following near bifurcation, Comm. Pure Appl. Math. XXXIV (1981) 149-175.

[77] M. Degiovanni, Homotopical properties of a class of nonsmooth functions, Ann. Math. pur. appl. CLVI (1990) 37-71.

[78] K. Deimling, *Nichtlineare Gleichungen und Abbildungsgrade* (Springer-Verlag, 1974).

[79] F. Demengel and J.M. Ghidaglia, Some remarks on the smoothness of inertial manifolds, Nonl. Anal., T.M.A. 16 (1991) 79-87.

[80] G. Dinca and D. Mateescu, Well-posed variational problems and Friedrichs extension, Math. Nachr. 148 (1990) 277-291.

[81] A.L. Dontchev and H.Th. Jongen, On regularity of the Kuhn-Tucker curve, SIAM J. Contr. Opt. 24 (1986) 169-176.

[82] A. Dress, E. Ruch and E. Triesch, Extremals in the cone of normals and cohomology of c^3– stratification, Preprint, Adv. Appl. Math., to appear.

[83] Y. Du, A deformation lemma and some critical point theorems, Bull. Austral. Math. Soc. 43 (1991) 161-168.

[84] J.J. Duistermaat, On the Morse index in variational calculus, Adv. Math. 21 (1976) 173-195.

[85] K.L. Duggal, Lorentzian geometry of CR submanifolds, Acta Appl. Math. 17 (1989) 171-193.

[86] S. Eilenberg and N. Steenrod, *Foundations of Algebraic Topology* (Princeton University Press, 1952).

[87] I. Ekeland, On the variational principle, J. Math. Anal. Appl. 47 (1974) 324-353.

[88] I. Ekeland, Une théorie de Morse pour les systèmes hamiltoniens, C. R. Acad. Sc. Paris, Série I, t.296 (1983) 117-120.

[89] I. Ekeland, A Morse theory for Hamiltonian systems, in: *Differential Equations*, I.W. Knowles and R.T. Lewis, eds. (North-Holland, 1984) 165-172.

[90] I. Ekeland and R. Temam, *Convex Analysis and Variational Problems* (North-Holland, 1976).

[91] E.R. Fadell and P.H. Rabinowitz, Bifurcation for odd potential operators and an alternative topological index, J. Funct. Anal. 26 (1977) 48-67.

[92] E.R. Fadell and P.H. Rabinowitz, Generalized cohomological index theories for Lie group actions with an application to bifurcation questions for Hamiltonian systems, Invent. Math. 45 (1978) 139-174.

[93] G. Faltings, Formale Geometrie und homogene Räume, Invent. math. 64 (1981) 123-165.

[94] G. Faltings, Neuere Entwicklungen in der arithmetischen algebraischen Geometrie, in: *Proc. Int. Congr. Math.* I, A.M. Gleason, ed. (AMS, Berkeley 1986) 55-61.

[95] V.A. Fiacco, *Introduction to Sensitivity and Stability Analysis in Nonlinear Programming* (Academic Press, 1983).

[96] A.V. Fiacco and Y. Ishizuwa, Suggested research topics in sensitivity and stability analysis for semi-infinite programming problems, appeared in Annals of Operations Research 27 (1990).

[97] E. Fontich and C. Simó, Invariant manifolds for near identity differentiable maps and splitting of separatrices, Ergod. Th. & Dyn. Syst. 10 (1990) 319-346.

[98] J.M. Franks, Morse-Smale flows and homotopy theory, Topology 18 (1979) 199-215.

[99] D.S. Freed and K.K. Uhlenbeck, *Instantons and Four-Manifolds* (Springer-Verlag, 1984).

[100] S. Fučik, J. Nečas, J.Souček and V. Souček, *Spectral Analysis of Nonlinear Operators*, Lect. N. Math. 346 (Springer-Verlag, 1973).

[101] T. Fujisawa and E.S. Kuh, Piecewise-linear theory and nonlinear networks, SIAM J. Appl. Math. 22 (1972) 307-328.

[102] O. Fujiwara, A note on differentiability of global optimal values, Math. Oper. Res. 10 (1985) 612-618.

[103] H. Furstenberg, *Recurrence in Ergodic Theory and Combinatorical Number Theory* (Princeton University Press, 1981).

[104] R.E. Gaines, Existence of solutions to Hamiltonian dynamical systems of optimal growth, J. Econ. Th. 12 (1976).

[105] R.E. Gaines and J. Mawhin, *Coincidence Degree and Nonlinear Differential Equations*, Lect. N. Math. 568 (Springer-Verlag, 1977).

[106] R.E. Gaines and J.K. Peterson, Periodic solutions to differential inclusions, Nonl. Anal., T.M.A. 5 (1981) 1109-1131.

[107] R.E. Gaines and J.K. Peterson, Degree theoretic methods in optimal control, J. Math. Anal. Appl. 94 (1983) 44-77.

[108] T. Gal, H.-J. Kruse and P. Zörnig, Survey of solved and open problems in the degeneracy phenomenon, Math. Progr. 42 (1988) 125-133.

[109] J. Gauvin, A necessary and sufficient regularity condition to have bounded multipliers nonconvex programming, Math. Progr. 12 (1977) 136-138.

[110] H. Gfrerer, J. Guddat, Hj. Wacker and W. Zulehner, Path-following for Kuhn-Tucker curves by an active index set strategy, in: *System and Optimization*, A. Bagchi and H.Th. Jongen, eds., Lect. Notes Contr. Inform. Sc. 66 (1985) 111-132.

[111] K. Glashoff, Duality theory of semi-infinite programming, in: *Semi-Infinite Programming*, R. Hettich, ed., Lect. Notes Contr. Inf. Sc. (Springer-Verlag, 1979) 1-16.

[112] M.A. Goberna and M.A. López, Conditions for the closedness of the characteristic cone associated with an infinite linear system, in: *Infinite Programming*, E.J. Anderson and A.B. Philpott, eds., Lect. Notes Econ. Math. Syst. 259 (Springer-Verlag,) 16-28.

[113] M. Golubitsky and J. Marsden, The Morse lemma in infinite dimensions via singularity theory, SIAM J. Math. Anal. 14 (1983) 1037-1044.

[114] M. Golubitsky and D.G. Schaeffer, *Singularities and Groups in Bifurcation Theory I* (Springer-Verlag, 1985).

[115] M. Goretsky and R. Mac Pherson, *Stratified Morse Theory* (Springer-Verlag, 1988).

[116] B.F. Gray and R.A. Thuraisingham, The cubic autocatalator: the influence of degenerate singularities in a closed system, J. Engin. Math. 23 (1989) 283-293.

[117] A.O. Griewank, Generalized descent for global optimization, J. Opt. Th. Appl. 34 (1981) 11-39.

[118] D. Gromoll and W. Meyer, On differentiable functions with isolated critical points, Topology 8 (1969) 361-369.

[119] J. Guckenheimer, Solving a single conservation law, in: *Geometry and Topology*, J. Palis and M. do Carmo, eds., Lect. N. Math. 597 (1977) 108-134.

[120] J. Guddat, F. Guerra Vasquez and H.Th. Jongen, *Parametric Optimization: Singularities, Pathfollowing and Jumps* (John Wiley, 1990).

[121] J. Guddat and H.Th. Jongen, Structural stability and nonlinear optimization, Optimization 18 (1987) 617-631.

[122] J. Guddat, H.Th. Jongen, B. Kummer and F. Nožička (eds.) *Parametric Optimization and Related Topics*, Mathematical Research 35 (Akademie-Verlag, Berlin 1987).

[123] J. Guddat, H.Th. Jongen, B. Kummer and F. Nožička (eds.) *Parametric Optimization and Related Topics* II, Mathematical Research 62 (Akademie-Verlag, Berlin 1991).

[124] J. Guddat, H.Th. Jongen and D. Nowack, Parametric optimization: pathfollowing with jumps, in: *Approximation and Optimization*, A. Gómez et al., eds., Lect. Notes Math. 1354 (Springer-Verlag, 1988) 43-53.

[125] J. Guddat, H.Th. Jongen and J. Rückmann, On stability and stationary points in nonlinear optimization, J. Austral. Math. Soc., Ser.B 28 (1986) 36-56.

[126] S.-Å. Gustafson and K.O. Kortanek, Semi-infinite programming and applications, in: *Mathematical Programming, the State of the Art*, A. Bachem et al., eds. (Springer-Verlag, 1984) 132-157.

[127] E. Gutkin and R. Johnson, Intersection theory for linear eigenvalue problems, J. reine angew. Math. 401 (1989) 1-24.

[128] J. Gwinner and W. Oettli, Duality and regularization for inf-sup problems, Results Math. 15 (1989) 227-237.

[129] W.W. Hager, Lipschitz continuity for continuity for constrained processes, SIAM J. Contr. Opt. 17 (1979) 321-338.

[130] S.-P. Han, On the Hessian of Lagrangian and second order optimality conditions, SIAM J. Contr. Opt. 24 (1986) 339-345.

[131] R.M. Hardt, Some analytic bounds for subanalytic sets, in: *Differential Geometric Control Theory*, Proceedings of the Conference held at Michigan Technological University, 1982, R.W. Brockett et al., eds. (Birkhäuser-Verlag, Boston-Basel-Stuttgart 1983) 259-267.

[132] R. Hartshorne, *Algebraic Geometry* (Springer-Verlag, 1977).

[133] E. Heinz, An elementary analytic theory of the degree of mappings in n-dimensional space, J. Math. Mech. 8 (1959) 519-574.

[134] D. Henry, *Geometric Theory of Semilinear Parabolic Equations*, Lect. Notes Math. 840 (Springer-Verlag, 1981).

[135] R. Hettich, A Comparison of some numerical methods for semi-infinite programming, in : *Semi-Infinite Optimization*, R. Hettich, ed., Lect. Notes Contr. Inf. Sc. (Springer-Verlag, 1979) 112-125.

[136] R. Hettich and G. Gramlich, A note on an implementation of a method for quadratic semi-infinite programming, Math. Progr. 46 (1990) 249-254.

[137] R. Hettich and W. van Honstede, On quadratically convergent methods for semi-infinite programming, in : *Semi-Infinite Optimization*, R. Hettich, ed., Lect. Notes Contr. Inf. Sc. (Springer-Verlag, 1979) 97-111.

[138] R. Hettich and H.Th. Jongen, Semi-infinite programming: conditions of optimality and applications, in: *Optimization Techniques 2*, J. Stoer, ed., Lect. Notes Contr. Inf. Sc. (Springer-Verlag, 1978) 1-11.

[139] R. Hettich and P. Zencke, *Numerische Methoden der Approximation und semi-infiniten Optimierung* (Teubner Studienbücher, Stuttgart 1982).

[140] G. Hetzer, A diffusion system for climate modeling, in: *Differential Equations and Applications* I, A.R. Aftabizadeh, ed., Proc. Int. Conf. Theory Appl. Diff. Equat. (1989) 441-444.

[141] G. Hetzer, H. Jarausch and W. Mackens, A multiparameter sensitivity analysis of a $2D$ diffusive climate model, Impact of Computing in Science and Engineering 1 (1989) 327-393.

[142] G. Hetzer and P.G. Schmidt, A global attractor and stationary solutions for a reaction-diffusion system arising from climate modelling, Nonl. Anal., T.M.A. 14 (1990) 915-926.

[143] G. Hetzer and P.G. Schmidt, Global existence and asymptotic behaviour for a quasilinear reaction-diffusion system from climate modelling, J. Math. An. Appl., to appear.

[144] J.-B. Hiriart-Urruty and A. Seeger, The second order subdifferential and the Dupin indicatrices of a non-differentiable convex function, Proc. London Math. Soc. 58 (1989) 351-365.

[145] J.-B. Hiriart-Urruty and A. Seeger, Calculus rules on a new set-valued second order derivative for convex functions, Nonl. Anal., T.M.A. 13 (1989) 721-738.

[146] M.W. Hirsch, *Differential Topology* (Springer-Verlag, 1976).

[147] M.W. Hirsch and C.C. Pugh, Stable manifolds and hyperbolic sets , Proc. Symp. Pure Math. 14, Amer. Math. Soc. (1970) 133-164.

[148] W.W. Hogan, Point-to-set maps in mathematical programming, SIAM Rev. 15 (1973) 591-603.

[149] W. van Honstede, An approximation method for semi-infinite problems, in : *Semi-Infinite Optimization*, R. Hettich, ed., Lecture Notes in Control and Information Sciences (Springer-Verlag,1979) 126-136.

[150] W.-Y. Hsiang, R.S. Palais and Ch.-L. Terng, The topology of isoparametric submanifolds, J. Diff. Geom. 27 (1988) 423-460.

[151] H. Hu, A one-phase algorithm for semi-infinite linear programming, Math. Progr. 46 (1990) 85-103.

[152] A.D. Ioffe, Regular points of Lipschitz functions, Trans. Amer. Math. Soc. 251 (1979) 61-69.

[153] A.D. Ioffe, Approximate subdifferentials and applications I: the finite dimensional theory, Trans. Amer. Math. Soc. 281 (1984) 389-416.

[154] A.D. Ioffe, Calculus of Dini subdifferentials of functions and contingent coderivatives of set-valued maps, Nonl. Anal., T.M.A. 8 (1984) 517-539.

[155] H.Th. Jongen, Three lectures on nonlinear optimization, Mathematical Methods in Operations Research, P. Kenderov, ed. (Sofia, Bulgaria 1985) 60-69.

[156] H.Th. Jongen, Parametric optimization: critical points and local minima, Lect. Appl. Math. 26 (1990) 317-335.

[157] H.Th. Jongen, P. Jonker and F. Twilt, On Newton flows in optimization, Meth. Oper. Res. 31 (1979) 345-359.

[158] H.Th. Jongen, P. Jonker and F. Twilt, *Nonlinear Optimization in $\mathbb{R}^n$, I: Morse Theory, Chebychev Approximation* (Peter Lang Verlag, Frankfurt a.M.-Bern-New York, 1983).

[159] H.Th. Jongen, P. Jonker and F. Twilt, *Nonlinear Optimization in $\mathbb{R}^n$, II: Transversality , Flows, Parametric Aspects* (Peter Lang Verlag, Frankfurt a.M.-Bern-New York, 1986).

[160] H.Th. Jongen, P. Jonker and F. Twilt, One-parameter families of optimization problems: equality constraints, J. Optim. Th. Appl. 48 (1986) 141-161.

[161] H.Th. Jongen, P. Jonker and F. Twilt, Critical sets in parametric optimization, Math. Progr. 34 (1986) 333-353.

[162] H.Th. Jongen, P. Jonker and F. Twilt, Parametric optimization: the Kuhn-Tucker set, in: *Parametric Optimization and Related Topics*, J.Guddat et al., eds. (Akademie-Verlag, Berlin 1987) 196-208.

[163] H.Th. Jongen, P. Jonker and F. Twilt, A note on Branin's method for finding the critical points of smooth functions, in: *Parametric Optimization and Related Topics*, J. Guddat et al., eds. (Akademie-Verlag, Berlin 1987) 209-228.

[164] H.Th. Jongen, D. Klatte and K. Tammer, Implicit functions and sensitivity of stationary points, Math. Progr. 49 (1991) 123-138.

[165] H.Th. Jongen and D. Pallaschke, On linearization and continuous selections of functions, Optimization 19 (1988) 343-353.

[166] H.Th. Jongen and J. Rückmann, Nonlinear optimization: on connected components of level sets, Preprint No. 24, Lehrstuhl C für Mathematik, RWTH Aachen (1990), submitted for publication.

[167] H.Th. Jongen, J. Rückmann and G.-W. Weber, One-parametric nonlinear optimization: on the stability of the feasible set (in preparation).

[168] H.Th. Jongen and J. Sprekels, The index k-stabilizing differential equation, OR-Spektrum 2 (1981) 223-225.

[169] H.Th. Jongen, F. Twilt and G.-W. Weber, Semi-infinite optimization: structure and stability of the feasible set, J. Opt. Th. Appl., to appear.

[170] H.Th. Jongen and G.-W. Weber, On parametric nonlinear programming, Annals of Operations Research 27 (1990) 253-284.

[171] H.Th. Jongen and G.-W. Weber, Nonlinear optimization : characterization of structural stability, J. Global Optimization, to appear.

[172] H.Th. Jongen and G. Zwier, On the local structure of the feasible set in semi-infinite optimization, in: *Parametric Optimization and Approximation*, B. Brosowski and F. Deutsch, eds., ISNM 72 (Birkhäuser-Verlag, Basel-Boston-Stuttgart, 1985) 185-202.

[173] H.Th. Jongen and G. Zwier, On regular semi-infinite optimization, in: *Infinite Programming*, E.J. Anderson and A.B. Philpott, eds., Lect. Notes Econ. Math. Syst. 259 (Springer-Verlag, 1985) 53-64.

[174] H.Th. Jongen and G. Zwier, On regular minimax optimization, J. Opt. Th. Appl. 62 (1989) 419-447.

[175] P. Kall, *Mathematische Methoden des Operations Research* (Teubner-Verlag, Stuttgart 1976).

[176] H. Kawasaki, Second order necessary optimality conditions for minimizing a sup-type function, Math. Progr. 49 (1991) 213-229.

[177] K.C. Kiwiel, *Methods of Descent for Nondifferentiable Optimization*, Lect. Notes Math. 1133 (Springer-Verlag, 1985).

[178] D. Klatte and K. Tammer, Strong stability of stationary solutions and Karush-Kuhn-Tucker points in nonlinear optimization, Annals of Operations Research 27 (1990) 285-308.

[179] F. Klein, *Vorlesung über die Entwicklung der Mathematik im 19. Jahrhundert, II: Die Grundbegriffe der Invariantentheorie und ihr Eindringen in die mathematische Physik* (Springer-Verlag, 1927).

[180] M. Kojima, Strongly stable stationary solutions in nonlinear programs, in: *Analysis and Computation of Fixed Points*, S.M. Robinson, ed. (Academic Press, 1980) 93-138.

[181] M. Kojima and R. Hirabayashi, Contiuuous deformations of nonlinear programs, Math. Progr. Study 21 (1984) 150-198.

[182] M. Kojima, S. Mizuno and T. Noma, A new continuation method for complementarity problems with uniform P-functions, Math. Prog. 43 (1989) 107-113.

[183] J. Kollár, The structure of algebraic threefolds: an introduction to Mori's program, Bull. (New Series) Am. Math. Soc. 17 (1987) 211-273.

[184] A.N. Kolmogorov, On the representation of continuous functions of several variables, Am. Math. Soc. Transl. 17 (1961) 369-373.

[185] A.I. Korablev, $\epsilon-$ subgradient method for the solution of nonlinear extremal problems, J. Sov. Math. 43 (1988) 2419-2425.

[186] A.I. Korablev, Relaxation method of minimization of pseudo-convex functions, J. Sov. Math. 44 (1989) 1-5.

[187] K.O. Kortanek, Semi-infinite programming and continuum physics, in : *Infinite Programming*, E.J. Anderson and A.B. Philpott, eds., Lect. N. Econ. Math. Syst. 259 (Springer-Verlag, 1985) 65-78.

[188] W. Kotarski, On some specification of the Dubovicki-Milutin theorem for Pareto optimal problems, Nonl. Anal., T.M.A. 14 (1990) 287-291.

[189] W. Krabs, *Optimization and Approximation* (John Wiley, 1979).

[190] D. Kuhn and R. Löwen, Piecewise affine bijections of $\mathbb{R}^n$, and the equation $Sx^+ - Tx^- = y$, Lin. Alg. Appl. 96 (1987) 109-129.

[191] B. Kummer, Newton's method for non-differentiable functions, in: *Advances in Mathematical Optimization* 45, J. Guddat et al., eds. (Akademie-Verlag, Berlin 1988) 114-125.

[192] B. Kummer, Some pathological Lipschitzian functions, Preprint, Humboldt-University, Berlin, Department of Mathematics (1988).

[193] B. Kummer, An implicit-function theorem for $C^{0,1}-$ equations and parametric $C^{1,1}-$ optimization, Math. Anal. Appl., to appear.

[194] B. Kummer, The inverse of a Lipschitz function in $\mathbb{R}^n$: complete characterization by directional derivatives, Preprint, Humboldt-University, Berlin, Department of Mathematics (1989).

[195] J. Kyparisis, Solution differentiability for variational. inequalities, Math. Progr. 48 (1990) 285-301.

[196] A. Langenbach, Über lipschitzstetige Funktionen, Z. Anal. Anwendungen 8 (1989) 289-292.

[197] A.C. Lazer and S. Solimini, Nontrivial solutions of operator equations and Morse indices of critical points of minmax type, Nonl. Anal., T.M.A. 12 (1988) 761-775.

[198] E.B. Lee and L. Markus, *Foundations of Optimal Control Theory* (John Wiley, 1969).

[199] T.Y. Li, T. Sauer and J.A. Yorke, The Cheater's homotopy: an efficient procedure for solving systems of polynomial equations, SIAM J. Numer. Anal. 26 (1989) 1241-1251.

[200] P.L. Lions, Bifurcation and optimal stochastic control, Nonl. Anal., T.M.A. 7 (1983) 79-90.

[201] L.A. Ljusternik and L.G. Schnirelman, *Topological Methods in the Calculus of Variations* (Hermann, Paris 1934).

[202] L.A. Ljusternik and W.I. Sobolew, *Elemente der Funktionalanalysis* (Akademie-Verlag, Berlin 1960).

[203] N.G. Lloyd, *Degree Theory* (Cambridge University Press, 1978).

[204] B. Luderer and R. Rösiger, On Shapiro's result in quasidifferential calculus, Math. Progr. 46 (1990) 403-407.

[205] O.L. Mangasarian and S. Fromovitz, The Fritz John necessary optimality conditions in the presence of equality and inequality constraints, J. Math. Anal. Appl. 17 (1967) 37-47.

[206] L. Markus, Optimal control of limit cycles or what control theory can do to cure a heart attack or to cause one, in: *Symposium on Ordinary Differential Equations*, Minneapolis, W.A. Harris and Y. Sibuya, eds. (Springer-Verlag, 1973) 108-134.

[207] A. Marino, The calculus of variations and some semilinear variational inequalities of elliptic and parabolic type, in : *Partial Differential Equations and the Calculus of Variations* II , F. Colombini, et al. , eds. (Birkhäuser-Verlag, Boston-Basel-Berlin 1989) 787-822.

[208] J.E. Marsden and M. Mc Cracken, *The Hopf Bifurcation and Its Applications* (Springer-Verlag, 1976).

[209] L. Martein, Lagrange multipliers and generalized differentiable functions in vector extremum problems, J. Opt. Th. Appl. 63 (1989) 281-297.

[210] J. Mather, Stability of C^∞ mappings, III: finitely determined map-germs, Publ. Math. I.H.E.S. 35 (1968) 127-156.

[211] T. Matsumoto, Change of stationary index of multi-parametrized nonlinear programs, I, Preprint, Fish. Manag. Econ. Div., Nat. Res. Inst. Fish. Sc., Tokyo (1990).

[212] H. Maurer and J. Zowe, Second-order necessary and sufficient optimality conditions for infinite-dimensional programming problems, in: *Optimization Techniques* 2, J. Stoer, ed., Lect. Notes Contr. Inf. Sc. (Springer-Verlag, 1978) 13-21.

[213] J. Mawhin, *Topological Degree Methods in Nonlinear Boundary Value Problems*, Regional Conference Series in Mathematics 40 (1979).

[214] M.C. Memory, Stable and unstable manifolds for partial functional differential equations, Nonl. Anal., T.M.A. 16 (1991) 131-142.

[215] H.-G. Meier, *Diskrete und kontinuierliche Newton-Systeme im Komplexen*, Thesis, RWTH Aachen (1991).

[216] R. Mifflin, Semismooth and semiconvex functions in constrained optimization, SIAM J. Contr. Opt. 15 (1977) 959-972.

[217] J. Milnor, *Morse Theory*, Annals of Math. Studies 51 (Princeton University Press, 1963).

[218] J. Milnor, On the Betti numbers of real varieties, Proc. Amer. Math. Soc. 15 (1964) 275-280.

[219] J. Milnor, *Lectures on the h-Cobordism Theorem*, Math. Notes 1 (Princeton University Press, 1965).

[220] J. Montgomery, Cohomology of isolated invariant sets under perturbation, J. Diff. Eq. 13 (1973) 257-299.

[221] J. Moser, Periodic orbits near an equilibrium and a theorem of A. Weinstein, Comm. Pure Appl. Math. XXIX (1976) 727-747.

[222] D. Motreanu, Existence in optimization with nonconvex constraints, J. Math. Anal. Appl. 117 (1986) 128-137.

[223] D. Motreanu, Embeddings of C^∞ -subcarţesian spaces, Analele ştienţifice ale Universităţii "Al.I. Cuza" din Iaşi, T. XXV, S.Ia, f.1 (1979) 65-70.

[224] D. Motreanu, Optimization problems on complete Riemannian manifolds, Colloq. Math. LIII (1987) 229-238.

[225] D. Motreanu, Tangent vectors to sets in the theory of geodesics, Nagoya Math. J. 106 (1987) 29-47.

[226] D. Motreanu, Morse functions arising in elliptic problems, Preprint (1990).

[227] D. Motreanu and N.H. Pavel, Quasi-tangent vectors in flow-invariance and optimization problems in Banach manifolds, J. Math. Anal. Appl. 88 (1982) 116-132.

[228] M. Mrozek, The Conley index on compact ANR's is of finite type, Results in Mathematics 18 (1990) 306-313.

[229] J.R. Munkres, *Elements of Algebraic Topology* (Addison-Wesley, 1984).

[230] Z. Naniewicz, On some nonconvex variational problems related to hemivariational inequalities, Nonl. Anal., T.M.A. 13 (1989) 87-100.

[231] O. Náther, On certain characterization of generalized continuity of multifunctions, Univ. Comeniana, Acta Math. Univ. Comenianae L-LI (1987) 75-88.

[232] L. Nirenberg, Variational and topological methods in nonlinear problems, Bull. (New Series) Amer. Math. Soc. 4 (1981) 267-302.

[233] T. Nishizeki and N. Chiba, *Planar Graphs: Theory and Algorithms* (North-Holland, 1988).

[234] P. Nistri, Periodic control problems for a class of nonlinear periodic differential systems, Nonl. Anal., T.M.A. 7 (1983) 79-90.

[235] J.L. Noakes, Fitting maps of the plane to experimental data, Math. Progr. (1989) 225-234.

[236] R.S. Palais, Morse theory on Hilbert manifolds, Topology 2 (1963) 299-340.

[237] R.S. Palais, *Seminar on the Atiyah-Singer Index Theorem* (Princeton University Press, 1965).

[238] R.S. Palais, Homotopy theory of infinite dimensional manifolds, Topology 5 (1966) 1-16.

[239] R.S. Palais, Lusternik-Schnirelman theory on Banach manifolds, Topology 5 (1966) 115-132.

[240] R.S. Palais and C.-l. Terng, *Critical Point Theory and Submanifold Geometry*, Lect. N. Math. 1353 (Springer-Verlag, 1988).

[241] R.S. Palais and S. Smale, A generalized Morse theory, Bull. Amer. Math. Soc. 70 (1964) 165-172.

[242] D. Pallaschke and P. Recht, On extensions of the second-order derivative, in: *Trends in Mathematical Optimization*, K.-H. Hoffmann et al., eds., ISNM 84 (Birkhäuser-Verlag, Basel 1988) 247-258.

[243] J.-P. Penot, Metric regularity, openess and Lipschitzian behaviour of multifunctions, Nonl. Anal., T.M.A. 13 (1989) 629-643.

[244] J.K. Peterson, *Degree Theoretic Methods in Optimal Control*, Thesis, State University, Fort Collins, Colorado (1980).

[245] R. Pignoni, Density and stability of Morse functions on a stratified space, Ann. Scuola Norm. Sup Pisa C1. Sci 4 (1979) 592-608.

[246] A. Pommelet, *Analyse convex et théorie de Morse*, Dissertation, Université de Paris-Dauphine (1982).

[247] A.B. Poore, Bifurcations in parametric nonlinear programming, Preprint, appeared in Annals of Operations Research 27 (1990).

[248] A.B. Poore and C.A. Tiahrt, Bifurcation problems in nonlinear parametric programming, Math. Progr. 39 (1987) 189-205.

[249] M. Prüfer, Calculating global bifurcation, in: *Continuation Methods*, Hj. Wacker, ed. (Academic Press, 1978) 187-213.

[250] P. Quittner, An instability criterion for variational inequalities, Nonl. Anal., T.M.A. 15 (1990) 1167-1180.

[251] P.H. Rabinowitz, Periodic solutions of Hamiltonian systems, Comm. Pure Appl. Math. XXXI (1978) 157-184.

[252] P.H. Rabinowitz, *Minimax Methods in Critical Point Theory with Applications to Differential Equations*, AMS-Regional Conference Series in Mathematics 65 (American Mathematical Society, Providence, 1986).

[253] P. Recht, On generalized gradients, Preprint, Inst. Stat. Math. Econ., University of Karlsruhe (1990).

[254] J. Renegar, Rudiments of an averaging case complexity theory for piecewise-linear path following algorithms, Math. Progr. 40 (1988) 113-163.

[255] R. Rennie, Geometry and topology of chiral anomalies in gauge theories, Adv. Phys. 39 (1990) 617-779.

[256] W.C. Rheinboldt and J.S. Vandergraft, On piecewise affine mappings in $\mathbb{R}^n$, SIAM J. Appl. Math. 29 (1975) 680-689.

[257] S.M. Robinson, Extension of Newton's method to nonlinear function with values in a cone, Num. Math. 19 (1972) 341-347.

[258] S.M. Robinson, Normed convex processes, Trans. Amer. Math. Soc. 174 (1972) 127-140.

[259] S.M. Robinson, Stability theory for systems of inequalities, part I: linear systems, SIAM J. Num. Anal. 12 (1975) 754-769.

[260] S.M. Robinson, Stability theory for systems of inequalities, part II: differentiable nonlinear systems, SIAM J. Numer. Anal. 13 (1976) 497-513.

[261] S.M. Robinson, Generalized equations and their solutions, part I: basic theory, Math. Progr. Study 10 (1979) 128-141.

[262] S.M. Robinson, Strongly regular generalized equations, Math. Oper. Res. 5 (1980) 43-62.

[263] S.M. Robinson, Generalized equations and their solutions, part II: applications to nonlinear programming, Math. Progr. Study 19 (1982) 200-221.

[264] S.M. Robinson, Generalized equations, in: *Math. Progr., The State of the Art*, A. Bachem et al., eds. (Springer-Verlag, 1983) 346-367.

[265] S.M. Robinson, Local structure of feasible sets in nonlinear programming, part I: regularity, in: *Numerical Methods*, V. Pereyra and A. Reinoza, eds., Lect. N. Math. 1005 (Springer-Verlag, 1983) 240-251.

[266] S.M. Robinson, Local structure of feasible sets in nonlinear programming, part III: stability and sensitivity, Math. Progr. Study 30 (1987) 45-66.

[267] S.M. Robinson, Mathematical foundations of nonsmooth embedding methods, Math. Progr. 48 (1990) 221-229.

[268] S.M. Robinson, Normal maps induced by linear transformations, Math. Progr., to appear.

[269] R.T. Rockafellar, *Convex Analysis* (Princeton University Press, 1970).

[270] R.T. Rockafellar, Saddle points of Hamiltonian systems in convex Lagrange problems having a nonzero discount rate, J. Econ. Theory 12 (1976) 71-113.

[271] R.T. Rockafellar, Integral functionals, normal integrands and measurable selections, Lect. N. Math. 543 (Springer-Verlag, 1977) 157-207.

[272] R.T. Rockafellar, Lipschitzian stability in optimization: the role of nonsmooth analysis, in: *Nondifferentiable Optimization: Motivations and Applications*, V.F. Demyanov and D. Pallaschke, eds., Lect. N. Econ. Math. Syst. 255 (Springer-Verlag, 1985) 55-73.

[273] R.T. Rockafellar, First- and second-order epi-differentiability in nonlinear programming, Trans. Amer. Math. Soc. 307 (1988) 75-108.

[274] J. Roe, An index theorem on open manifolds. I, J. Diff. Geom. 27 (1988) 87-113.

[275] J. Roe, An index theorem on open manifolds. II, J. Diff. Geom. 27 (1988) 115-136.

[276] J. Roe, Exotic cohomology and index theory, Bull. (New Series) Amer. Math. Soc. 23 (1990) 447-453.

[277] H. Rosenberg, A generalization of Morse-Smale inequalities, Bull. Amer. Math. Soc. 70 (1964) 422-427.

[278] J. Rückmann and G.-W. Weber, A characterization of excisional stability in semi-infinite optimization (in preparation, 1991).

[279] A. Ruiz, personal communication (1990).

[280] Th. Rupp, *Kontinuitätsmethoden zur Lösung einparametrischer semi-infiniter Optimierungsprobleme*, Thesis, University of Trier, Germany (1988).

[281] A. Ruske, Bifurcation in case of one-parametric nonlinear optimization problems, Preprint, Humboldt-University, Berlin, Department of Mathematics (1989), submitted for publication.

[282] K.P. Rybakowski, On the homotopy index for infinite-dimensional semiflows, Trans. Amer. Math. Soc. 269 (1982) 351-382.

[283] B.P. Rynne, Bifurcation from eigenvalues in nonlinear multiparameter problems, Nonl. Anal., T.M.A. 15 (1990) 185-198.

[284] R.S. Sadyrkhanov, Local and global implicit function theorem and equations with singularities , Sov. Math. Dokl. 40 (1990) 410-412.

[285] H. Samuelson, R.M. Thrall and J. Wesler, A partition theorem for Euclidean n-space, Proc. Amer. Math. Soc. 9 (1958) 805-807.

[286] L.A. Santalo, *Integral geometry and geometric probability* (Addison-Wesley, 1976).
[287] R. Schoen and S.-T. Yau, Existence in incompressible minimal surfaces and the topology of three dimensional manifolds with non-negative scalar curvature, Ann. Math. 110 (1979) 127-142.
[288] R. Schoen and S.-T. Yau, On the structure of manifolds with positive scalar curvature, manuscripta math. 28 (1979) 159-183.
[289] J.T. Schwartz, Generalizing the Lusternik-Schnirelman theory of critical points, Comm. Pure Appl. Math. XVII (1964) 307-315.
[290] J.T. Schwartz, *Nonlinear Functional Analysis* (Gordon and Breach, New York 1969).
[291] S. Schwartzman, Asymptotic cycles, Ann. Math. 66 (1957) 270-284.
[292] J.-P. Serre, Homologie singulière des espaces fibrés, Ann. Math. 54 (1951) 425-505.
[293] I.R. Shafarevich, *Basic Algebraic Geometry* (Springer-Verlag, 1974).
[294] I.R. Shafarevich, Abelian and nonabelian mathematics, Math. Intell. 13 (1991) 67-75.
[295] A. Shapiro, Second-order derivatives of extremal-value functions and optimality conditions for semi-infinite programs, Math. Oper. Res. 10 (1985) 207-219.
[296] S. Shindoh, R. Hirabayashi and T. Matsumoto, Structure of solution set to nonlinear programs with two parameters: I. change of stationary indices, Res. Rep. Inf. Sc. B-224, Series B: Oper. Res, Tokyo Inst. Tech. (1989).
[297] N.Z. Shor, Generalized gradient methods of nondifferentiable optimization employing space dilatation operations, in : *Math. Progr., The State of the Art* (Springer, 1983) 501-529.
[298] N.Z. Shor, *Minimization Methods for Non-Differentiable Functions* (Springer-Verlag, 1985).
[299] M. Shub, Dynamical systems, filtrations and entropy, Bull. Amer. Math. Soc. 80 (1974) 27-41.
[300] M. Shub and D. Sullivan, Homology theory and dynamical systems, Topology 14 (1975) 109-132.
[301] S. Smale, Morse inequalities for a dynamical system, Bull. Amer. Math. Soc, 66 (1960) 43-49.
[302] S. Smale, Morse theory and a non-linear generalization of the Dirichlet problem, Ann. Math. 80 (1964) 382-396.
[303] S. Smale, Differentiable dynamical systems, Bull. Amer. Math. Soc. 73 (1967) 747-817.
[304] S. Smale, Global analysis and economics I : Pareto optimum and a generalization of Morse theory, in : *Dynamic Systems*, M. Peixoto, ed. (Academic Press, 1973) 531-544.
[305] S. Smale, A convergence process in price adjustment and global Newton methods, J. Math. Econ. 3 (1976) 107-120.

[306] S. Smale, On the efficiency of algorithms of analysis, Bull. (New Series) Amer. Math. Soc. 13 (1985) 87-121.

[307] S. Smale, Algorithms for solving equations, in: *Proc. Int. Congr. Math.* I, M.S. Gleason, ed. (AMS Berkeley ,1986) 172-195.

[308] J.Q. Smith, P.J. Harrison and E.C. Zeeman, The analysis of some discontinuous decision processes, Europ. J. Oper. Res. 7 (1981) 30-43.

[309] J.A. Smoller, *Shock Waves and Reaction-Diffusion Equations* (Springer-Verlag, 1983).

[310] D. Solow, A constrained optimization approach to solving certain convex systems of equations, Preprint, Weatherhead School of Management, Cleveland (1989), submitted for publication.

[311] G. Sonnevend, An "analytic center" for polyhedrons and new classes of global algorithms for linear (smooth, convex) programming, in: *System Modelling and Optimization*, A. Prékopa et al. , eds., Lect. Notes Contr. Inf. Sci. 84 (Springer-Verlag, 1985) 866-876.

[312] G. Sonnevend, On some algorithm for constructing feedback controls in linear games; optimal rejection of bounded disturbances, Schwerpunktprogr. der Deutschen Forschungsgem. "Anwendungsbezogene Optimierung und Steuerung", Preprint 107 (1988).

[313] G. Sonnevend and J. Stoer, Global ellipsoidal approximations and homotopy methods for solving convex analytic programs, Appl. Math. Opt. 21 (1990) 139-165.

[314] E.H. Spanier, *Algebraic Topology* (Mac Graw-Hill, New York 1966).

[315] L.E. Stanfel, A Lagrangian treatment of certain nonlinear clustering problems, Europ. J. Oper. Res. 7 (1981) 121-132.

[316] J. Stoer and C. Witzgall, *Convexity and Optimization in Finite Dimensions* I (Springer-Verlag, 1970).

[317] P. Szilági, Characterization of the Kuhn-Tucker vector set by constraint qualifications, Preprint, Eötvös L. University, Dep. Oper. Res., Budapest (1990).

[318] F. Takens, A note on sufficiency of jets, Invent. math. 13 (1971) 225-231.

[319] E. Tarafdar and S.K. Teo, On the existence of solutions of the equation $Lx \in Nx$ and a coincidence degree theory, J. Austr. Math. Soc., Ser.A) 28 (1979) 139-173.

[320] R. Temam, *Infinite-Dimensional Dynamical Systems in Mechanics and Physics* (Springer-Verlag, 1988).

[321] R. Thom, *Stabilité Structurelle et Morphogénèse: Essai d'une Théorie Générale des Modèles* (W.A. Benjamin, Reading, MA, 1972).

[322] L. Thibault, On generalized differentials and subdifferentials of Lipschitz vector-valued functions, Nonl. Anal., T.M.A. 6 (1982) 1037-1053.

[323] C.A. Tiahrt and A.B. Poore, A bifurcation analysis of the nonlinear parametric programming problem, Math. Prog. 47 (1990) 117-141.

[324] M.J. Todd, On convergence properties of algorithms for unconstrained minimization, IMA J. Num. Anal. 9 (1989) 435-441.

[325] P.Tracqui, J.F. Staub and A.M. Perault-Staub, Analysis of degenerate Hopf bifurcations for a nonlinear model of calcium metabolism, Nonl. Anal., T.M.A. 13 (1989) 429-457.

[326] E. Triesch, *Über die Komplexität von Grapheneigenschaften*, Thesis, RWTH Aachen (1984).

[327] C. Udrişte, Kuhn-Tucker theorem on Riemannian manifolds, in: *Topics in Differential Geometry* II, J. Szenthe and L. Tamássy, eds., Colloq. Math. Soc. János Bolyai 46 (North-Holland, 1988) 1247-1259.

[328] Hj. Wacker (ed.) *Continuation Methods* (Academic Press, 1978).

[329] C.T.C. Wall, Lectures on C^∞ -stability and classification, in: *Proc. Liverpool Singul.-Symp.* I, C.T.C. Wall, ed., Lect. N. Math. 192 (Springer-Verlag, 1971) 178-206.

[330] R. Walter, *Differentialgeometrie* (BI-Verlag, Zürich 1978).

[331] J. Wang, Continuity of the feasible solution sets of probabilistic constrained programs, J. Opt. Th. Appl. 63 (1989) 79-89.

[332] T. Wang, A minimax principle without differentiability, Nanjing Daxue Xuebao Shuxue Bannian Kan 6 (1989) 46-52.

[333] T. Wang, A static bifurcation theorem for critical points, Nonl. Anal., T.M.A., 15 (1990) 1181-1186.

[334] J.R.L. Webb, On degree theory for multivalued mappings and applications, Bolletino unione matematica italiana 99 (1974) 137-158.

[335] J.R.L. Webb, On uniqueness of topological degree of set-valued mappings, Poc. Royal Soc. Edinburgh 74 A (1974/75) 225-229.

[336] A. Weinstein, Periodic orbits for convex Hamiltonian systems, Ann. Math. 108 (1978) 507-518.

[337] W. Wetterling, Definitheitsbedingungen für relative Extrema bei Optimierungs- und Approximationsaufgaben, Num. Math. 15 (1970) 122-136.

[338] E. Witten, A new proof of the positive energy theorem, Comm. Math. Phys. 80 (1981) 381-402.

[339] E. Witten, Supersymmetry and Morse theory, J. Diff. Geom. 17 (1982) 661-692.

[340] E. Witten, Physics and geometry, in: *Proc. Int. Congr. Math.* I, A.M. Gleason, ed. (AMS Berkeley 1986) 267-303.

[341] S. Wright, An inexact algorithm for composite nondifferentiable optimization, Math. Progr. 44 (1989) 221-234.

[342] Y. Yomdin, On functions representable as a supremum of a family of smooth functions, SIAM J. Math. Anal. 14 (1983) 239-246.

[343] Y. Yomdin, On representability of convex functions as maxima of linear families, Preprint, Max-Planck Institut für Mathematik, Bonn (1984).

[344] Y. Yomdin, Global bounds for the Betti numbers of regular fibers of differentiable mappings, Topology 24 (1985) 145-152.

[345] Y. Yomdin, Metric semialgebraic geometry with applications in smooth analysis, Preprint (1988).

[346] E.C. Zeeman, *Catastrophe Theory* (Addison-Wesley Publ.Comp., 1977).

[347] J. Zowe, Nondifferentiable optimization - a motivation and a short introduction into the subgradient - and the bundle concept, in: *Computational Mathematical Programming*, K. Schittkowski, ed. (Springer-Verlag, 1985) 321-356.

[348] J. Zowe, Optimization and nonsmooth data, OR Spektrum 9 (1987) 195-201.

NONLINEAR OPTIMIZATION PROBLEMS UNDER DATA PERTURBATIONS

Diethard Klatte
Fachbereich Mathematik, Pädagogische Hochschule Halle-Köthen
Kröllwitzer Straße 44, D-O 4050 Halle (Saale)

Abstract: In this paper, we survey results on the stability of local minimizers and stationary solutions to a nonlinear optimization problem under data perturbations. The main purpose is to present local optimality and stability conditions, avoiding the assumption of twice differentiability of the functions which enter into the problem. As essential tools we use arguments from the analysis of Lipschitzian mappings. Some motivations and applications are sketched.

1. Introduction

Motivations for the study of sensitivity and stability of optimization problems come from different areas in mathematical theory, numerical methods and applications, for example, from the convergence analysis of optimization procedures, the study of incorrect models, the solution of multi-level optimization problems, the equilibrium theory, postoptimal analysis, input-output-modelling. For recent surveys of the field of stability and parametric analysis in optimization we refer, e.g., to the monographs or collections of papers by Bank, Guddat, Klatte, Kummer, Tammer [4], Fiacco [9,10,11], Jongen, Jonker and Twilt [19], Malanowski [36], Guddat, Jongen, Kummer, Nozicka [13, 14].

In the present paper, we are concerned with the following subject: the study of sufficient conditions for Hölder and Lipschitz continuity and strong stability of local minimizers and stationary solutions for nonlinear programs under perturbations. The particular aim is to generalize well-known second-order sufficient optimality and stability conditions developed for programs with C^2 data (cf., e.g., Fiacco and McCormick [12], Kojima [29], Robinson [41], Fiacco [9]) to the case of non-C^2 data. It will turn out that many weak and strong stability results for local minimizers given in [29] and [40,41] may be extended to certain classes of non-C^2 programs. Our main interest is devoted to programs with differentiable data functions whose gradients are locally Lipschitzian (so-called $C^{1,1}$ functions). Such functions arise quite naturally, cf., e.g., Hiriart-Urruty, Strodiot and Nguyen [17]. For example, the penalty function $(\max\{f,0\})^2$ is necessarily a $C^{1,1}$ function even if f

is C^2. Another example (interesting for two-level problems): Minimize z^2 s.t. $z \leq t$. The global optimal value is $v(t)=0$ if $t \geq 0$ and $v(t)=t^2$ if $t \leq 0$, hence, v is $C^{1,1}$ but not C^2. Typical classes of $C^{1,1}$ problems will be discussed below. Essential tools for handling these questions come from nonsmooth analysis, particularly, we use directional derivatives of Lipschitzian mappings in the sense of Clarke [7] and Thibault [47] - Kummer [32,33].
The paper is organized as follows. First we introduce notation and provide some preliminaries. Included are basic stability results for C^1 programs, cited from the literature. The Sections 3-5 present 2nd-order conditions for local optimality and stability for $C^{1,1}$ optimization problems. Interesting special cases, motivations and applications are discussed in the Sections 6 and 7.
Essential parts of the paper are based on a IIASA Working Paper [28] written together with Bernd Kummer and Ralf Walzebok.

2. Notation and Preliminaries

Throughout the paper we consider a parametric optimization problem of the following type:

$$f_0(x,t) \longrightarrow \min_x \quad \text{s.t.} \quad x \in M(t), \tag{2.1}$$

where t is a parameter varying in a metric space $(T , d(.,.))$, f_i: $R^n \times T, R$ $(i=0,1,\ldots,m)$ are given functions, and

$$M(t) := \{x \in R^n \mid f_i(x,t) = 0,\ i=1,\ldots,l;\ f_j(x,t) \leq 0,\ j=l+1,\ldots,m\}.$$

In particular, we study the following special classes of parametric problems. Suppose for each $t \in T$ and for $i=0,1,\ldots,m$ that

$$P^1(t): \quad \begin{array}{l} f_i(.,t) \text{ is continuously differentiable on } R^n, \\ f_i \text{ and } D_x f_i(.,.) \text{ are continuous on } R^n \times T. \end{array} \tag{2.2}$$

$$P^2(t): \quad \begin{array}{l} f_i(.,t) \text{ is twice continuously differentiable on } R^n, \\ f_i,\ D_x f_i(.,.),\ D^2_{xx} f_i(.,.) \text{ are continuous on } R^n \times T. \end{array} \tag{2.3}$$

$$P^{1,1}(t): \quad (2.2) \text{ holds, and } D_x f_i(.,t) \text{ is locally Lipschitzian on } R^n. \tag{2.4}$$

Here Df and D^2f ($D_x f(.,.)$ and $D^2_{xx} f_i(.,.)$) denote the first and second derivatives (partial derivatives w.r.t. to x) of a function f (resp. of f(.,.)).
In what follows, writing P(t), $t \in T$, we mean one of the problems $P^1(t)$, $P^2(t)$, $P^{1,1}(t)$. Denote by $(x,u) \mapsto l(x,u,t)$ the Lagrange function to P(t),

$$l(x,u,t) := f_o(x,t) + \sum_{i=1}^{m} u_i f_i(x,t).$$

A point x in M(t) which satisfies with some m-vector u

$$D_x l(x,u,t) = 0, \quad u_j \geqq 0, \quad u_j f_j(x,t) = 0 \quad (j=l+1,\dots,m) \qquad (2.5)$$

is said to be a <u>stationary solution</u> to P(t), in symbols

$$x \in S(t).$$

Further, the sets of <u>Kuhn-Tucker points</u> and <u>Lagrange multiplier vectors</u> are

$$KT(t) := \{(x,u) \mid (x,u) \text{ fulfils } (2.5)\}, \quad LM(x,t) := \{u \mid (x,u) \in KT(t)\}.$$

Let $\Psi_{loc}(t)$ symbolize the set of all local minimizers of P(t). Further, Argmin $(f(x) \mid x \in D)$ is the set of all global minimizers for f w.r. to D.

We continue with some general notation. B(z,r) denotes the closed r-neighborhood of z (for $z \in R^p$ w.r. to the Euclidean norm $||.||$, or w.r. to $z \in T$), B_p is the closed Euclidean unit ball in R^p, $||.||_1$ is the sum norm in R^p; the distance of $x \in R^p$ and $Y \subset R^p$ is $\text{dist}(x,Y) := \inf\{||x-y|| \ / \ y \in Y\}$. Vectors in R^p are usually considered as column vectors, A^T denotes the transpose of a matrix A, thus x^Ty means the scalar product of $x,y \in R^p$. For $X \subset R^p$ let conv X, bd X, cl X denote the convex hull, the boundary, the closure of X, and ext X is the set of extreme points of X if X is convex. If $z \in R^p$ and $X \subset R^p$, let $z+X := \{z+x \mid x \in X\}$ and $z^TX := \{z^Tx \mid x \in X\}$. Denote by a^+ and a^- the numbers $a^+ := \max\{a,0\}$ and $a^- := \min\{a,0\}$ for a from R. Analogously, for x in R^p we write x^+ for the vector $(x_1^+,\dots,x_p^+)^T$.

A function g from a metric space (T, d(.,.)) to R^p is called Lipschitzian on T if for some $\beta \geqq 0$, $||g(t')-g(t'')|| \leqq \beta\, d(t',t'')$ holds for all $t',t'' \in T$. Given an open subset Q of R^s, $C^1(Q,R^p)$ and $C^2(Q,R^p)$ denote the classes of once and twice continuously differentiable functions from Q to R^p, respectively, where $C^{0,1}(Q,R^p)$ will be the class of all functions f from Q to R^p being locally Lipschitzian on Q (i.e., for each x in Q there is some neighborhood V of x such that f is Lipschitzian on V). $C^{1,1}(Q,R^p)$ denotes the class of all functions f in $C^1(Q,R^p)$ whose gradient mappinng Df(.) is locally Lipschitzian on Q. We shall make also use of the Landau symbols O(.) and o(.).

Now we make available some basic definitions and results concerning regularity and stability. By <u>LICQ</u> and <u>MFCQ</u> we abbreviate the well-known <u>Linear Independence Constraint Qualification</u> and the <u>Mangasarian-Fromovitz Constraint Qualification</u>, respectively, for dicussions of these CQ we refer, e.g., to [9,29,41].

For fixed $t=t^o$, a local minimizer x^o of $P(t^o)$ is said to be strongly stable with order 2 (concerning $P(t)$, $t \in T$), if there are positive real numbers δ, r and c as well as some mapping x(.) from $B(t^o, \delta)$ to $B(x^o, r)$ such that

$$x(.) \text{ is continuous at } t^o \text{ and } x(t^o)=x^o, \tag{2.6}$$

$$S(t) \cap B(x^o,r) = \{x(t)\} \text{ for all } t \in B(t^o, \delta), \tag{2.7}$$

$$f_o(x,t) - f_o(x(t),t) \geq c\, ||x - x(t)||^2 \quad \forall\, t \in B(t^o, \delta)\ \forall\, x \in M(t) \cap B(x(t),2r). \tag{2.8}$$

(2.8) in particular says that x(t) is a strict local minimizer with order 2 w.r. to P(t) (in Auslender's [3] terminology). The properties (2.6) and (2.7) together give, in our situation, the specialization of Kojima's [29] definition of strongly stable stationary solutions. Regarding the problem $P^2(t)$, the properties (2.6)-(2.8) hold under MFCQ and the so-called strong second-order sufficient optimality condition, cf. [29].

We emphasize that all results noticed in the remainder of this section concern the class of programs of type $P^1(t)$, and so, they hold for the other two classes, too. First we recall conditions for the Hölder continuity (of order 1/2) of solutions and for Lipschitz continuity of the infima.

Proposition 2.1: Let $t^o \in T$, and let x^o be a local minimizer of $P(t^o)$ satisfying MFCQ. Suppose that for some neighborhood Q of x^o and for i=0,1,...,m, the functions $f_o, f_1, \ldots, f_m$ are Lipschitzian on Q x T. Then one has:

(i) If x^o is a strict local minimizer with order 2, then there are positive real numbers δ_1, r_1 and L_1 such that for each element t of $B(t^o, \delta_1)$, $\emptyset \neq X(t) := \text{Argmin}_x(f_o(x,t) \mid x \in M(t) \cap B(x^o,2r_1)) \subset \Psi_{loc}(t) \cap B(x^o,r_1)$ and $||x - x^o||^2 \leq L_1\, d(t,t^o)$ for all $x \in X(t)$. (2.9)

(ii) If x^o is strongly stable with order 2, then there are δ_2, $L_2 > 0$ with

$$||x(t') - x(t")||^2 \leq L_2\, d(t',t") \quad \text{for all } t',t" \in B(t^o, \delta_2). \tag{2.10}$$

Proof: (i) cf., e.g., Alt [1, Thm. 4.6, Cor. 4.7], Klatte [24, Thm. 1]; (ii) cf. Klatte [25, Proof of Thm. 2.5].

A similar version of the preceding proposition is also true in the absense of differentiability but under some extended MFCQ, cf. Auslender [3]. Generally, (2.9) may not be improved to a linear rate - even if x^o is strongly stable in Kojima's sense (i.e. (2.6) and (2.7) hold but (2.8) may fail). In the case (ii) above, differentiability of the value function $f_o(x(.),.)$ does not automatically follow. Both facts mentioned last will now be illustrated.

Example 2.2: The problem $x_2 - tx_1 \longrightarrow \min_x$ s.t. $x_2 \geqq x_1^2$, $x_2 \geqq t$ $(t \in R)$ has for each t near 0 the unique global optimal solution $x(t) = ((t^+)^{1/2}, t^+)$. //

Example 2.3: The problem $x_1^2 + x_2 \longrightarrow \min_x$ s.t. $x_2 \geqq x_1^2$, $x_2 \geqq t$ $(t \in R)$ has the unique global minimizer $x(t) = (0, t^+)$, $x(0)$ is strongly stable with order 2, but $t \longmapsto f_o(x(t),t) = t^+$ is not differentiable at t=0. Note that there is no continuous selection of the multifunction LM(x(.),.), since $LM(x(0),0) = \{(u,v) | u+v=1,\ u \geqq 0,\ v \geqq 0\}$, $LM(x(t),t) = \{(1,0)\}$ if $t < 0$, but $= \{(0,1)\}$ if $t > 0$. //

The existence of a continuous selection of LM(x(.),.), together with some Lipschitz property of x(.) (where x(.) is a selection of S(.)), ensure the differentiability of $f_o(x(.),.)$, cf. Jongen, Möbert and Tammer [21] and Section 7 of our paper. However, there holds

Proposition 2.4: [24, § 4, Cor. 1] Under the assumptions of Proposition 2.1, Part (i), the local infimum function $t \longmapsto f_o(x(t),t)$ is Lipschitzian on some neighborhood U of t^o.

The constraint qualifications MFCQ and LICQ guarantee certain stability properties of the stationary solution set mapping S(.) and of the Langrange multiplier set mapping LM(.,.).

Proposition 2.5: (Robinson [41, Thm. 23]) Let $t^o \in T$, and let x^o be a feasible point of $P^1(t^o)$ which fulfils MFCQ. Then there exists neighborhoods U of t^o and V of x^o such that for each (x,t) in V x U, x satisfies MFCQ w.r. to M(t). Moreover there holds:
If $x^o \in S(t^o)$ then there exist neighborhoods U' of t^o and V' of x^o such that the multifunctions $t \in U' \rightrightarrows S(t) \cap V'$ and $(x,t) \in V' \times U' \rightrightarrows LM(x,t)$ are closed and locally bounded (and hence upper semicontinuous in Berge's sense) on U' and V' x U', respectively.

Recall that the validity of MFCQ at $x^o \in S(t^o)$ is equivalent to the compactness of $LM(x^o,t^o)$. The upper semicontinuity property for LM(.,.) may be strengthened if each of the functions f_i (i=0,1,...,m) satisfies with some positive real numbers β_i and β_i':

$$| f_i(x,t) - f_i(x^o,t^o) | \leqq \beta_i (||x-x^o|| + d(t,t^o)), \tag{2.11}$$

$$||D_x f_i(x,t) - D_x f_i(x^o,t^o)|| \leqq \beta_i' (||x-x^o|| + d(t,t^o)) \tag{2.12}$$

for all x and t belonging to some neighborhoods V_o of x^o and U_o of t^o.

Proposition 2.6: [25, Thm. 2.2] Let $t^o \in T$ and $x^o \in S(t^o)$. Suppose that MFCQ holds at x^o, and let (2.11) and (2.12) be satisfied. Then there exist neighborhoods U of t^o and V of x^o and some $L > 0$ so that for all $(x,t) \in V \times U$ and all $u \in LM(x,t)$, one has $\text{dist}\,(u, LM(x^o,t^o)) \leq L\,(||x - x^o|| + d(t,t^o))$.

Note that the assumptions of the preceding proposition do not ensure that LM(x,t) is nonempty for (x,t) near (x^o,t^o), cf., e.g., [24, Part II, Ex. 1]. If (x^o,t^o) is a strongly stable stationary solution to $P(t^o)$ this property persists under perturbations, we have

Proposition 2.7: ([27, Thm. 2.4], [28, Thm. 2.4]) Let $t^o \in T$ and $x^o \in S(t^o)$. Suppose that x^o satisfies MFCQ, and that x^o is strongly stable in Kojima's sense, i.e., (2.6), (2.7) hold with some mapping x(.). Then there is some neighborhood U of t^o such that x(.) is continuous on U, and for each t in U, x(t) is a stationary solution of $P^1(t)$ which is strongly (Kojima-)stable.

For the class of problems of type $P^2(t)$ the assertion of the former proposition is known from [29].
It is well-known that if $x^o \in S(t^o)$ satisfies LICQ then LM(x,t) is single-valued for all (x,t) near (x^o,t^o) which belong to the domain of LM(.,.), and LM(.,.) is continuous there. A refinement of this property provides

Proposition 2.8: [27, Lemma 2.1] Let $t \in T$ be fixed, and let x be in M(t). Denote by A the (m,n)-matrix whose rows are the gradients of $f_i(.,t)$ at x and by F the diagonal matrix with diagonal elements $f_i(x,t)$, i=1,...,m. Then the following hold:
(i) x satisfies LICQ if and only if the matrix $(AA^T - F)$ is nonsingular.
(ii) If x satisfies LICQ and belongs to S(t) then the unique Lagrange multiplier u in LM(x,t) has the representation $u = -(AA^T-F)^{-1} A\, D_x f_o(x,t)$.

3. Generalized Directional Derivatives

In what follows we shall use two concepts of generalized directional derivatives for vector-valued functions: that of going back to the generalized (set-valued) Jacobian matrix in Clarke's sense [7] and that of the limit set introduced by Thibault [47] and extensively studied by Kummer [32,33]. Given $G \in C^{0,1}(R^p,R^s)$ and $z \in R^p$, the set of (s,p)-matrices

$$\partial G(z) := \text{conv}\left\{ M \mid \exists\ z^k \to z:\ z^k \in E_G\ (\forall k) \text{ and } DG(z^k) \to M \right\}$$

is called the generalized Jacobian of G at z, where E_G is the set of all points y for which the Jacobian DG(y) exists. This definition is justified by the fact that each locally Lipschitzian mapping from R^p to R^s is almost everywhere differentiable (cf., e.g., [8]). As usual, the set $\partial G(z)$ will be considered as a subset of the linear space $R^{s\times p}$ of (s,p)-matrices, which is endowed with the associated matrix norm. For the following properties we refer the reader to [7]. Let $G,H\in C^{0,1}(R^p,R^s)$ and $z\in R^p$, then one has

$\partial G(z)$ is a nonempty, convex and compact subset of $R^{s\times p}$, (3.1a)
the multifunction $\partial G(.)$ is closed and locally bounded on R^p, (3.1b)
if G is continuously differentiable at z then $\partial G(z) = \{DG(z)\}$, (3.1c)
$\partial(G+H)(z)h \subset \partial G(z)h + \partial H(z)h$, where $\partial G(z)h := \{Mh \mid M\in \partial G(z)\}$. (3.1d)

Let G as before. According to [32,33] the set (first considered in |47|)

$$\Delta G(z;h) := \left\{v\in R^s \mid \exists\ z^k \longrightarrow z\ \exists\ q_k \downarrow 0:\ q_k^{-1}(G(z^k+q_k h) - G(z^k)) \longrightarrow v\right\}$$

is called the <u>directional derivative of G at z in direction h</u>. This concept does not require the use of Rademacher's theorem (except for deriving (3.2f) below). Now we summarize several properties of this directional derivative, the proofs can be found in [32]. Let any $G,H\in C^{0,1}(R^p,R^s)$, $z,h\in R^p$ be given. Then one has:

$\Delta G(z;\beta h) = \beta\,\Delta G(z;h)$ for all real numbers β, (3.2a)
$\Delta(G+H)(z;h) \subset \Delta G(z;h) + \Delta H(z;h)$, (3.2b)
$\Delta G(z;h)$ is a nonempty, compact and connected set, (3.2c)
$\Delta G(.,.)$ is a closed and locally bounded multifunction, (3.2d)
if G is continuously differentiable at z then $\Delta G(z;h) = \{DG(z)h\}$, (3.2e)
$(\mathrm{ext}\,\partial G(z))\,h \subset \Delta G(z;h) \subset \partial G(z)\,h$, (3.2f)
if L is the Lipschitz modulus for G on some neighborhood V of z then $\Delta G(z;h') \subset \Delta G(z;h'') + L\,\|h'-h''\|\,B_p$ for all h',h" in R^p. (3.2g)

Proposition 3.1: Let $G\in C^{0,1}(R^p,R^s)$. Then for all z,h in R^p and all y in R^s one has $y^T\Delta G(z;h) = y^T\partial G(z)\,h$.

Proof: By (3.1a), (3.2c) and (3.2f), $y^T\Delta G(z;h)$ and $y^T\partial G(z)\,h$ are compact intervals with $y^T\Delta G(z;h) \subset y^T\partial G(z)\,h$. Since, obviously, ext $(y^T\partial G(z)\,h)$ is a subset of $y^T(\mathrm{ext}\,\partial G(z))\,h$, (3.2f) then implies the desired equality. //

If g is a $C^{1,1}$ function from R^p to R, then $(\partial Dg(x))h$ and $\Delta Dg(x;h)$ define directional derivatives of second-order. In the form which is essential in deriving 2nd-order optimality conditions, both sets coincide, by the preceding proposition: $y^T(\partial Dg(x))\,h = y^T\Delta Dg(x;h)$ for all x,y,h in R^p.

Based on a mean-value theorem for $C^{0,1}$ mappings, a <u>second-order Taylor expansion</u> for $C^{1,1}$ functions holds, namely

Proposition 3.2: (Hiriart-Urruty et al. [17,Thm.2.3], Kummer [32,Prop.5.1]) Suppose $g \in C^{1,1}(R^p,R)$ and $z,h \in R^p$. Then there is some q: $0 < q < 1$ such that

$$g(z+h) \in g(z) + h^T Dg(z) + (1/2)\, h^T \Delta\, Dg(z+qh;h). \tag{3.3}$$

We conclude this section by providing two technical lemmas which will be helpful in what follows. Consider the parametric problem $P^{1,1}(t)$, $t \in T$. We put for $(x,u,t) \in R^n \times R^m \times T$, with $J_1 := \{1,\dots,l\}$, $J_2 := \{l+1,\dots,m\}$,

$$\begin{aligned} I_2(x,t) &:= \{ j \in J_2 \mid f_j(x,t)=0 \}, & I(x,t) &:= J_1 \cup I_2(x,t), \\ I_2^+(u) &:= \{ j \in J_2 \mid u_j > 0 \}, & I^+(u) &:= J_1 \cup I_2^+(u), \\ W^+(x,u,t) &:= \{ h \in R^n \mid h^T D_x f_i(x,t) = 0 ,\ i \in I^+(u) \}, \\ W(x,u,t) &:= \{ h \in W^+(x,u,t) \mid h^T D_x f_j(x,t) \le 0,\ j \in I_2(x,t) \setminus I_2^+(u) \}. \end{aligned} \tag{3.4}$$

Lemma 3.3: Suppose that

(i) $t^0 \in T$, $x^0 \in S(t^0)$ and $u^0 \in LM(x^0,t^0)$, and that t^k, x^k, y^k and u^k are any sequences satisfying for all k, $x^k \in S(t^k)$, $y^k \in M(t^k)$ and $u^k \in LM(x^k,t^k)$,

(ii) $(x^k,u^k,t^k) \to (x^0,u^0,t^0)$ and $y^k \to x^0$, and

(iii) there is some function $\varphi(.)$: $R_+ \to R$ of the type $\varphi(.) = o(.)$ so that

(a) $f_0(y^k) - f_0(x^k) \le \varphi(\|y^k - x^k\|)$ $(\forall k)$, <u>or</u>

(b) $(y^k - x^k)^T D_x f_i(y^k,t^k) \ge \varphi(\|y^k - x^k\|)$ $(\forall k\ \forall i \in I^+(u^0))$

hold.

Then the sequence $h^k := \|y^k - x^k\|^{-1} (y^k - x^k)$ has an accumulation point $h \in W^+(x^0,u^0,t^0) \cap \text{bd}\, B_n$.

Further, if $(x^k,t^k) \equiv (x^0,t^0)$ (or $(y^k,t^k) \equiv (x^0,t^0)$, respectively) then h^k (or $-h^k$, respectively) even has an accumulation point in $W(x^0,u^0,t^0) \cap \text{bd}\, B_n$.

Proof: To show the assertion of the lemma, one may follow standard arguments as those used by Robinson [41, Thm 2.2]. The details are omitted here. //

In view of the later use, we note that (i) and (ii) imply

$$I^+(u^0) \subset I^+(u^k) \subset I(x^k,t^k) \subset I(x^0,t^0) \quad \text{for k large.} \tag{3.5}$$

Let $\Delta\, (D_x l(.,u,t))(x^0;h)$ denote the directional derivative of the function $D_x l(.,u,t)$ at x^0 in direction h (for fixed (u,t)).

Lemma 3.4: Let assumption (i) of Lemma 3.3 be satisfied, and suppose that for some real number $c \geq 0$ and for all k,

$$f_o(y^k,t^k) - f_o(x^k,t^k) < \frac{c}{4} ||y^k - x^k||^2 \tag{3.6}$$

holds. Let $h^k := ||y^k - x^k||^{-1} (y^k - x^k)$. Then, there are sequences $q_k \downarrow 0$ and $\{v^k\} \subset R^n$ such that for each k, one has

$$v^k \in \Delta (D_x l(.,u^k,t^k))(x^k + q_k h^k; h^k) \qquad \text{and} \qquad h^{k^T} v^k < \frac{c}{2}. \tag{3.7}$$

Proof: Let k be fixed. For simplicity, we put $l_k := l(.,u^k,t^k)$ and denote by $H_k(x;h)$ the set $\Delta (Dl_k)(x;h)$. Since $l_k \in C^{1,1}(R^n,R)$, Proposition 3.2 implies the existence of some $\bar{q}_k \in (0,1)$ and of some $\bar{v}^k \in H_k(x^k + \bar{q}_k(y^k - x^k); y^k - x^k)$ such that, together with (3.6) and $x^k \in S(t^k)$,

$$\frac{1}{2} (y^k-x^k)^T \bar{v}^k = (y^k-x^k)^T Dl_k(x^k) + \frac{1}{2} (y^k-x^k)^T \bar{v}^k \leq$$

$$\leq l_k(y^k) - l_k(x^k) \leq f_o(y^k,t^k) - f_o(x^k,t^k) < \frac{c}{4} ||y^k - x^k||^2.$$

With $v^k := ||y^k - x^k||^{-1} \bar{v}^k$ it follows $(h^k)^T v^k < (c/2)$, where, by (3.2a), v^k is an element of $H(x^k + q_k h^k; h^k)$ with $q_k := \bar{q}_k \; ||y^k - x^k|| \downarrow 0$. //

Lemma 3.4 already suggests the kind of second-order condition which has to be required to ensure existence and stability of local minimizers in $C^{1,1}$ optimization problems. This will be discussed in what follows.

4. Optimality Conditions and Weak Stability Results

In this section, we present - for $C^{1,1}$ optimization problems - second-order conditions which correspond to the standard second-order condition for C^2 problems. Such as in the C^2 case, this condition implies, together with MFCQ, that a stationary solution is a strict and isolated local minimizer, and so, by Proposition 2.1 and Proposition 2.5, a local minimizer (and a stationary solution) of a slightly perturbed problem exists, and the multifunctions S(.) and LM(.,.) behave upper semicontinuous. Generalizing a result given by Robinson [41] in the C^2 case, we may even show upper Lipschitz continuity of the Kuhn-Tucker point mapping.

Throughout this section, we consider $C^{1,1}$ programs of the type $P^{1,1}(t)$ as introduced in Section 2. We start by presenting local optimality conditions. For simplicity, let us agree that for fixed $t=t^o$ we sometimes delete the variable t^o and write $f_i(x)$, M, LM(x), W(x,u) and so on instead of $f_i(x,t^o)$, $M(t^o)$, $LM(x,t^o)$, $W(x,u,t^o)$ and so on.

First we cite a necessary second-order optimality condition for $C^{1,1}$ programs given by Hiriart-Urruty, Strodiot and Nguyen [17]. Consider for fixed $t=t^o$ the problem $P^{1,1} = P^{1,1}(t^o)$. Put for (x,u) in KT,

$$Z := \{z \in R^n \mid f_i(z) = 0,\ i \in I^+(u),\ f_j(z) \leq 0,\ j \in I_2(x,u) \setminus I_2^{\,+}(u)\}.$$

Let $T_Z(x)$ denote the Bouligand tangent cone to Z at x. It is well-known that for a Kuhn-Tucker point (x,u) of $P^{1,1}$, $T_Z(x)$ is a subset of the cone $W(x,u)$ defined in (3.4), where both cones coincide if an additional regularity condition (such as LICQ) holds at x. Let CQ be any constraint qualification which guarantees that a local minimizer of $P^{1,1}$ is a stationary solution. As before, let $\Delta(D_x l(.,u))(x^o;h)$ denote the directional derivative of $D_x l(.,u)$ at x^o in direction h.

Proposition 4.1: (Second-order necessary condition [17, Thm.3. 2]) Let x^o be a local minimizer to $P^{1,1}$ which satisfies CQ. Then for each $u \in LM(x^o)$ and for each $h \in T_Z(x^o)$, one has

$$\exists\ v \in \Delta\,(D_x l(.,u))(x^o;h) : \qquad h^T v \geq 0. \tag{4.1}$$

Note that, by Proposition 3.1, (4.1) is equivalent to the formulation in the original paper [17].

The following condition naturally extends the standard (C^2-) second-order condition (cf., e.g., [9,12,29,41]) to problems of the type $P^{1,1}$.

In view of the application to stability studies below we formulate the next definition in parametric form.

Definition 4.2: [26,28] Let $t^o \in T$ and $(x^o,u^o) \in KT(t^o)$. We shall say that (x^o,u^o) satisfies the <u>generalized</u> (standard) <u>second-order sufficient condition</u> with modulus $c>0$ (symbolized by <u>GSOSC</u>(x^o,u^o,t^o,c)) if $h^T v \geq c$ holds for all $h \in W(x^o,u^o,t^o) \cap \text{bd}\, B_n$ and all $v \in \Delta\,(D_x l(.,u^o,t^o))(x^o;h)$.

Theorem 4.3: (Second-order sufficient optimality condition [26,28]) Let t^o be a fixed element of T, consider $P^{1,1} = P^{1,1}(t^o)$, and let $x^o \in S = S(t^o)$.

(i) If for some $u^o \in LM(x^o)$ and some $c>0$, GSOSC(x^o,u^o,c) is satisfied, then x^o is a strict local minimizer with order 2 of $P^{1,1}$.

(ii) If, moreover, x^o fulfils MFCQ, and for all $u \in LM(x^o)$ and some $c>0$, GSOSC(x^o,u,c) is satisfied, then there exists a neighborhood V of x^o such that $S \cap V = \Psi_{loc} \cap V = \{x^o\}$.

Proof: (i) If there is a sequence y^k of feasible points satisfying

$$f_o(y^k) - f_o(x^o) < (c/4)\ \|y^k - x^o\|^2,$$

then, by Lemma 3.3. and Lemma 3.4 (put there $(x^k,u^k,t^k)\equiv(x^0,u^0,t^0)$), the sequence $h^k := ||y^k - x^0||^{-1}(y^k - x^0)$ has an accumulation point

$$h \in W(x^0,u^0) \cap \text{bd } B_n,$$

and there are sequences $q_k \downarrow 0$ and v^k such that for all k

$$v^k \in \Delta\,(D_x l(.,u^0))(x^0 + q_k h^k;h^k) \qquad \text{and} \qquad (h^k)^T v^k < c/2.$$

By Property (3.2d) of directional derivatives, v^k has an accumulation point

$$v \in \Delta\,(D_x l(.,u^0))(x^0;h) \qquad \text{and} \qquad h^T v \leq c/2 < c.$$

Now assertion (i) follows by contraposition.

(ii) To show $S \cap V = \{x^0\}$ for some neighborhood V of x^0, assume, on the contrary, that there is a sequence $\{z^k\} \subset S$ with $z^k \neq x^0$ ($\forall k$) and $z^k \to x^0$. By (i),

$$f_0(z^k) > f_0(x^0) \quad \text{for k sufficiently large.} \tag{4.2}$$

For each k, let u^k be an element of $LM(z^k)$. Since LM(.) is closed and locally bounded (Proposition 2.5), then by passing to a subsequence if necessary,

$$u^k \to u^0 \in LM(x^0)$$

holds. Setting $h^k := ||x^0-z^k||^{-1}(x^0-z^k)$ and using Lemma 3.3 with $t^k \equiv t^0$ and $(x^k,y^k) := (z^k,x^0)$, we have (at least for a subsequence)

$$-h^k \to h \in W(x^0,u^0) \cap \text{bd } B_n. \tag{4.3}$$

By (4.2), Property (3.6) in Lemma 3.4 holds, and therefore one may find a sequence $q_k \downarrow 0$ and a sequence of vectors v^k such that for large k,

$$v^k \in \Delta\,(D_x l(.,u^k))(z^k + q_k h^k;h^k) \qquad \text{and} \qquad (h^k)^T v^k < 0.$$

Again by (3.2d), v^k has an accumulation point v in $\Delta\,(D_x l(.,u^0))(x^0;-h)$ with $h^T v \geq 0$, hence, using (3.2a) and (4.3), we see that the pair (h,-v) contradicts $GSOSC(x^0,u^0,c)$. Consequently, x^0 is isolated as a stationary solution, and by Proposition 2.5 then also isolated as local minimizer. //

Note that Theorem 4.3 is a generalization of [41, Thms. 2.2, 2.4] to $C^{1,1}$ problems. In [41], Robinson gives a simple example of an C^2 program which shows that in the absence of MFCQ, GSOSC does not ensure isolatedness of the local minimizer, in general.

Applying Proposition 2.5 and Proposition 2.1, we immediately obtain the following stability results for a parametric problem of the type $P^{1,1}(t)$.

Corollary 4.4: Under the assumptions of Theorem 4.3 (ii), there are neighborhoods U of t^o and Q of x^o such that

(a) the multifunctions $t \in U \rightrightarrows S(t) \cap Q$ and $(x,t) \in Q \times U \rightrightarrows LM(x,t)$ are upper semicontinuous,

(b) the multifunction $t \rightrightarrows S(t) \cap Q$ is continuous at t^o, and

(c) for each neighborhood $Q' \subset Q$ of x^o there is a neighborhood U' of t^o with $\emptyset \neq \Psi_{loc}(t) \cap Q' \subset S(t)$ for all $t \in U'$.

Having more information about the structure of the data functions, we can improve the Hölder continuity result of Proposition 2.1 (i): The following theorem establishes local upper Lipschitz continuity of the Kuhn-Tucker set mapping.

Theorem 4.5: (Upper Lipschitz continuity of KT(.)) Consider $P^{1,1}(t)$, $t \in T$, let $t^o \in T$ and $x^o \in S(t^o)$. Suppose that x^o satisfies MFCQ, and that with some positive real number c and for each $u \in LM(x^o,t^o)$, GSOSC$(x^o,u,t^o c)$ holds. Further, suppose that the functions f_i $(i=0,1,\dots,m)$ obey (2.11) and (2.12). Then one has:

a) For each neighborhood Q of x^o there is a neighborhood U_Q of t^o such that $KT(t) \cap (Q \times R^r)$ is nonempty for each t in U_Q.

b) There are neighborhoods U of t^o and V of x^o and a constant $\eta > 0$ such that $\text{dist}((x,u),KT(t^o)) \leq \eta\, d(t,t^o) \quad \forall t \in U \quad \forall (x,u) \in KT(t) \cap (V \times R^m)$.

Proof: a) follows from Corollary 4.4, Part c).

b) By Theorem 4.3 (ii) and Proposition 2.5 there are neighborhoods U' of t^o and V' of x^o so small that for all $(x,t) \in V' \times U'$ with $x \in S(t)$ and for all $u \in LM(x,t)$, one has $\text{dist}((x,u), KT(t^o)) = \text{dist}((x,u), \{x^o\} \times LM(x^o,t^o))$. Let U',V' be chosen in such a way that the estimation of Proposition 2.6 is also true with some $L > 0$, i.e., for all $t \in U'$ and all $(x,u) \in KT(t) \cap (V' \times R^m)$,

$$\begin{aligned} \text{dist}((x,u),KT(t^o)) &\leq ||x-x^o|| + \text{dist}(u, LM(x^o,t^o), \\ &\leq ||x-x^o|| + L(||x-x^o||+d(t,t^o)) \\ &\leq (1+L)\, ||x-x^o|| + L\, d(t,t^o), \end{aligned} \tag{4.4}$$

hence, it is sufficient to show that for some constant $\mu > 0$ and for certain neighborhoods $U \subset U'$ of t^o and $V \subset V'$ of x^o, one has

$$||x - x^o|| \leq \mu\, d(t,t^o) \quad \forall t \in U \quad \forall x \in S(t) \cap V. \tag{4.5}$$

The proof of (4.5) will be indirect. We assume, on the contrary, that for some sequences

$$t^k \to t^o \quad \text{and} \quad x^k \to x^o \text{ with } x^k \in S(t^k)$$

we have

$$||x^k - x^o|| > k\, d(t^k,t^o) \quad \text{for all } k. \tag{4.6}$$

Then MFCQ entails regularity of the constraints, cf. Robinson [39, Thm. 1], and with (2.11) and (2.12) we obtain for some $\beta' > 0$ and some index k' that $\text{dist}\,(x^o, M(t^k)) \leq \beta'\, d(t^k,t^o)$ and $\text{dist}\,(x^k, M(t^o)) \leq \beta'\, d(t^k,t^o)$ for all $k \geq k'$. For simplicity put

$$\beta := \max\{\beta_o,\dots,\beta_m,\beta_o',\dots,\beta_m',\beta'\}$$

with β_i, β_i' according to (2.11) and (2.12). Let z^k and $z^{o,k}$ be points with

$$z^k \in M(t^k) \text{ and } ||x^o - z^k|| = \text{dist}(x^o,M(t^k)) \leq \beta\, d(t^k,t^o) \quad \text{for } k \geq k', \tag{4.7}$$

$$z^{o,k} \in M(t^o) \text{ and } ||x^k - z^{o,k}|| = \text{dist}(x^k,M(t^o)) \leq \beta\, d(t^k,t^o) \quad \text{for } k \geq k'. \tag{4.8}$$

Hence

$$z^k \to x^o \quad \text{and} \quad z^{o,k} \to x^o.$$

We may assume that for some $k'' \geq k'$,

$$x^o \neq z^{o,k} \quad \text{for all } k \geq k'',$$

otherwise, if $x^o = z^{o,k}$ for some subsequence holds then (4.8) contradicts (4.6). For each k, let $u^k \in LM(x^k,t^k)$ be arbitrarily chosen. Since $LM(.,.)$ is locally bounded and closed, we may assume without loss of generality that

$$u^k \to u^o \in LM(x^o,t^o). \tag{4.9}$$

Now the proof runs as follows. We consider the sequence

$$h^k := ||z^{o,k} - x^o||^{-1}(z^{o,k} - x^o)$$

which has an accumulation point h^o in $\text{bd}\, B_n$. Let us assume again $h^k \to h^o$, and we shall first show

$$h^o \in W(x^o,u^o,t^o) \cap \text{bd}\, B_n. \tag{4.10}$$

After that, applying the condition GSOSC, we shall construct a contradiction to (4.6).

To avoid indices, we think k being sufficiently large but fixed, so that (2.11), (2.12), (4.7) and (4.8) hold, and we write

$$t\,,\ \tilde{x}\,,\ z\,,\ z^o\,,\ u \quad \text{instead of} \quad t^k\,,\ x^k\,,\ z^k\,,\ z^{o,k}\,,\ u^k.$$

Applying (2.11), (4.7), (4.8) and $z \in M(t)$, we obtain for any $i \in I^+(u^o)$,

$$\begin{aligned}
&\beta(\beta+1)d(t,t^o) \\
&\geq f_i(z,t) + \beta(\beta+1)d(t,t^o) \\
&\geq f_i(z,t) + \beta(\|z-z^o\|+d(t,t^o)) \\
&\geq f_i(x^o,t^o) \\
&= f_i(z^o,t^o) + (x^o-z^o)^T D_x f_i(z^o,t^o) + o(\|x^o-z^o\|) \\
&\geq f_i(\tilde{x},t) - \beta(\|\tilde{x}-z^o\|+d(t,t^o)) + (x^o-z^o)^T D_x f_i(z^o,t^o) + o(\|x^o-z^o\|) \\
&\geq f_i(\tilde{x},t) - \quad \beta(\beta+1)d(t,t^o) \quad + (x^o-z^o)^T D_x f_i(z^o,t^o) + o(\|x^o-z^o\|).
\end{aligned}$$

Due to (4.9) we may assume for all $i \in I^+(u^o)$

$$u_i > 0 \quad , \qquad \text{hence} \quad f_i(\tilde{x},t) = 0,$$

and so,

$$(x^o - z^o)^T D_x f_i(z^o,t^o) + o(\|x^o - z^o\|) \leq 2\,\beta(\beta+1)d(t,t^o).$$

By (4.6) - (4.8),

$$k\, d(t,t^o) < \|\tilde{x}-x^o\| \leq \|\tilde{x}-z^o\| + \|z^o-x^o\| \leq \beta d(t,t^o) + \|z^o-x^o\|,$$

hence, assuming without loss of generality $\beta \leq k/2$, we obtain

$$d(t,t^o) \leq (2/k)\, \|z^o-x^o\|, \tag{4.11}$$

and so it follows that for all $i \in I^+(u^o)$,

$$(z^o-x^o)^T D_x f_i(z^o,t^o) \geq o(\|z^o-x^o\|) - (4/k)\beta(\beta+1)\, \|z^o-x^o\|.$$

This means that the assumptions of Lemma 3.3 are satisfied, in particular, condition (iii)/(b) (put in Lemma 3.3 $(x^k,t^k) \equiv (x^o,t^o)$, $y^k := z^o$ $(= z^{o,k})$). Thus,

$$h^o \in \bar{W}(x^o,u^o,t^o) \cap \text{bd } B_n.$$

Now we apply that $\text{GSOSC}(x^o,y,t^o,c)$ holds for each $y \in \text{LM}(x^o,t^o)$. Since the function $(x,y) \longmapsto D_x l(x,y,t^o)$ is locally Lipschitzian, the multifunction

$$(x,y,h) \ \rightrightarrows\ \Delta\,(D_x l(.,y,t^o))(x;h)$$

is upper semicontinuous. Using the compactness of $\text{LM}(x^o,t^o)$ (due to MFCQ) and the second-order condition $\text{GSOSC}(x^o,y,t^o,c)$ $\forall y \in \text{LM}(x^o,t^o)$, we have for some $\delta > 0$, with $W_\delta := (\bigcup_y (W(x^o,y,t^o) \cap \text{bd } B_n)) + \delta B_n$, where the union is taken over $y \in \text{LM}(x^o,t^o)$,

$$h^T \Delta\,(D_x l(.,y,t^o))(x;h) \geq c/2 \quad \forall x \in B(x^o,\delta) \quad \forall y \in \text{LM}(x^o,t^o)+\delta B_n \quad \forall h \in W_\delta. \tag{4.12}$$

Here $h^T G \geq b$ for $G \subset R^n$ and $b \in R$ reads as $h^T g \geq b$ for all $g \in G$.
(4.12) applies to our situation: We may assume that, by construction (recall that $(\tilde{x}, z^0, u) = (x^k, z^{0,k}, u^k)$),

$$\tilde{x}, z^0 \in B(x^0, \delta),\; u \in LM(x^0,t^0) + \delta B_m \quad \text{and} \quad ||z^0 - x^0||^{-1}(z^0 - x^0) \in W_\delta . \qquad (4.13)$$

Let $y^0 \in LM(x^0,t^0)$ be the vector satisfying

$$||y^0 - u|| = \text{dist}\,(u, LM(x^0,t^0)),$$

therefore, by Proposition 2.6, with some $L > 0$,

$$||y^0 - u||_1 \leq L\,(||x^0 - \tilde{x}|| + d(t,t^0)). \qquad (4.14)$$

Note that

$$z_q := q\, x^0 + (1-q)\, z^0 \in B(x^0, \delta) \quad \text{for} \quad q \in (0,1).$$

Thus, (4.12) and the second-order Taylor expansion for both $l(.,u,t^0)$ and $l(.,y^0,t^0)$ yield that for some $q' \in (0,1)$ and some $q'' \in (0,1)$,

$$\begin{aligned} l(x^0,u,t^0) &= l(z^0,u,t^0) + (x^0 - z^0)^T D_x l(z^0,u,t^0) + (1/2)\,(x^0 - z^0)^T v \\ &\geq l(z^0,u,t^0) + (x^0 - z^0)^T D_x l(z^0,u,t^0) + (c/4)\,||z^0 - x^0||^2, \end{aligned} \qquad (4.15)$$

where

$$v \in \Delta\,(D_x l(.,u,t^0))(z_{q'}; x^0 - z^0)$$

and , by using $y^0 \in LM(x^0,t^o)$,

$$\begin{aligned} l(z^0,y^0,t^0) &= l(x^0,y^0,t^0) + (z^0 - x^0)^T D_x l(x^0,y^0,t^0) + (1/2)\,(z^0 - x^0)^T w \\ &\geq f(x^0,t^0) \quad + (1/2)\,(z^0 - x^0)^T w \\ &\geq l(x^0,u,t^o), \end{aligned} \qquad (4.16)$$

where

$$w \in \Delta\,(D_x l(.,y^0,t^0))(z_{q''}; z^0 - x^0).$$

Therefore, with $c' := c/4$, taking (4.15) and (4.16) together, one has

$$\begin{aligned} c'\,||z^0 - x^0||^2 &\leq l(z^0,y^0,t^0) - l(z^0,u,t^0) + (z^0 - x^0)^T D_x l(z^0,u,t^0) \\ &\leq \sum_{i=1}^m (y_i^0 - u_i) f_i(z^0,t^0) \quad + \quad (z^0 - x^0)^T D_x l(z^0,u,t^0) \\ &\leq \sum_{i=1}^m (y_i^0 - u_i)(f_i(z^0,t^0) - f_i(\tilde{x},t)) \\ &\quad + (z^0 - x^0)^T (D_x l(z^0,u,t^0) - D_x l(\tilde{x},u,t)), \end{aligned} \qquad (4.17)$$

where $u \in LM(\tilde{x},t)$ and $y_i^0 f_i(\tilde{x},t) \leq 0$ $(i=1,\dots,m)$ were used. By (2.12) and (4.8),

$$||D_x L(z^o,u,t^o)-D_x L(\tilde{x},u,t)||$$
$$= ||D_x f_o(z^o,t^o)-D_x f_o(\tilde{x},t) + \sum_{i=1}^{m} u_i(D_x f_i(z^o,t^o)-D_x f_i(\tilde{x},t))||$$
$$\leq ß(ß+1)d(t,t^o) + ||u||_1 ß(ß+1)d(t,t^o). \tag{4.18}$$

Because of (4.13),

$$||u||_1 \leq ß'' := \max\left\{||y||_1 \mid y \in LM(x^o,t^o) + \delta B_m\right\}$$

Therefore, by (4.17) and (4.18),

$$c'||z^o-x^o||^2 \leq \sum_{i=1}^{m}|y_i{}^o-u_i||f_i(z^o,t^o)-f_i(\tilde{x},t)| + ß(ß+1)(ß''+1)||z^o-x^o||d(t,t^o)$$

and so, using (2.11), (4.14), (4.8) and (4.11), we obtain

$$c'\ ||z^o-x^o||^2 - ß(ß+1)(ß''+1)\ ||z^o-x^o||\ d(t,t^o)$$
$$\leq ß(ß+1)\ ||y^o-u||_1\ d(t,t^o)$$
$$\leq ß(ß+1)\ L\ (||x^o-\tilde{x}|| + d(t,t^o))\ d(t,t^o)$$
$$\leq ß(ß+1)\ L\ (||z^o-x^o|| + (ß+1)\ d(t,t^o))\ d(t,t^o)$$
$$\leq ß(ß+1)\ L\ (1 + (2/k)(ß+1))\ ||z^o-x^o||\ d(t,t^o).$$

Thus, because of $ß \leq k/2$, this implies

$$c'\ ||z^o-x^o|| \leq ß(ß+1)(L(1 + ß^{-1}(ß+1)) + ß'' +1)\ d(t,t^o).$$

By (4.6),

$$k\ d(t^k,t^o) < ||x^k - x^o|| \quad \text{for all } k,$$

on the other hand, using (4.8) and writing $z^{o,k}$ and t^k instead of z^o and t, we then have from the above estimation

$$||x^k-x^o|| \leq ||x^k-z^{o,k}|| + ||z^{o,k}-x^o|| \leq ß\ d(t^k,t^o) + ß^*\ d(t^k,t^o) \tag{4.19}$$

with

$$ß^* := (c')^{-1}ß(ß+1)(L(1 + ß^{-1}(ß+1)) + ß'' + 1),$$

and so a contradiction is found. This completes the proof of b). //

Under a stronger 2nd-order sufficient condition which even entails the strong stability of the local minimizer x^o, a version of the preceding theorem has been shown in [25]. The idea of the proof is partially based on an estimation trick used by Alt [2]. For the program $P^2(t)$, $t \in T$, Theorem 4.5 covers Corollary 4.3 in Robinson [41]. Note that the upper Lipschitz continuity result shown above is not necessarily true if one requires that GSOSC only holds for some (but not for all) u in $LM(x^o,t^o)$, cf. Example 2.2.

5. Strongly Stable Solutions for $C^{1,1}$ Programs

This section deals with conditions for a stationary solution of a $C^{1,1}$ program to be strongly stable. First we characterize under MFCQ and some generalized second-order sufficient condition local minimizers which are strongly stable with order 2. Thereby we extend results known from the C^2 case, cf., e.g., Kojima [29, Th.7.2, Cor. 7.8]. Then we present under LICQ some conditions for regularity (strong stability) of Kuhn-Tucker points for perturbed $C^{1,1}$ programs. This also corresponds to certain known facts in the C^2 case, cf., e.g., Kojima [29, Thms. 4.1, 4.2, 6.4], Robinson [40, Thm. 2.3].

Througout this section we consider a parametric program of the type $P^{1,1}(t)$, $t \in T$.

Definition 5.1: Let $t^o \in T$ and $x^o \in S(t^o)$. We shall say that the <u>uniform second-order sufficient condition</u> holds with modulus $c>0$ and w.r. to a neighborhood V of x^o, an open set $N \supset LM(x^o,t^o)$, a neighborhood U of t^o and some open set W including each of the sets $W^+(x^o,u,t^o) \cap \text{bd } B_n$, $u \in LM(x^o,t^o)$, symbolized by <u>USOSC</u>(V,N,U,W,c), if for all $(x,u,t,h) \in V \times N \times U \times W$ and all $v \in \Delta (D_x l(.,u,t))(x;h)$, one has $h^T v \geq c$.

The condition USOSC looks rather strong and hardly practicable, but we have to by-pass the difficulty that the partial directional derivative of a locally Lipschitzian mapping $(x,t) \mapsto g(x,t)$, $(x,t,h) \rightrightarrows \Delta_x g(.,t)(x;h)$, is not in general upper semicontinuous, we refer to Example 6.1. This property would be a suitable analogon to the requirement "$D_{xx}^2 g(.,.)$ continuous" which is usually imposed for parametric C^2 programs. A discussion in the next section will provide several specializations and simplifications which make more plausible and better usable the condition USOSC.

Theorem 5.2: (Strong stability of local minimizers) Consider $P^{1,1}(t)$, $t \in T$. Let $t^o \in T$ and $x^o \in S(t^o)$ be given. Suppose that x^o fulfils MFCQ and that with V, N, U, W, c (according to Definition 5.1) USOSC(V,N,U,W,c) holds.
Then x^o is a local minimizer of $P^{1,1}(t^o)$ which is strongly stable with order 2. Moreover, if the functions f_i, $i=0,1,\dots,m$, obey (2.11) and (2.12) then the stationary solution mapping $x(.)$ occuring in (2.6)-(2.8) satisfies locally a Lipschitz condition, i.e., there are a positive real number L and a neighborhood U of t^o such that $\|x(t) - x(t^o)\| \leq L\, d(t,t^o)$ holds for each $t \in U$.

Corollary 5.3: (Persistence of strong stability and Hölder continuity) Assume the hypotheses of Theorem 5.2, and let x(.) be the mapping occuring there. Then there exists a neighborhood U' of t^o such that x(t) is a strongly stable (with order 2) local minimizer of $P^{1,1}(t)$ if $t \in U'$. Moreover, if the functions f_i, i=0,1,...,m, additionally satisfy with some $\beta > 0$ and for neighborhoods U_o of t^o and V_o of x^o,

$$|f_i(x',t') - f_i(x'',t'')| \leq \beta(||x'-x''|| + d(t',t'')) \qquad (5.1)$$

$$||D_x f_i(x',t') - D_x f_i(x'',t'')|| \leq \beta(||x'-x''|| + d(t',t'')) \qquad \forall t',t'' \in U_o,\ \forall x',x'' \in V_o^o \qquad (5.2)$$

then x(.) is Hölder continuous with order 1/2 on some neighborhood U" of t^o, i.e., there is some $\nu > 0$ such that $||x(t')-x(t'')||^2 \leq \nu\, d(t',t'')$ $\forall t',t'' \in U''$.

Proof: The assertions follow from Theorem 5.2, by using additionally Proposition 2.7 and Proposition 2.1 (ii), respectively. //

Remark 5.4: Robinson [41, p.219] gives an example of a program with objective function $||x||^2$ which has to be minimized under parametric linear constraints and shows that the assumptions of Corollary 5.3 (Theorem 5.2) do not guarantee, in general, that the global optimal solution mapping x(.) is Lipschitzian "around" t^o, i.e., the rate of 1/2 may not be improved to the rate 1 in Corollary 5.3. In the case of C^2 programs in finite dimensions, Theorem 5.2 may be concluded from [29] and [41]. Alt has shown strong stability and Lipschitz continuity similar to Theorem 5.2 for C^2 optimization problems in Banach spaces, however, under the additional assumption that the multifunction LM(.,.) has a locally Lipschitzian selection, cf. [2].

Proof of Theorem 5.2: Taking Theorem 4.5 into account, we see that it suffices to show strong stability with order 2, the Lipschitz property then will follow. Note that, by Part a) of Theorem 4.5, we find for each neighborhood Q of x^o a neighborhood U_Q of t^o with

$$S(t) \cap Q \neq \emptyset \quad \forall t \in U_Q. \qquad (5.3)$$

We have to verify, whether there exist constants δ, r and ε and a mapping x(.) which satisfy (2.6), (2.7) and (2.8) (put in (2.8) $c := \varepsilon$). To get this, we shall show that for some $r > 0$ and for some $\delta > 0$ with $S(t) \cap B(x^o,r) \neq \emptyset$ $\forall t \in B(t^o,\delta)$ (such a number $\delta = \delta(r)$ exists due to (5.3)), one has

$$f_o(x,t) - f_o(z,t) \geq (c/2)\, ||x - z||^2$$
$$\forall t \in B(t^o,\delta)\ \forall z \in S(t) \cap B(x^o,r)\ \forall x \in M(t) \cap B(x^o,2r). \qquad (5.4)$$

Assume, for the moment, (5.4) is shown. Then for each $t \in B(t^o,\delta)$ and for any two points $x,z \in S(t) \cap B(x^o,r)$ with $x \neq z$, we have

$$f_o(x,t) - f_o(z,t) \geq (c/2)\ ||x-z||^2$$
$$f_o(z,t) - f_o(x,t) \geq (c/2)\ ||x-z||^2$$

which is impossible. Hence, there is some mapping x(.) with

$$S(t) \cap B(x^o,r) = \{x(t)\}\ \forall\ t \in B(t^o,\delta),$$

and so, (2.7) is shown. (2.8) is a special case of (5.4), and (2.6) follows from (5.3).

To show (5.4) indirectly, we assume that there are sequences

$$t^k \to t^o,\quad x^k \to x^o \quad \text{and} \quad y^k \to x^o$$

with

$$x^k \in S(t^k) \qquad \text{and} \qquad y^k \in M(t^k)$$

such that

$$f_o(y^k,t^k) - f_o(x^k,t^k) < (c/2)\ ||y^k-x^k||^2.$$

For each k let $u^k \in LM(x^k,t^k)$. Since LM(.,.,) is closed and locally bounded, we may assume without loss of generality that u^k converges to some u^o in $LM(x^o,t^o)$. Thus, the assumptions of the Lemmas 3.3 and 3.4 are satisfied, hence one finds a sequence $q_k \downarrow 0$ and obtains for k sufficiently large

$$h^k := ||y^k-x^k||^{-1}(y^k-x^k) \in W,\ z^k := x^k + q_k h^k \in V,\ u^k \in N,\ t^k \in U$$

with

$$(v^k)^T h^k < c \quad \text{for some} \quad v^k \in \Delta(D_x l(.,u^k,t^k))(z^k;h^k),$$

which contradicts USOSC(V,N,U,W,c). //

Corollary 5.5: (Strong stability of Kuhn-Tucker solutions under USOSC) Consider $P^{1,1}(t), t \in T$. Let $t^o \in T$ and $x^o \in S(t^o)$ be given. Suppose that x^o satisfies LICQ (hence $LM(x^o,t^o)$ consists of a single vector u^o), and that USOSC(V,N,U,W,c) holds with V, N, U, W, c according to Definition 5.1. Then there are neighborhoods $U' \subset U$ of t^o and $V' \subset V$ of x^o and some mapping $z(.) = (x(.),u(.))\colon U' \to V' \times R^m$ such that the following hold:

a) z(.) is continuous at t^o with $z(t^o) = (x^o,u^o)$.

b) $KT(t) \cap (V \times R^m) = \{z(t)\}$ for all $t \in U'$.

c) If the functions f_i and $D_x f_i$ (i=0,1,...,m) are Lipschitzian according to (5.1) and (5.2) then z(.) is Lipschitzian on some neighborhood of t^o.

Proof: Using Corollary 5.3, we immediately obtain the existence of some mapping z(.) which satisfies the properties a) and b). In order to show that z(.) is Lipschitzian on some neighborhood of t^o, one may follow the same line of proof as in Theorem 4.5. Applying Proposition 2.8 and the Lipschitz properties (5.1) and (5.2), we see that the unique Lagrange multiplier u associated with (x,t) (for (x,t) near (x^o,t^o) and $x \in S(t)$) depends Lipschitz continuously on (x,t), and so it sufficies to show the Lipschitz property for x(.). Assume, on the contrary, that for some sequences $t^{j,k} \to t^o$, j=1,2, one has

$$||x(t^{1,k})-x(t^{2,k})|| > k\, d(t^{1,k},t^{2,k}) \quad \text{for all } k. \tag{5.5}$$

Following the proof in Theorem 4.5, we may construct a contradiction similar to that between (4.6) and (4.19). However, we have to verify that the constants ß, ß", c', L appearing in (4.19) are uniform w.r. to some neighborhood of (t^o,x^o). This can be easily done, the details will be omitted. //

Definition 5.6: If a Kuhn-Tucker solution $(x^o u^o)$ of $P^{1,1}(t^o)$ satisfies the properties a), b) and c) of Corollary 5.5, then the mapping KT(.) will be called <u>regular</u> at (x^o,u^o,t^o).

If more information about the behavior of partial generalized second derivatives w.r. to the parameter is available, the uniform condition USOSC may be replaced by the following one:

Definition 5.7: Let $t^o \in T$ and $(x^o,u^o) \in KT(t^o)$. We shall say that (x^o,u^o) satisfies with modulus $c > 0$ the <u>strong second-order sufficient conditon</u>, symbolized by <u>SSOSC</u>(x^o,u^o,t^o,c), if $h^T v \geq c$ holds for all h in $W^+(x^o,u^o,t^o) \cap \text{bd } B_n$ and for all v in $\Delta(D_x l(.,u^o,t^o))(x^o;h)$.

In Section 6 we shall discuss several special cases in which the uniform condition USOSC may be replaced by SSOSC. In this place, let us assume that

$$T \text{ is an open subset of } R^p, \tag{5.6}$$

$$f_i(.,.) \in C^{1,1}(R^n \times T,R),\ i=0,1,\dots,m, \tag{5.7}$$

$$D_t Df_i(.,.) \text{ exists and is locally Lipschitzian, } i=0,1,\dots,m. \tag{5.8}$$

Then a chain role for directional derivatives of Lipschitz mappings, cf. Kummer [32], yields for a mapping $g \in C^{0,1}(R^n \times T,R^n)$:

$$\Delta g((x,t);(h,w)) = \Delta_x g((x,t);h) + D_t g(x,t)w,$$

where $\Delta_x g((x,t);h) := \Delta(g(.,t))(x;h)$. This implies that the directional derivative multifunctions of $D_x f_i(.,.)$ are closed. This anticipates the dicussion in 6.4 and 6.1 below, where we shall point out that in this situation the condition SSOSC entails the uniform condition USOSC. Hence, as a consequence of Corollary 5.5, we obtain regularity of a Kuhn-Tucker point associated with a local minimizer to a $C^{1,1}$ program:

Corollary 5.8: (Regularity of the Kuhn-Tucker mapping) Given $P^{1,1}(t)$, $t \in T$, let $t^o \in T$ and $x^o \in S(t^o)$. Suppose that (5.6)-(5.8) hold and that x^o satisfies LICQ (hence $LM(x^o,t^o) = \{u^o\}$). If $SSOSC(x^o,u^o,t^o,c)$ is fulfilled with some $c > 0$, then KT(.) is regular at (x^o,u^o,t^o).

Under (5.6)-(5.8), Kummer [33] has given an implicit function theorem which provides, as a special case, a result similar to that of Corollary 5.8 and which allows, moreover, to characterize regularity of Kuhn-Tucker solutions to $C^{1,1}$ programs also in the case when the stationary solution is <u>not</u> a local minimizer. The main idea is, by splitting the Lagrange multiplier vector into $v = (v_1,\dots,v_l)^T$ and $y = (y_{l+1},\dots,y_m)^T$, to replace the Kuhn-Tucker system (2.5) by

$$F(x,v,y,t) := \begin{bmatrix} D_x f_o(x,t) + \sum_{i=1}^{l} v_i D_x f_i(x,t) + \sum_{j=l+1}^{m} y_j^+ D_x f_j(x,t) \\ -f_i(x,t) \quad (i=1,\dots,l) \\ y_j^- - f_j(x,t) \quad (j=l+1,\dots,m) \end{bmatrix} = 0 \qquad (5.9)$$

and to handle $F(x,v,y,t) = 0$ as a Lipschitz equation. Note that $F(x,v,y,t)=0$ implies $(x,v,y^+) \in KT(t)$, and $(x,v,y) \in KT(t)$ implies $F(x,v,y+g(x,t),t)=0$, where $g(x,t) = (f_{l+1}(x,t),\dots,f_m(x,t))^T$. Both transformations are Lipschitz. A point (x,v,y) which satisfies (5.9) is called <u>critical</u> w.r. to $P^{1,1}(t)$, cf. Kummer [32,33], in symbols: $(x,v,y) \in CP(t)$.

Definition 5.9: A critical point (x,v,y) to $P^{1,1}(t)$ satisfies the second-order condition <u>SOC</u> if $\Delta(D_x l(.,v,y^+,t))(x;h) \cap K(x,y,t,h) = \emptyset$ holds for each $h \in W^+(x,v,y^+,t) \setminus \{0\}$, where

$$K(x,y,t,h) := \left\{ \sum_i \nu_i D_x f_i(x,t) + \sum_{j:\ y_j \geq 0} \eta_j D_x f_j(x,t) \;\middle|\; \begin{array}{l} \nu_i \in R,\ i=1,\dots,l \\ \eta_j h^T D_x f_j(x,t) \leq 0 \text{ if } y_j = 0 \end{array} \right\}$$

Proposition 5.10: (Kummer [33, Thm.2]) Consider $P^{1,1}(t)$, $t \in T$. Let $t^o \in T$, and let (x^o,v^o,y^o) be a critical point to $P^{1,1}(t^o)$. Suppose (5.6)-(5.8). If LICQ and SOC hold w.r. to (x^o,v^o,y^o) then the mapping CP(.) is regular at (x^o,v^o,y^o), and hence KT(.) is regular at $(x^o,v^o,(y^o)^+)$.

Kummer has shown that LICQ and SOC are also necessary in some sense for the regularity of the critical point mapping, for details see [33]. In the case of parametric optimization problems with twice differentiable data (of the type $P^2(t)$, $t\in T$), a broad discussion of equivalences between strong stability of stationary solutions, regularity of Kuhn-Tucker points and other stability/regularity-characterizations may be found in [27]. That paper is essentially based on implicit function theorems for locally Lipschitzian mappings, and it turns out that under LICQ, the regularity of KT(.) is, for example, equivalent to the Lipschitz invertibility of the mapping F (5.9) and also equivalent to nonsingularity of the generalized Jacobian of F at the critical point of interest.

6. A Discussion of the Uniform Second-Order Condition

In this section, we discuss how to replace USOSC (Definition 5.1) by more convenient conditions. In particular, we deal with a special class of $C^{1,1}$ problems for which the verification of USOSC or SSOSC reduces to checking whether finitely many matrices are positive definite.

Throughout this section we consider the parametric problem $P^{1,1}(t)$, $t\in T$.

6.1. Let $t^o\in T$, $x^o\in S(t^o)$ and suppose that for some open subset Q of R^n, some bounded open set N containing $LM(x^o,t^o)$, some open set W which contains each of the sets $W^+(x^o,u,t^o)\cap \text{bd } B_n$, $u\in LM(x^o,t^o)$ and some multifunction

$$H:\quad Q \times N \times T \times W \rightrightarrows R^n$$

the following hold:

$$H \text{ is closed and locally bounded on } Y_o := \{x^o\} \times LM(x^o,t^o) \times \{t^o\} \times \text{bd } B_n \tag{6.1}$$

and

$$\Delta(D_x l(.,u,t))(x;h)\subset H(x,u,t,h) \quad (\forall (x,u,t,h)\in Q \times N \times T \times W). \tag{6.2}$$

The ideal realization of (6.1) and (6.2) would be to choose $H = \Delta(D_x l)$, cf. 6.4 and the discussion before Corollary 5.8. Unfortunately, the partial directional derivative is not closed, in general, consider

Example 6.1.: (Kummer [32, Ex.1]) Take the function $g: R^2 \to R$ defined by $g(x,y) := y^+$ if $y\leq |x|$ and $g(x,y) := |x|$ if $|x|\leq y$. Hence $\Delta(g(.,y))(0;1)$ is $\{0\}$ if $y\leq 0$ and equal to the closed interval $[-1,1]$ if $y>0$.

Proposition 6.2: Assume (6.1) and (6.2). If SSOSC(x^o,u,t^o,c) holds for all $u \in LM(x^o,t^o)$ then USOSC($V,N',U,W',c/2$) is satisfied for some tuple of sets V,N',U,W' according to Definition 5.1.

Proof: Since N is bounded, $LM(x^o,t^o)$ is a compact set. By (3.5), the multifunction $u \rightrightarrows W^+(u) := W^+(x^o,u,t^o) \cap \text{bd } B_n$ is upper semicontinuous on $LM_o := LM(x^o,t^o)$, hence $W^+(LM_o)$, the image of LM_o under $W^+(.)$ is a compact set. By (6.1), H is upper semicontinuous on Y_o, thus $H(Y_o)$ is the image of the compact set Y_o and hence compact, too. Using SSOSC, we find open sets $W'' \supset W^+(LM_o)$ and $H'' \supset H(Y_o)$ such that

$$h^T v \geq c/2 \quad \text{for all } h \in W'' \text{ and all } v \in H''. \tag{6.3}$$

Again by (6.1), there are neighborhoods V of x^o and U of t^o and open sets N': $LM_o \subset N' \subset N$ and W': $W^+(LM_o) \subset W' \subset W \cap W''$ such that

$$H(x,u,t,h) \subset H'' \quad \text{for all } (x,u,t,h) \text{ in } V \times N' \times U \times W'.$$

Hence, (6.3) and (6.2) imply that (read as set-valued inequality)

$$h^T \Delta (D_x l(.,u,t))(x;h) \geq c/2$$

holds for each (x,u,t,h) in $V \times N' \times U \times W'$. //

6.2. Now consider a problem of the type $P^2(t)$, $t \in T$, cf. Section 2. By property (3.2e), $h^T \Delta (D_x l(.,u,t))(x;h) = \{h^T D_{xx}{}^2 l(x,u,t)\, h\}$ holds, which immediately implies that our condition GSOSC (Definition 4.2) reduces to the standard 2nd-order sufficient condition used, e.g., in [9,12,41]. Moreover, by assumption (2.3) on $P^2(t)$, (6.1) and (6.2) are automatically fulfiled with $H(x,u,t,h) := \{h^T D_{xx}{}^2 l(x,u,t) h\}$ and with any open set $N \supset LM_o$ (provided LM_o is bounded, which is equivalent to the validity of MFCQ) and $W = R^n$. Thus, USOSC passes to a special version of the condition in Proposition 6.2, cf., e.g., [29, Condition 7.3].

6.3. The previous remarks immediately allow to specify USOSC in the case that a $C^{1,1}$ optimization problem is perturbed by C^2 functions. For the given parametric program, consider the case that for each i, f_i has the representation $f_i(x,t) = g_{o,i}(x) + g_{1,i}(x,t)$, where $g_{o,i}$ is a $C^{1,1}$ function and $g_{1,i}$ satisfies the C^2 assumptions (2.3). It is easy to show, cf. [28], that the multifunction $H = \Delta (D_x l)$ satisfies (6.1) and (6.2), therefore, Proposition 6.2 applies to this situation.

We note that literature on decomposition methods pays a special attention to optimization problems in which the objective function is separable w.r. to two groups of variables (cf., e.g., Beer [5]), where of particular interest is the case $g_{0,i} \equiv 0$ ($i \geq 1$), but $f_0(x,t) = g_{0,0}(x) + g_{1,0}(t)$. Obviously, $\Delta(D_x l(.,u,t))(x;h)$ then has a very simple form.

6.4. The following assumptions on f_i and $D_x f_i$ ($i=0,1,\dots,m$) ensure the closedness of the partial directional derivative $\Delta(D_x l)$: Let $t^o \in T$ and $x^o \in S(t^o)$, and let U_o and V_o be neighborhoods of t^o and x^o, respectively, so that with some constant $\beta > 0$,

$$||D_x f_i(x',t) - D_x f_i(x'',t)|| \leq \beta\, ||x'-x''|| \qquad (\forall x',x'' \in V_o\ \forall t \in U_o), \tag{6.4}$$

$$\limsup_{\substack{t \to t^o \\ x \to x^o}} \Delta(D_x f_i(.,t) - D_x f_i(.,t^o))(x;h) = \{0\} \quad (\forall h \in \mathrm{bd}\, B_n) \tag{6.5}$$

($i=0,1,\dots,m$) hold. In the case of C^2 data, (6.5) corresponds to the continuity of $D_{xx}{}^2 f_i(.,.)$.

Proposition 6.3: Let for $x^o \in S(t^o)$ MFCQ be satisfied, and suppose (2.2), (6.4) and (6.5). Then $(x,u,t,h) \rightrightarrows \Delta(D_x l(.,u,t))(x;h)$ is locally bounded and closed on $\{x^o\} \times LM(x^o,t^o) \times \{t^o\} \times \mathrm{bd}\, B_n$.

For the proof we refer to [28]. One has essentially to use the properties (3.2d) and (3.2g) for directional derivatives of Lipschitzian mappings.

6.5. Now we recall a broad class of $C^{1,1}$ functions g for which a simple representation of Clarke's generalized Jacobian ∂g is possible, and which is of interest in several applications, see also Section 7.
Given an open subset Q of R^n, functions $g_i \in C^2(Q,R)$, let g be a selection from $\{g_1,\dots,g_s\}$ satisfying

(a) For each $x \in Q$ there is some $i=i(x) \in \{1,\dots,s\}$ such that $g(x) = g_i(x)$,
(b) g is continuous on Q,
(c) for each pair $i,j \in \{1,\dots,s\}$ and each $x \in Q_i \cap Q_j$ one has $Dg_i(x) = Dg_j(x)$, where $Q_i := \{x \in Q \mid g(x) = g_i(x)\}$.

Proposition 6.4: ([26, Thm. 4]) The function g belongs to $C^{1,1}(Q,R)$, and for each x in Q there is an index set $J=J(x) \subset \{i \mid g(x)=g_i(x)\}$ such that $\partial(Dg)(x) = \mathrm{conv}\{D^2 g_i(x) \mid i \in J\}$.

A function g satisfying (a)-(c) is called a $C^{1,1}$ selection of $\{g_1,\dots,g_s\}$. Studying $P^{1,1}(t)$ in the case $T=R^p$ and assuming that the functions f_i are $C^{1,1}$ selections from $\{g_1,\dots,g_s\}$, $g_j \in C^2(R^n \times R^p, R)$, $j=1,\dots,s$, $i=0,1,\dots,m$, the imbedding multifunction H (6.1),(6.2) could be chosen as follows:

For $(x,u,t,h) \in Q \times R^m \times T \times R^n$ put

$$H(x,u,t,h) := H_0(x,t,h) + u_1 H_1(x,t,h) + \ldots + u_m H_m(x,t,h)$$

where

$$H_i(x,t,h) \quad := \quad \text{conv} \left\{ D_{xx}{}^2 g_j(x,t)\, h \mid j \in I(f_i,x,t) \right\},$$

$$I(f_i,x,t) \quad := \quad \left\{ j \in \{1,\ldots,s\} \mid f_i(x,t) = g_j(x,t) \right\}.$$

Another way to choose H is to represent the Lagrange function $l(.,.,.)$ as a $C^{1,1}$ selection of functions being of the same structure and being constructed from $g_1,\ldots,g_m$. How to use this approach in semi-infinite programming will be discussed in 7.2. below.

7. Related Work, Motivations and Applications

The present paper reflects only a little part of subjects being of interest in parametric analysis of nonsmooth optimization problems. The approach is restricted to the use of the generalized directional derivatives introduced in Section 3, other concepts of generalized derivatives are left out of consideration. More about parametric optimization via nonsmooth analysis may be found, e.g., in the recent papers of Kummer [32,33,34], Robinson [43], King and Rockafellar [23], Rockafellar [44], Jongen, Klatte and Tammer [20]. In this last section we shall outline some motivations and applications of nonsmooth parametric optimization, in particular, with regard to the framework given in the sections before.

7.1. One of the subjects for which optimality conditions and stability analysis under nonsmoothness apply is that of multi-level optimization. We only refer to decomposition procedures (cf. e.g. Beer [5], Jongen, Möbert and Tammer [21]), local reduction or aggregation of infinitely many constraints in semi-infinite programming (cf. e.g. Hettich and Jongen [16], Ben- Tal, Teboulle and Zowe [6], Ioffe [18], Jongen, Wetterling and Zwier [22], Shapiro [46]), two-stage problems in stochastic programming (cf., e.g., Römisch and Schultz [45]) and Stackelberg problems (cf.,e.g., Loridan and Morgan [35], Outrata [37]).

In our context, iterated local minimization for nonconvex programs is of particular interest ([21,25]). Let $f_i \in C^1(R^n \times R^p)$, $i=0,1,\ldots,m$, be given functions, define

$$X := \left\{ (x,t) \in R^n \times R^p \mid f_i(x,t)=0,\ i=1,\ldots,l;\ f_j(x,t) \leq 0,\ j=l+1,\ldots,m \right\}$$

and consider the (fixed) nonlinear optimization problem

$$(P) \qquad f_o(x,t) \longrightarrow \min_{(x,t)} \qquad \text{s.t.} \qquad (x,t)\in X.$$

Defining $M(t) := \{x \mid (x,t)\in X\}$, the associated parametric program

$$P(t): \qquad f_o(x,t) \longrightarrow \min_x \qquad \text{s.t.} \qquad x\in M(t)$$

(with parameter $t\in T := R^p$) is even the parametric problem $P^1(t)$, $t\in T$, introduced in Section 2; (2.2) is fulfiled. Now suppose that for some t^o and some neighborhood U of t^o, x(.) and u(.) are vector functions satisfying $(x(t),u(t))\in KT(t)$, $t\in U$. Let

$$f^*(t) := f_o(x(t),t) \;, \quad t\in U,$$

i.e., f^* is the value function which assigns to each $t\in U$ the value of the sationary solution x(t) (in particular cases, of course, this is the global minimal value or a local minimal value). Let $x^o := x(t^o)$. Trivially, (x^o,t^o) is a global minimizer of (P) if and only if x^o is a global minimizer for $P(t^o)$ and t^o is a global minimizer for the function $t \mapsto \inf_{x\in M(t)} f_o(x,t)$ s.t. the effective domain of M(.). This suggests to conjecture that a local minimizer x^o of $P(t^o)$, a selection $x(.)\in \Psi_{loc}(.)$ which is continuous at t^o and a local minimizer t^o for f^* lead to a local minimizer (x^o,t^o) of (P). Jongen, Möbert and Tammer [21] give an example of an unconstrained optimization problem with a polynomial objective function, showing that this conjecturture is false. To overcome this difficulty, one has to require some isolateness of the local minima x(t), cf. Klatte [25, Thm. 1.1].

In this place, let us cite two typical results on decomposition for nonconvex problems, where x(.), u(.), t^o, x^o are as above:

(i) [28, Thm. 2.5; 25, Thm. 1.2] Let $x(.)\in S(.)$ be continuous at t^o, and suppose that x^o is a strongly stable local minimizer of $P(t^o)$ and satisfies MFCQ. Then (x^o,t^o) is a local minimizer of (P) if t^o is a local minimizer for f^*.

(ii) [21, Lemma 2.1, Thm. 2.1] Let $J(t^o) := \{i \mid D_x f_i(x,t)\equiv 0$ for (x,t) near $(x^o,t^o)\}$. Write $g_i(t)$ instead of $f_i(x,t)$ for $i\in J(t^o)$, let $J_1(t^o)$ be the intersection of $J(t^o)$ and $\{1,\dots,1\}$. If u(.) is continuous at t^o, and for some positive ß, x(.) satisfies $||x(t)-x^o|| \leq ß\, ||t-t^o||$ for all $t\in T$, then f^* is differentiable at t^o, and one has: (x^o,t^o) is a stationary solution of (P) if t^o is a stationary solution of the problem $f^*(t) \longrightarrow \min$ s.t. $g_i(t)=0$, $i\in J_1(t^o)$, $g_j(t)\leq 0$, $j\in J(t^o)\setminus J_1(t^o)$.

To satisfy the hypotheses of (i) or (ii), one may use, for example, the results of Theorem 4.5, Theorem 5.2 and Corollary 5.5. w.r. to $C^{1,1}$ programs.

7.2. As announced in 7.1., the studies of our paper are also motivated by several questions arising in semi-infinite programming. Given an open subset T of R^p, twice continuously differentiable functions $f_0: R^n \times T \to R$, $f: R^n \times R^s \to R$ and $g_j: R^s \to R$, $j=1,\dots,d$, we consider a semi-infinite program depending on a parameter vector t from T which only appears in the objective function:

$$\text{SIP(t):} \quad f_0(x,t) \longrightarrow \min_x \quad \text{s.t.} \quad f(x,z) \leq 0 \quad \forall z \in Z,$$

where

$$Z := \{ z \in R^s \mid g_j(z) \leq 0, \; j=1,\dots,d \}$$

is supposed to be compact.
In order to study stability, it suggests itself to transform SIP(t) into a program with finitely many constraints by taking $f(x,.) := \max_{z \in Z} f(x,z)$ (and handling SIP(t) as a nonsmooth program, as done, e.g., in [6,18,46]), or by using the reduction ansatz of Hettich and Jongen [16].
Following the second approach, we present here an exemplary result. Let x^0 be a feasible point of interest and suppose that

$$E(x^0) := \{ z \in Z \mid f(x^0,z) = 0 \} \text{ is nonempty.}$$

Since every z in $E(x^0)$ is a global maximizer for $f(x^0,.)$ on Z, the idea is to study the parametric auxiliary program (with x near x^0 as parameter)

$$P(x): \quad f(x,z) \longrightarrow \max_z \quad \text{s.t.} \quad z \in Z.$$

Note that P(x), x near x^0, is a parametric problem of the type P^2 (2.3).
Suppose that for each z in $E(x^0)$, LICQ holds w.r. to Z, and (z,y_z) satisfies $\text{SSOSC}(z,y_z,x^0,c)$ with some $c>0$ (cf. Definition 5.7), where y_z is the unique multiplier vector associated with z and SSOSC has to be adapted for a maximum type problem. Recall that SSOSC reduces to a known 2nd-order condition (cf. the discussion in 6.2.). We emphasize that SSOSC does not include the strict complementarity slackness condition.
Under these assumptions, each z in $E(x^0)$ is a strongly stable local maximizer, and, in particular, $E(x^0)$ is finite, say $E(x^0) = \{ z^1,\dots,z^m \}$. The local reduction theorem and the structural analysis of P(x) [22] yield that there are a neighborhood U of x^0 and mappings $z^i(.)$ from U to R^s $(i=1,\dots,m)$ such that for each $i \in \{1,\dots,m\}$:

a) $z^i(x)$ is a local maximizer to $P(x)$, $x \in U$, $z^i(.)$ is Lipschitzian on U with $z^i(x^o) = z^i$, and $f_i(.)$ defined by $f_i(x) := f(x,z^i(x))$, $x \in U$, is continuously differentiable on U,

b) there exist continuously differentiable on U mappings $z^{ij}(.)$ from U to R^s, $j=1,\dots,k_i$, having the property that for each $x \in U$ there is some j with $z^i(x) = z^{ij}(x)$, and so $f_i(x) = f_{ij}(x) := f(x,z^{ij}(x))$, and each f_{ij} is twice continuously differentiable,

c) if $x \in U$ then one has $f(x,z) \leq 0 \quad \forall z \in Z$ iff $f_i(x) \leq 0$, $i=1,\dots,m$.

Hence, under the assumptions imposed on $P(x)$, the infinitely many constraints reduce in some neighborhood of x^o to finitely many one, and we have the reduced parametric program

$$\text{RED}(t): \quad f_o(x,t) \longrightarrow \min_x \quad \text{s.t.} \quad f_i(x) \leq 0, \quad i=1,\dots,m.$$

By Proposition 6.4, the functions $f_1(.),\dots,f_m(.)$ are $C^{1,1}$ selections of a finite set of C^2 functions, this property carries over to the Lagrange function of RED(t),

$$L(x,u,t) := f_o(x,t) + u_1 f_1(x) + \dots + u_m f_m(x), \quad (x,u,t) \in R^n \times R^m \times T.$$

More exactly, for each (x,u,t) there is an index tuple $J=(j_1,\dots,j_m)$, where $j_i \in \{1,\dots,k_i\}$ $(i=1,\dots,m)$, such that

$$L(x,u,t) = L^J(x,u,t) := f_o(x,t) + u_1 f_{1j_1}(x) + \dots + u_m f_{mj_m}(x),$$

provided that $x \in U$. In this case, J will be called <u>feasible for (x,u,t)</u>. Since RED(t) is of the type $P^{1,1}(t)$ (without equality constraints), the notation introduced for $P^{1,1}(t)$ can be used, and the propositions in the Sections 4, 5 and 6 apply to the present situation. The second-order conditions now read as follows.

<u>GSOSC(x^o,u^o,t^o,c) w.r. to RED(t^o):</u> For each index tuple J being feasible for (x^o,u^o,t^o) and for each $h \in W(x^o,u^o,t^o) \cap \text{bd}\, B_n$, there holds that $h^T D^2_{xx} L^J(x^o,u^o,t^o)\, h \geq c$.

<u>SSOSC(x^o,u^o,t^o,c) w.r. to RED(t^o):</u> replace in GSOSC the cone $W(x^o,u^o,t^o)$ by the linear space $W^+(x^o,u^o,t^o)$.

Without using the tool of nonsmooth analysis, a condition similar to GSOSC was given by Jongen, Wetterling and Zwier in [22] for characterizing local minima of semi-infinite programs.

7.3. Even in the case of smooth initial data, optimality conditions provide a more complicated structure such as nonlinear complementarity problems, nonsmooth equations, generalized equations etc. This is one of the reasons why implicit function theorems for these more complicated problems are of great interest for stability studies in optimization. In this paper, we mentioned several times this fact, let us refer once more, in this place, to the papers [20,23,29,32-34,40,41,43,44]. A similar interest concerns the creation of Newton and Newton-like methods for these classes of problems.

However, smooth optimization and complementarity problems often produce a rather special structure. Consider, for example, a nonlinear complementarity problem (NLCP) with continuously differentiable function f from R^n to R^n:

$$x \geqq 0 \ , \quad f(x) \geqq 0 \ , \quad x^T f(x) = 0. \tag{7.1}$$

Of course, x is a solution of (7.1) iff x is a zero of the equation

$$\min \{ x \ , \ f(x) \} = 0,$$

where the minimum is taken component-wise. Via the so-called Minty map

$$h(x) = f(x^+) + x^- \tag{7.2}$$

one has two other equivalent characterizations: x is a zero of (7.2) if and only if x^+ solves (7.1), and this is equivalent to minimizing $h(\xi)^T h(\xi)$. In all cases, to the problem (7.1) with smooth data is assigned a problem defined via special nonsmooth functions. A similar phenomenon was studied in Section 5 with respect to Kuhn-Tucker sets. Newton-type methods for solving this kind of problems were, e.g., given in the papers of Pang [38], Harker and Xiao [15], Kojima and Shindo [30], Kummer [31].

Actually, the development of Newton methods for equations and generalized equations under local Lipschitz continuity assumptions plays an essential role in the literature, we also refer to recent papers of Robinson [42] and Kummer [34]. The papers differ a) in using the one or the other concept of generalized derivatives (which replaces the classical term Df(x) or Df(x)h) or b) in using a certain approximation requirement (which replaces the classical linear approximation properties for C^1 functions). Newton's method is extended, e.g., in the case of PC^1 functions [30,31], for multivalued derivatives (including that of Clarke type) [31], for point-based approximations and B-derivatives [42], for homogenious approximations by strong B-derivatives [15,38], for classical directional derivatives [33].

Acknowledgement: The author is indebted to Bernd Kummer and Ralf Walzebok for being allowed to use some material from the common Working Paper [28]. He wishes to thank Bernd Kummer also for recent stimulating discussions on the subject of this paper.

References

[1] Alt, W.: Lipschitzian perturbations of infinite optimization problems. In: A.V. Fiacco, ed., Mathematical Programming with Data Perturbations. M. Dekker, New York and Basel, 1983.

[2] Alt, W.: Stability of solutions for a class of nonlinear cone constrained optimization problems, Part 1: Basic theory. Numer. Funct. Anal. and Optimiz. 10 (1989), 1053-1064.

[3] Auslender, A.: Stability in mathematical programming with nondifferentiable data. SIAM J. Control Optim. 22 (1984), 239-254.

[4] Bank, B., J. Guddat, D. Klatte, B. Kummer, K. Tammer: Non-Linear Parametric Optimization. Akademie-Verlag, Berlin, 1982.

[5] Beer, K.: Lösung großer linearer Optimierungsaufgaben. VEB Deutscher Verlag der Wissenschaften, Berlin, 1977.

[6] Ben-Tal, A., M. Teboulle and J. Zowe: Second order necessary optimality conditions for semi-infinite optimization problems. In: R. Hettich, ed., Semi-Infinite Programming, Springer, Berlin-Heidelberg-New York, 1979.

[7] Clarke, F.: Optimization and Nonsmooth Analysis. Wiley, New York, 1983.

[8] Federer, H.: Geometric Measure Theory. Springer, Berlin-Heidelberg-New York, 1969.

[9] Fiacco, A.V.: Introduction to Sensitivity and Stability Analysis in Nonlinear Programming. Academic Press, New York-London, 1983.

[10] Fiacco, A.V., ed.: Sensitivity, Stability and Parametric Analysis. Math. Programming Study 21 (1984).

[11] Fiacco, A.V., ed.: Optimization with Data Perturbations. Annals of Operations Research 27 (1990).

[12] Fiacco, A.V. and G.P. McCormick: Nonlinear Programming: Sequential Unconstrained Minimization Techniques. Wiley, New York, 1968.

[13] Guddat. J., H.Th. Jongen, B. Kummer, F. Nozicka, eds.: Parametric Optimization and Related Topics. Akademie-Verlag, Berlin, 1987.

[14] Guddat, J., H.Th. Jongen, B. Kummer, F. Nozicka, eds.: Parametric Optimization and Related Topics II. Akademie-Verlag, Berlin, 1991.

[15] Harker, P.T. and B. Xiao: Newton's method for the nonlinear complementarity problem: a B-differentiable equation approach. Math. Programming 48 (1990), 339-357.

[16] Hettich, R. and H.Th. Jongen: Semi-infinite programming: Conditions of optimality and applications. In: J. Stoer, ed., Optimization Techniques, Springer, Berlin-Heidelberg-New York, 1978.

[17] Hiriart-Urruty, J.-B., J.J. Strodiot and V. Hien Nguyen: Generalized Hessian matrix and second-order optimality conditions for problems with C 1,1 - data. Appl. Math. Optim. 11 (1984), 43-56.

[18] Ioffe, A.D.: Second order conditions in nonlinear nonsmooth problems of semi-infinite programming. In: A.V. Fiacco and K.O. Kortanek, eds., Semi-Infinite Programming and Applications. Springer, Berlin-Heidelberg-New York, 1983.

[19] Jongen, H.Th., P. Jonker and F. Twilt: Nonlinear Optimization in R^n, II: Transversality, Flows, Parametric Aspects. P. Lang, Frankfurt/Main, 1986.

[20] Jongen, H.Th., D. Klatte and K. Tammer: Implicit functions and sensitivity of stationary points. Math. Programming 49 (1990), 123-138.

[21] Jongen, H.Th., T. Möbert and K. Tammer: On iterated minimization in nonconvex optimization. Math. Operations Res. 11 (1986), 679-691.

[22] Jongen, H.Th., W. Wetterling and G. Zwier: On sufficient conditions for local optimality in semi-infinite programming. optimization 18 (1987), 165-178.

[23] King, A.J. and R.T. Rockafellar: Sensitivity analysis for nonsmooth generalized equations. Research Report RC 14639 (#65641), IBM Research Division, Yorktown Heights, NY, 1989.

[24] Klatte, D.: On the stability of local and global optimal solutions in parametric problems of nonlinear programming. Seminarbericht Nr. 75, Sektion Mathematik, Humboldt-Universität, Berlin, 1985.

[25] Klatte, D.: Strong stability of stationary solutions and iterated local minimization. In: |14|.

[26] Klatte, D. and K. Tammer: On second-order sufficient optimality conditions for C 1,1 optimization problems. optimization 19 (1988), 169-179.

[27] Klatte, D. and K. Tammer: Strong stability of stationary solutions and Karush-Kuhn-Tucker points in nonlinear optimization. Annals Oper. Res. 27 (1990), 285-308.

[28] Klatte, D., B. Kummer and R. Walzebok: Conditions for optimality and strong stability in nonlinear programs without assuming twice differentiability of data. IIASA Working Paper WP-89-089, Laxenburg, Austria, 1989.

[29] Kojima, M.: Strongly stable stationary solutions in nonlinear programs. In: S.M. Robinson, ed., Analysis and Computation of Fixed Points. Academic Press, New York, 1980.

[30] Kojima, M. and S. Shindo: Extensions of Newton and Quasi-Newton methods to systems of PC 1 equations. J. Oper. Res. Soc. Japan 29 (1987), 352-374.

[31] Kummer, B.: Newton's method for non-differentiable functions. In: J. Guddat et al., eds., Advances in Mathematical Optimization. Akademie-Verlag, Berlin, 1988.

[32] Kummer, B.: The inverse of a Lipschitz function in R^n: Complete characterization by directional derivatives. Preprint Nr. 195 (Neue Folge), Sektion Mathematik, Humboldt-Universität, Berlin, 1988.

[33] Kummer, B.: An implicit-function theorem for C 0,1 - equations and parametric C 1,1 - optimization. Manuscript, Sektion Mathematik, Humboldt-Universität, Berlin, 1989, to appear in J. Math. Anal. Appl.

[34] Kummer, B.: Newton's method based on generalized derivatives: Convergence analysis. Manuscript, Fachbereich Mathematik, Humboldt-Universität, Berlin, 1991.

[35] Loridan, P. and J. Morgan: Approximate solutions for two-level optimization problems. International Series of Numerical Mathematics Vol. 84 (1988), 181-196, Birkhäuser, Basel.

[36] Malanowski, K.: Stability of Solutions to Convex Problems of Optimization, Springer, Berlin-Heidelberg-New York, 1987.

[37] Outrata, J.V.: On the numerical solution of a class of Stackelberg problems. ZOR - Methods and Models of Operations Research 34 (1990), 255-277.

[38] Pang, J.-S.: Newton's method for B-differentiable equations. Math. Operations Res. 15 (1990), 311-341.

[39] Robinson, S.M.: Stability theory for systems of inequalities, Part II: Differentiable nonlinear systems. SIAM J. Numer. Anal. 13 (1976), 497-513.

[40] Robinson, S.M.: Strongly regular generalized equations. Math. Operations Res. 5 (1980), 43-62.

[41] Robinson, S.M.: Generalized equations and their solutions, Part II: Applications to nonlinear programming. Math. Programming Study 19 (1982), 200-221.

[42] Robinson, S.M.: Newton's method for a class of nonsmooth functions. Manuscript, Department of Industrial Engineering, University of Wisconsin-Madison, Madison, 1988.

[43] Robinson, S.M.: An implicit-function theorem for a class of nonsmooth functions. Manuscript, Department of Industrial Engineering, University of Wisconsin-Madison, Madison, 1989, to appear in Math. Operation Res.

[44] Rockafellar, R.T.: Perturbation of generalized Kuhn-Tucker points in finite-dimensional optimization. In: F. H. Clarke et al., eds., Nonsmooth Analysis and Related Topics, 393-402.Plenum Press, 1989.

[45] Römisch, W. and R. Schultz: Stability analysis for stochastic programs. Preprint Nr. 242, Sektion Mathematik, Humboldt-Universität, Berlin, 1989.

[46] Shapiro, A.: Second-order derivatives of extremal value functions and optimality conditions for semi-infinite programs. Math. Operations Res. 10 (1985), 207-219.

[47] Thibault, L.: On generalized differentials and subdifferentials of Lipschitz vector-valued functions. Nonlinear Anal. Theory Methods Appl. 6 (1982), 1037-1053.

Difference Methods for Differential Inclusions

Frank Lempio

Lehrstuhl für Angewandte Mathematik, Universität Bayreuth,
Postfach 10 12 51, D-8580 Bayreuth, Federal Republic of Germany

Abstract. First we introduce differential inclusions by means of several model problems. These model problems shall illustrate the significance of differential inclusions for a wide range of applications, e.g. dynamic systems with discontinuous state equations, nonlinear programming, and optimal control.

Then we concentrate on difference methods for initial value problems. The basic convergence proof for linear multistep methods is given. Main emphasis is laid on the fundamental ideas behind the proof techniques in order to clarify the meaning of all relevant assumptions. Especially, instead of global boundedness of the right-hand side we prefer imposing a growth condition, and moreover examine the influence of errors thoroughly.

Finally, we outline order of convergence proofs for differential inclusions satisfying a one-sided Lipschitz condition. For higher dimensional problems the underlying difference methods must satisfy consistency and stability properties familiar from ordinary stiff differential equations and not shared by explicit methods. Nevertheless, we can clarify the proof structure already by the classical explicit Euler method for one-dimensional problems. Thus, by the way we prove first order convergence of Euler method for special problems not necessarily satisfying the Lipschitz condition.

1 Introduction

We introduce differential inclusions by means of several model problems, thus illustrating their significance for a wide range of applications, e.g. dynamic systems with discontinuous state equations, nonlinear programming, and optimal control.

Naturally, every ordinary differential equation can obviously be considered as a differential inclusion. This is an adequate point of view especially for differential equations not satisfying the classical Lipschitz condition.

1.1. Example. Let $I = [0, T]$ with $T > 0$. Find an absolutely continuous function

$$y(\cdot) : I \longrightarrow \mathbb{R}$$

with

$$\begin{aligned} y'(t) &= 3y(t)^{2/3} \qquad (t \in I)\,, \\ y(0) &= 0\,. \end{aligned} \tag{1.1}$$

Obviously, this initial value problem is solved by any function $y_\tau(\cdot)$ of the following type

$$y_\tau(t) = \begin{cases} 0 & (0 \le t \le \tau) \\ (t-\tau)^3 & (\tau \le t \le T) \end{cases} ,$$

where τ can be chosen arbitrarily in I. Since the right-hand side

$$f : I \times \mathbb{R} \longrightarrow \mathbb{R}$$

of differential equation (1.1), defined by

$$f(t,x) = 3x^{2/3} \qquad (t \in I, x \in \mathbb{R}) ,$$

does not satisfy the Lipschitz condition standard convergence proofs for classical difference methods do not apply. In fact, not even the sequence of Euler approximations will converge if perturbations have to be taken into account. Nevertheless, using compactness arguments familiar from Peano's existence proof at least convergence of a subsequence could be proved. Convergence properties of this type are typical for broad classes of general differential inclusions, as we will see in Section 2. ∎

We continue with an ordinary differential equation with right-hand side discontinuous with respect to the state variable. Contrary to Ex. 1.1, now the existence of global classical solutions of the corresponding initial value problem cannot longer be expected. Typically, subintervals occur where a trajectory can be reasonably defined by the underlying model, but its derivative does not coincide with the original right-hand side evaluated at that trajectory, but rather belongs to a certain set-valued extension of the right-hand side.

Following the pioneering work of A. F. Filippov [10], such differential equations have to be reformulated as special differential inclusions to get a mathematically sound notion of solution and satisfactory existence results. This transition from differential equations with discontinuous right-hand sides to differential inclusions can be done in several ways, cp. the detailed presentation in [11]. We confine ourselves to the special extension described in the following Ex. 1.2.

1.2. Example. Consider the following second order differential equation modelling a mechanical system with forced vibrations and combined dry and viscous damping

$$z''(t) + 0.2z'(t) + 4\mathrm{sgn}(z'(t)) + z(t) = 2\cos \pi t \tag{1.2}$$

on a time interval $I = [0, T]$ with $T > 0$ and initial conditions

$$\begin{aligned} z(0) &= 3 , \\ z'(0) &= 4 . \end{aligned}$$

In a standard way, the differential equation (1.2) can be written as the two-dimensional system

$$\begin{aligned} y_1'(t) &= y_2(t) \\ y_2'(t) &= -y_1(t) - 0.2y_2(t) - 4\mathrm{sgn}(y_2(t)) + 2\cos\pi t \end{aligned} \tag{1.3}$$

for $t \in I$ with initial conditions

$$\begin{aligned} y_1(0) &= 3\,, \\ y_2(0) &= 4\,. \end{aligned}$$

Here,

$$\mathrm{sgn}(\xi) = \begin{cases} -1 & (\xi < 0) \\ 0 & (\xi = 0) \\ 1 & (\xi > 0) \end{cases} \qquad (\xi \in \mathbf{R}) \tag{1.4}$$

denotes the usual sgn-function.

A thorough analysis shows that this initial value problem has no global classical solution, e.g. on the interval $[0, 6]$. Following [10], [11], to get a solution in a generalized sense the right-hand side

$$f : I \times \mathbf{R}^2 \longrightarrow \mathbf{R}^2$$

of the system (1.3) of ordinary differential equations has to be replaced by the set-valued mapping

$$F : I \times \mathbf{R}^2 \Longrightarrow \mathbf{R}^2$$

with

$$F(t,x) = \bigcap_{\delta>0} \bigcap_{\mu(N)=0} \mathrm{cl}\left(\mathrm{conv}\left(f\left(t, \left\{\xi \in \mathbf{R}^2 : \|\xi - x\|_2 \le \delta\right\} \setminus N\right)\right)\right)$$

for all $t \in I$, $x \in \mathbf{R}^2$, where $\mathrm{cl}(\mathrm{conv}(A))$ denotes the closed convex hull of a set A, $\|\cdot\|_2$ the Euclidean norm, and $\mu(\cdot)$ Lebesgue measure.

This process results in the following differential inclusion

$$\begin{aligned} y_1'(t) &= y_2(t) \\ y_2'(t) &\in -y_1(t) - 0.2y_2(t) - 4\mathrm{Sgn}(y_2(t)) + 2\cos\pi t \end{aligned} \tag{1.5}$$

for almost all $t \in I$, where now

$$\mathrm{Sgn}(\xi) = \begin{cases} -1 & (\xi < 0) \\ [-1,1] & (\xi = 0) \\ 1 & (\xi > 0) \end{cases} \qquad (\xi \in \mathbf{R}) \tag{1.6}$$

denotes the set-valued analogue of the usual single-valued sign-function (1.4).

The existence of an absolutely continuous function

$$y : I \longrightarrow \mathbf{R}^2$$

satisfying (1.5) for almost all $t \in I$ with $y_1(0) = 3$, $y_2(0) = 4$ can now be proved, cp. [2] or [11]. In fact, this solution is unique. Its structure is revealed by the plots of $y_1(\cdot)$ and $y_2(\cdot)$ in Fig. 1 and of the corresponding phase portrait in Fig. 2.

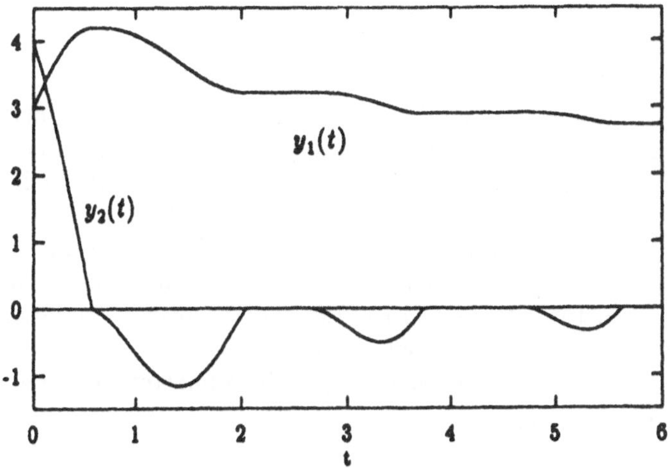

Figure 1: Solution $y(\cdot)$ of differential inclusion (1.5)

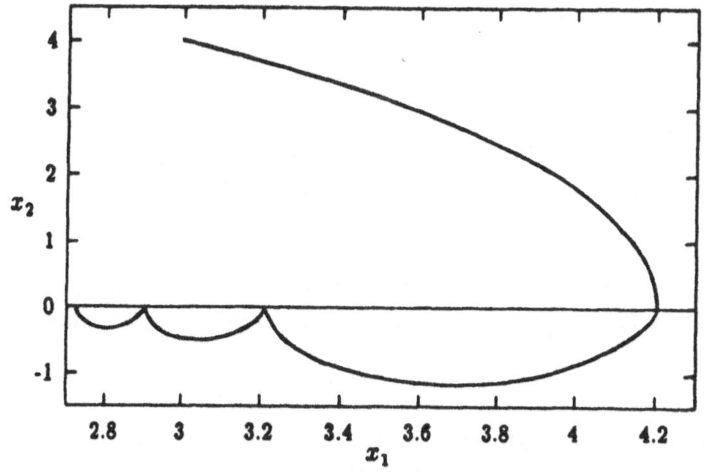

Figure 2: Phase portrait of $y(\cdot)$

Typically, several states of rest

$$y_1(t) \equiv \text{ const}, \; y_2(t) \equiv 0 \qquad (\tau_0 \leq t \leq \tau_1)$$

occur on certain subintervals corresponding to sharp peaks in the phase portrait. On these subintervals the solution is necessarily a generalized one in the sense of Filippov, since otherwise the driving force had to be constant on these subintervals. ∎

The situation described in Ex. 1.2 is typical for a broad class of differential equations with discontinuous right-hand side. In fact, modelling optimal control systems by feedback controls with state dependent switching processes inevitably leads to systems of differential equations with discontinuous right-hand sides. In all these cases, the Filippov extension of the single-valued right-hand side to a set-valued mapping results in a differential inclusion, and the question arises how to solve such differential inclusions numerically.

The approximations in Fig. 1 and Fig. 2 were calculated by Euler's method with small stepsize. An outline of the convergence proof for general linear multistep methods is the central subject of Section 2, which is based on the original results of K. Taubert [37], [36], and [35]. But instead of global boundedness of the right-hand side of the differential inclusion we merely impose a growth condition, and moreover examine the influence of rounding and truncation errors thoroughly. The resulting Convergence Theorem 2.2 does not exploit the full structure inherent in Ex. 1.2, consequently it can be applied to much broader classes of general differential inclusions.

Another class of differential inclusions is closely connected with finite dimensional convex optimization problems.

1.3. Example. Let $\varphi : \mathbf{R}^n \to \mathbf{R}$ be convex, $I = [t_0, T]$ a fixed interval, and $y_0 \in \mathbf{R}^n$. Find an absolutely continuous function $y(\cdot) : I \to \mathbf{R}^n$ with

$$\begin{aligned} y(t_0) &= y_0 , \\ y'(t) &\in -\partial\varphi(y(t)) \end{aligned} \tag{1.7}$$

for almost all $t \in I$.

Here, for all $x_0 \in \mathbf{R}^n$

$$\partial\varphi(x_0) = \{l \in \mathbf{R}^n : \varphi(x_0) + l^\star(x - x_0) \leq \varphi(x) \qquad (x \in \mathbf{R}^n)\}$$

is the subdifferential of φ at x_0, where $(\cdot)^\star$ denotes transposition.

It is well-known that the initial value problem (1.7) has a unique global solution for all $y_0 \in \mathbf{R}^n$, cp. [2]. Naturally, every stationary solution $y(t) \equiv y_0$ satisfies

$$0_{\mathbf{R}^n} \in -\partial\varphi(y_0) ,$$

i.e. minimizes the functional φ. Moreover, if φ achieves a minimum at all, then for every initial value $y_0 \in \mathbf{R}^n$ the solution $y(t)$ of (1.7) as $t \to \infty$ converges to a point which minimizes φ. Thus, we encounter an interesting connection between convex optimization and differential inclusions. In fact, some standard minimization techniques for convex functionals can directly be related to difference methods for the differential inclusion (1.7), e.g. the subgradient method, cp. [27], can be regarded as Euler's method, applied to the differential inclusion (1.7) with unbounded interval $[t_0, \infty)$, on a specially selected infinite grid. ·

The set-valued mapping

$$-\partial\varphi : \mathbf{R}^n \Longrightarrow \mathbf{R}^n$$

satisfies

$$\begin{aligned}(l_1 - l_2)^*(x_1 - x_2) &= l_1^*(x_1 - x_2) - l_2^*(x_1 - x_2) \\ &\le \varphi(x_2) - \varphi(x_1) + \varphi(x_1) - \varphi(x_2) \\ &= 0\end{aligned} \tag{1.8}$$

for all $l_1 \in -\partial\varphi(x_1)$, $l_2 \in -\partial\varphi(x_2)$ and all $x_1, x_2 \in \mathbf{R}^n$, i.e. $-\partial\varphi$ is antitone and $\partial\varphi$ monotone. In fact, $\partial\varphi$ is even maximal monotone, and this is the basis for the nice theoretical properties of the differential inclusion (1.7).

The relation (1.8) can be interpreted in the following way: There exists an inner product $(\cdot|\cdot)$ in $\mathbf{R}^n$, in fact the usual Euclidean inner product, with induced norm $\|\cdot\|$ and a constant $\lambda = 0$ with

$$(l_1 - l_2 | x_1 - x_2) \le \lambda \|x_1 - x_2\|^2$$

for all $l_1 \in -\partial\varphi(x_1)$, $l_2 \in -\partial\varphi(x_2)$ and all $x_1, x_2 \in \mathbf{R}^n$, i.e. $-\partial\varphi$ satisfies a so-called one-sided Lipschitz condition with one-sided Lipschitz constant $\lambda = 0$. ■

For antitone set-valued mappings the one-sided Lipschitz constant can always be chosen equal to 0. But there are broader classes of set-valued mappings with one-sided Lipschitz constant $\lambda \ge 0$, e.g. all single-valued mappings satisfying the classical Lipschitz condition. But everybody familiar with stiff differential equations knows that the classical Lipschitz condition is not the adequate tool, that rather the one-sided Lipschitz condition is essential for this type of equations, cp. e.g. [6]. Inspection of Ex. 1.2, originating from a differential equation with discontinuous right-hand side, shows equally well that the classical Lipschitz condition is not adequate whereas the one-sided Lipschitz condition applies. Especially, negative one-sided Lipschitz constants are most agreeable. We give an example which was used by A. Kastner-Maresch [18], [19] as a model problem for the numerical test of difference methods for differential inclusions. It is a special example of a well-known model equation of A. Prothero and A. Robinson, cp. [6], with added discontinuity.

1.4. Example. Let $I = [0, 2]$ and the function

$$h(t) = -\frac{4}{\pi}\arctan(t-1) \qquad (t \in I)$$

be fixed, and define the set-valued mapping

$$F(t, x) = L(h(t) - x) + h'(t) + L - L\,\mathrm{Sgn}(x) \tag{1.9}$$

for all $t \in I$, $x \in \mathbf{R}$, where L is a positive parameter.

Find an absolutely continuous function $y(\cdot)$ on I with

$$\begin{aligned} y(0) &= y_0\,, \\ y'(t) &\in F(t, y(t)) \end{aligned} \tag{1.10}$$

for almost all $t \in I$.

For initial values y_0 near $h(0) = 1$ the solution $y(\cdot)$ is positive on some time interval $[0, \tau)$ and equal to

$$y(t) = h(t) + e^{-Lt}(y_0 - h(0)) \qquad (0 \le t < \tau)\,.$$

At the point τ the solution vanishes for the first time and from then on

$$y(t) = 0 \qquad (\tau \le t \le T)\,.$$

For large parameter values $L \gg 0$, the solution has a slowly decreasing part $h(t)$ and a very fast decreasing part $e^{-Lt}(y_0 - 1)$ for $0 \le t < \tau$. This behaviour is typical for stiff differential equations. Together with the additional discontinuity of the derivative there is no hope to solve this test problem for large L numerically with a classical explicit difference method. The plots of the direction fields for $L = 2$ in Fig. 3 and $L = 200$ in Fig. 4 together with the special solution

$$y(t) = \begin{cases} h(t) & (0 \le t \le 1) \\ 0 & (1 \le t \le 2) \end{cases}$$

for initial value

$$y_0 = h(0) = 1$$

show the typical difficulties encountered by stiffness and simultaneous discontinuities of the right-hand side.

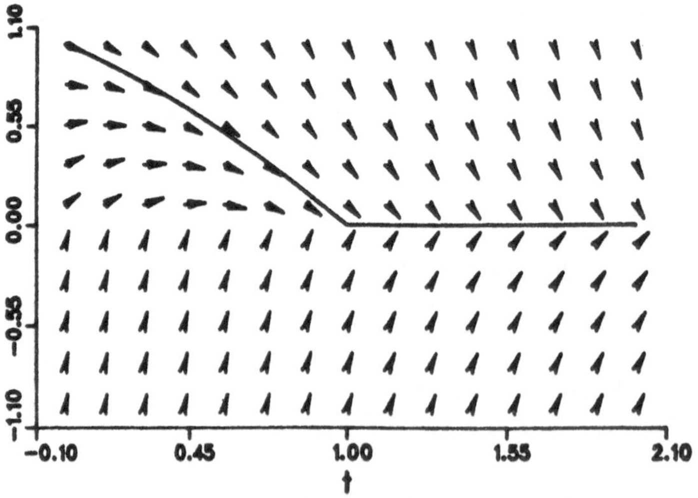

Figure 3: Direction field for differential inclusion (1.10) with $L = 2$

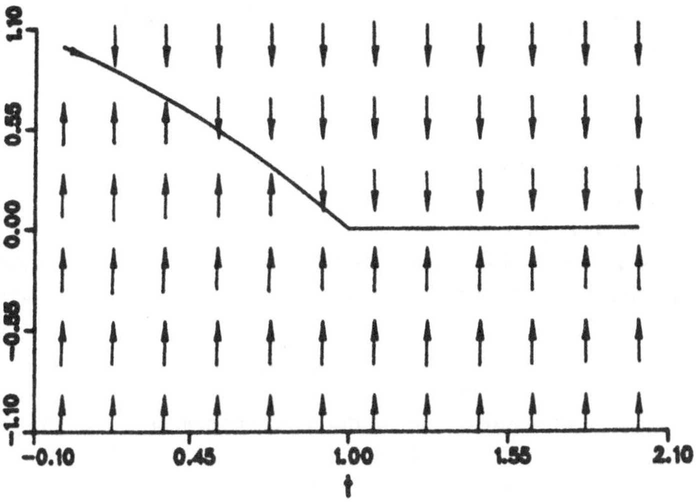

Figure 4: Direction field for differential inclusion (1.10) with $L = 200$

Nevertheless, the set-valued mapping $F(\cdot,\cdot)$ satisfies

$$(l_1 - l_2)^*(x_1 - x_2) \leq -L\|x_1 - x_2\|_2^2$$

for all $x_1, x_2 \in \mathbf{R}$ and all $l_1 \in F(t, x_1)$, $l_2 \in F(t, x_2)$ uniformly for all $t \in I$.

This is just the one-sided Lipschitz condition for non-autonomous differential inclusions with negative one-sided Lipschitz constant $-L$. ∎

Most interesting is the fact that this one-sided Lipschitz condition is an essential tool for the proof of order of convergence results for difference methods. Naturally, these methods must have appropriate consistency and stability properties not shared by classical explicit difference methods. Nevertheless, for one-dimensional problems the proof structure can already be clarified by the classical explicit Euler method. We choose this approach in Section 3 more or less for didactical reasons, and by the way, thus prove first order convergence of the classical Euler method for a special class of differential inclusions not necessarily satisfying the Lipschitz condition. We sketched these results in [22]. They can be extended to higher dimensional problems satisfying a one-sided Lipschitz condition of a special kind. If only the usual one-sided Lipschitz condition according to Definition 3.1 holds or if additional stiffness phenomena occur as in Ex. 1.4 for large L, then one has to switch over to implicit methods. We outline the basic order of convergence results for implicit s-stage Runge-Kutta methods in Theorem 3.6 following [18], [19].

Until now, all presented examples resulted in initial value problems for differential inclusions. Finally, we describe a boundary value problem for a differential inclusion originating from optimal control.

1.5. Example. Let $I = [0, T]$ with $T > 0$ and $x_0 \in \mathbf{R}^n$ be fixed,

$$\varphi : I \times \mathbf{R}^n \times \mathbf{R}^m \longrightarrow \mathbf{R} ,$$
$$\psi : I \times \mathbf{R}^n \times \mathbf{R}^m \longrightarrow \mathbf{R}^n$$

be continuous mappings with continuous partial derivatives with respect to the state variables, i.e. with respect to their second up to their $(n+1)$-st argument. The control variables, i.e. the last m arguments, are allowed to vary in a given arbitrary set $U \subset \mathbf{R}^m$.

These data determine the following simple control problem with fixed time interval and free end condition as a prototype for more general control problems:

Minimize

$$\int_0^T \varphi(\tau, x(\tau), u(\tau)) \, d\tau$$

on the set of all pairs of absolutely continuous trajectories $x(\cdot) : I \to \mathbf{R}^n$ and essentially bounded control functions $u(\cdot) : I \to \mathbf{R}^m$ which satisfy

$$\begin{aligned} x(0) &= x_0 , \\ u(t) &\in U , \\ x'(t) &= \psi(t, x(t), u(t)) \end{aligned}$$

for almost all $t \in I$. Define the function

$$H : I \times \mathbf{R}^n \times \mathbf{R}^m \times \mathbf{R}^n \longrightarrow \mathbf{R}$$

by

$$H(t, x, u, p) = p^\star \psi(t, x, u) - \varphi(t, x, u)$$

for all $t \in I$, $x \in \mathbf{R}^n$, $u \in \mathbf{R}^m$, $p \in \mathbf{R}^n$, where p corresponds to additional so-called adjoint state variables, and use the notations

$$\frac{\partial H}{\partial x} \quad \text{resp.} \quad \frac{\partial H}{\partial p}$$

for the functional matrix of H with respect to the state variables resp. to the adjoint state variables. Then, as is well-known from the theory of optimal control, cp. e.g. the monographs [28], [15], [16], the following maximum principle holds:

Let $\bar{x}(\cdot)$ be an optimal trajectory and $\bar{u}(\cdot)$ a corresponding optimal control. Then there exists an absolutely continuous adjoint trajectory $p(\cdot) : I \to \mathbf{R}^n$ satisfying

$$\begin{aligned} \bar{x}(0) &= x_0, \qquad p(T) = 0_{\mathbf{R}^n} , \\ \bar{u}(t) &\in U \\ \bar{x}'(t) &= \frac{\partial H}{\partial p}(t, \bar{x}(t), \bar{u}(t), p(t))^\star , \\ p'(t) &= -\frac{\partial H}{\partial x}(t, \bar{x}(t), \bar{u}(t), p(t))^\star , \\ H(t, \bar{x}(t), \bar{u}(t), p(t)) &= \max_{w \in U} H(t, \bar{x}(t), w, p(t)) \end{aligned}$$

for almost all $t \in I$.

Indirect methods for optimal control problems consist in the calculation of trajectories, corresponding control functions and adjoint trajectories satisfying the set of necessary optimality conditions given by the maximum principle. In a second step the optimal solution has to be singled out by means of sufficient optimality conditions, or existence and uniqueness theorems, or convexity arguments.

Attempts to eliminate the control functions lead to the set-valued mapping

$$R : I \times \mathbf{R}^n \times \mathbf{R}^n \Longrightarrow \mathbf{R}^m ,$$

defined by

$$R(t,x,p) = \left\{ u \in U : H(t,x,u,p) = \sup_{w \in U} H(t,x,w,p) \right\}$$

for all $t \in I$, $x \in \mathbf{R}^n$, $p \in \mathbf{R}^n$.

Introducing this mapping into the maximum principle leads to the following boundary value problem for a $2n$-dimensional system of differential inclusions:

Find absolutely continuous functions $x(\cdot) : I \to \mathbf{R}^n$ and $p(\cdot) : I \to \mathbf{R}^n$ with

$$\begin{aligned} x(0) &= x_0, \qquad p(T) = 0_{\mathbf{R}^n} , \\ x'(t) &\in \frac{\partial H}{\partial p}(t, x(t), R(t, x(t), p(t)), p(t))^* , \\ p'(t) &\in -\frac{\partial H}{\partial x}(t, x(t), R(t, x(t), p(t)), p(t))^* \end{aligned}$$

for almost all $t \in I$.

Naturally, only in simple cases the mapping R can be calculated explicitly. If R turns out to be single-valued, then this mapping leads to a feedback representation of the controls as functions of states and adjoint states. But generally this dependence is discontinuous, thus the transition to set-valued right-hand sides would be necessary even in this case. If the mapping R is really set-valued then in general the Hamiltonian

$$H(\cdot, \bar{x}(\cdot), \bar{u}(\cdot), p(\cdot))$$

does not determine the optimal control $\bar{u}(\cdot)$ uniquely on the whole time interval I, i.e. so-called singular subarcs have to be reckoned with. ∎

Solving boundary value problems for differential inclusions is a difficult task. Attempts have been made in [30], [29] to apply simplicial fixed point algorithms. Another approach consists in the application of general discretization theories to difference methods for boundary value problems yielding convergence results which are analogous to K. Taubert's results for initial value problems, cp. [26], [25]. Eventually, algorithms for initial value problems with discontinuous right-hand sides, cp. [34], [33], will lead to more efficient methods for boundary value problems with discontinuous right-hand sides by shooting techniques. Here, we cannot go into the details of these algorithms, but have to restrict our presentation to difference methods for initial value problems of the following type.

1.6. Initial Value Problem. *Let $I = [t_0, T]$ with $T > t_0$ be a real interval,*

$$F : I \times \mathbf{R}^n \Longrightarrow \mathbf{R}^n$$

be a set-valued mapping of $I \times \mathbf{R}^n$ into the set of all subsets of $\mathbf{R}^n$, and the initial vector $y_0 \in \mathbf{R}^n$ be fixed.

Find an absolutely continuous function $y(\cdot) : I \to \mathbf{R}^n$ which satisfies the initial condition

$$y(t_0) = y_0$$

and the differential inclusion

$$y'(t) \in F(t, y(t))$$

for almost all $t \in I$. ■

In the next section we study convergence of linear multistep methods for Initial Value Problem 1.6, afterwards we will investigate order of convergence properties of s-stage Runge-Kutta methods. But before entering the presentation of the underlying ideas, the author wants to express his thanks to the organizers and participants of the summer school at Thurnau for the stimulating atmosphere. Moreover, the author is indebted to A. Kastner-Maresch, who contributed several numerical tests on the VAX 8600 of the Computer Centre of the University of Bayreuth, and to A. Dontchev for fruitful discussions during the preparation of the survey [8]. Last but not least, he appreciates the indefatigable dedication of Mrs. K. Dulleck to the preparation of the manuscript with LATEX.

2 Convergent Difference Approximations

In this section, we outline the basic structure of convergence proofs for difference methods for Initial Value Problem 1.6. Because of lack of space we restrict ourselves to the class of linear multistep methods with equidistant grids, but we present the underlying ideas in such a way that possible generalizations to multistep multistage methods can be carried out on similar lines.

Every such method consists in a replacement of the original differential inclusion on the whole interval $I = [t_0, T]$ by a sequence of discrete inclusions on a sequence of grids

$$t_0 = t_0^N < t_1^N < \ldots < t_N^N = T \qquad (N \in \mathbf{N}')$$

with stepsize

$$h = \frac{T - t_0}{N} = t_j^N - t_{j-1}^N \qquad (j = 1, \ldots, N)$$

and in the subsequent numerical solution of these discrete inclusions, resulting in a sequence

$$\left(\eta^N\right)_{N \in \mathbf{N}'}$$

of grid functions

$$\eta^N = \left(\eta_0^N, \eta_1^N, \ldots, \eta_N^N\right) \qquad (N \in \mathbf{N}') .$$

Here, as usual, $\mathbf{N}'$ denotes a subsequence of $\mathbf{N}$ converging to infinity, and only where necessary we indicate explicitely the dependence of the grid points, discrete approximations, discrete selections and errors on the discretization parameter N resp. on the stepsize h. For simplicity we always identify the grid function η with its piecewise linear continuous interpolant $\eta(\cdot)$ on the whole interval I, using the notation

$$\eta_j = \eta(t_j) \qquad (j = 0, \ldots, N) ,$$

where appropriate.

With these conventions in mind we consider the following linear multistep method as a model for more general types of difference methods.

2.1. Linear Multistep Method. *Fix $r \in \mathbf{N}$ and*

$$a_i,\ b_i \in \mathbf{R} \qquad (i = 0, \ldots, r)$$

with $a_r \neq 0$, $|a_0| + |b_0| > 0$. Choose a stepsize

$$h = \frac{T - t_0}{N} \qquad (N \in \mathbf{N}' \subset \mathbf{N}) ,$$

starting approximations

$$\eta_j \in \mathbf{R}^n$$

and corresponding starting selections

$$\zeta_j \in F(t_j, \eta_j) \qquad (j = 0, \ldots, r - 1) .$$

For $j = r, \ldots, N$ compute approximations η_j and selections ζ_j from the difference inclusions

$$\frac{1}{h}\sum_{i=0}^{r} a_i \eta_{j-r+i} = \sum_{i=0}^{r} b_i \zeta_{j-r+i} + \epsilon_j , \tag{2.1}$$

$$\zeta_j \in F(t_j, \eta_j) . \tag{2.2}$$

■

Starting approximations and starting selections can be computed by a linear $\hat{r}$-step method with $\hat{r} < r$ or by a one-step method, formally we define the corresponding errors

$$\epsilon_j = \epsilon_j^N = \eta_j - y_0 \qquad (j = 0, \ldots, r-1) .$$

At each step we have to solve the finite dimensional inclusion (2.2) under the additional constraint (2.1). Hence, even for explicit methods ($b_r = 0$) generally some numerical procedure for the solution of nonlinear systems of equations and inequalities has to be applied, e.g. the starting phase of some nonlinear programming code [31]. The error terms

$$\epsilon_j = \epsilon_j^N \qquad (j = r, \ldots, N)$$

reflect all rounding and truncation errors made during these calculations.

In the following we outline several steps of a convergence analysis of Linear Multistep Method 2.1, based essentially on K. Taubert's papers [37], [36], [35]. But we prefer imposing a growth condition on F instead of global boundedness, and in addition examine the influence of all errors $\epsilon_0, \ldots, \epsilon_N$ closely. To clarify the essential ideas we do not specify the relevant assumptions until they are really needed.

Step 1. The existence of global solutions of Initial Value Problems 1.6 on the whole interval I can only be expected if an appropriate growth condition is satisfied. Hence we assume $F(t, x) \neq \emptyset$ and

$$\|\zeta\|_\infty \leq a\|x\|_\infty + b \tag{2.3}$$

for all $\zeta \in F(t, x)$ and all $t \in I$, $x \in \mathbf{R}^n$ with constants $a \geq 0$, $b \geq 0$, where $\|\cdot\|_\infty$ is the maximum norm in $\mathbf{R}^n$. Moreover, we assume for a while that the root condition is satisfied, i.e. that all zeros λ of the polynomial

$$\sum_{i=0}^{r} a_i \lambda^i \tag{2.4}$$

have absolute value $|\lambda| \leq 1$, and those with $|\lambda| = 1$ are simple. It is well-known that this condition is equivalent to stability of the linear difference equation given by the left-hand side of (2.1), when the right-hand side of (2.1) is considered as perturbation, in the following sense, cp. [32], p. 205:

$$\|\eta_j\|_\infty \leq \Gamma \left[\max_{\mu=0,\ldots,r-1} \|\eta_\mu\|_\infty + h \sum_{\mu=r}^{j} \| \sum_{i=0}^{r} b_i \zeta_{\mu-r+i} + \epsilon_\mu \|_\infty \right] \tag{2.5}$$

for all $j = r, \ldots, N$ and all $N \in \mathbf{N}'$.

Introducing the growth condition yields

$$\begin{aligned}\|\eta_j\|_\infty &\le \Gamma\left[\max_{\mu=0,\dots,r-1}\|\eta_\mu\|_\infty + h\sum_{\mu=r}^{j}\left(\sum_{i=0}^{r}|b_i|\,(a\|\eta_{\mu-r+i}\|_\infty + b) + \|\epsilon_\mu\|_\infty\right)\right]\\ &\le \Gamma\max_{\mu=0,\dots,r-1}\|\eta_\mu\|_\infty + h\Gamma\sum_{\mu=r}^{j}\left(\|\epsilon_\mu\|_\infty + \sum_{i=0}^{r}|b_i|b\right)\\ &\qquad + h\Gamma\sum_{i=0}^{r}|b_i|a\sum_{\mu=r}^{j}\max_{\nu=\mu-r,\dots,\mu}\|\eta_\nu\|_\infty\end{aligned}$$

for all $j = r,\dots,N$ and all $N \in \mathbf{N}'$.

All solutions η^N of this difference inequality are uniformly bounded for $N \in \mathbf{N}'$, cp. [32], pp. 71–72,

$$\|\eta_j^N\|_\infty \le C_1 \qquad (j = 0,\dots,N;\ N \in \mathbf{N}')\,, \tag{2.6}$$

if the starting approximations are uniformly bounded,

$$\|\eta_j^N\|_\infty \le C_2 \qquad (j = 0,\dots,r-1;\ N \in \mathbf{N}')\,, \tag{2.7}$$

and if the mean error

$$\frac{T-t_0}{N}\sum_{\mu=r}^{N}\|\epsilon_\mu^N\|_\infty \le C_3 \qquad (N \in \mathbf{N}') \tag{2.8}$$

is uniformly bounded as well.

Due to the growth condition (2.3), under these assumptions all selections ζ_j^N are likewise uniformly bounded,

$$\|\zeta_j^N\|_\infty \le C_4 \qquad (j = 0,\dots,N;\ N \in \mathbf{N}')\,. \tag{2.9}$$

Step 2. The stability inequality (2.5) can be improved using Spijker's norm, if the strong root condition is satisfied, i.e. if all zeros λ of the polynomial (2.4) have absolute value $|\lambda| < 1$ except the simple zero $\lambda = 1$. This condition is equivalent to stability of the left-hand side of (2.1), when the right-hand side is considered again as perturbation, in the following sense, cp. [32], p. 205:

There exists a constant Γ_0 such that

$$\|\eta_j\|_\infty \le \Gamma_0\left[\max_{\mu=0,\dots,r-1}\|\eta_\mu\|_\infty + h\max_{\mu=r,\dots,j}\left\|\sum_{\nu=r}^{\mu}\left(\sum_{i=0}^{r}b_i\zeta_{\nu-r+i} + \epsilon_\nu\right)\right\|_\infty\right] \tag{2.10}$$

for all $j = r,\dots,N$ and all $N \in \mathbf{N}'$.

This stability inequality suggests a cancellation of perturbations if one would estimate the differences

$$\eta_{j+1} - \eta_j \qquad (j = r,\dots,N-1)$$

regarding

$$\eta_1 - \eta_0,\dots,\eta_r - \eta_{r-1}$$

as starting approximations and

$$\sum_{i=0}^{r} b_i\zeta_{j+1-r+i} + \epsilon_{j+1} - \sum_{i=0}^{r} b_i\zeta_{j-r+i} - \epsilon_j \qquad (j = r, \ldots, N-1)$$

as perturbations. This trick results in the estimate

$$\begin{aligned}\|\eta_{j+1} - \eta_j\|_\infty &\leq \Gamma_0 \max_{\mu=0,\ldots,r-1} \|\eta_{\mu+1} - \eta_\mu\|_\infty \\ &\quad + h\Gamma_0 \max_{\mu=r,\ldots,j} \Big\| \sum_{\nu=r+1}^{\mu+1} \Big(\sum_{i=0}^{r} b_i\zeta_{\nu-r+i} + \epsilon_\nu\Big) - \sum_{\nu=r}^{\mu} \Big(\sum_{i=0}^{r} b_i\zeta_{\nu-r+i} + \epsilon_\nu\Big) \Big\|_\infty \\ &= \Gamma_0 \max_{\mu=0,\ldots,r-1} \|\eta_{\mu+1} - \eta_\mu\|_\infty \\ &\quad + h\Gamma_0 \max_{\mu=r,\ldots,j} \Big\| \sum_{i=0}^{r} b_i \left(\zeta_{\mu+1-r+i} - \zeta_i\right) + \epsilon_{\mu+1} - \epsilon_r \Big\|_\infty\end{aligned}$$

for all $j = r, \ldots, N-1$ and all $N \in \mathbf{N}'$.

Remember that we have already proved (2.9) and that it will do not much harm to sharpen the error behaviour (2.8) to

$$\|\epsilon_j^N\|_\infty \leq C_5 \qquad (j = r, \ldots, N;\ N \in \mathbf{N}') \,, \tag{2.11}$$

then we get the estimate

$$\|\eta_{j+1}^N - \eta_j^N\|_\infty \leq \Gamma_0 \max_{\mu=0,\ldots,r-1} \|\eta_{\mu+1}^N - \eta_\mu^N\|_\infty + hC_6 \tag{2.12}$$

for all $j = r, \ldots, N-1$ and all $N \in \mathbf{N}'$.

Assume the existence of a constant C_7 with

$$\|\eta_{j+1}^N - \eta_j^N\|_\infty \leq C_7 h \qquad (j = 0, \ldots, r-2) \tag{2.13}$$

uniformly for $N \in \mathbf{N}'$. In fact, this condition can be realized recursively by $\hat{r}$-step methods with $\hat{r} < r$, since it is void for one-step methods. Hence, it remains to estimate $\eta_r^N - \eta_{r-1}^N$.

Because of

$$\sum_{i=0}^{r} a_i = 0 \,, \tag{2.14}$$

which is part of the consistency condition (2.28) used later anyway, the difference equation (2.1) yields

$$\frac{1}{h}\sum_{i=0}^{r} a_i \left(\eta_i - \eta_{r-1}\right) = \sum_{i=0}^{r} b_i\zeta_i + \epsilon_r \,,$$

whence it follows directly

$$\eta_r - \eta_{r-1} = -\sum_{i=0}^{r-1} \frac{a_i}{a_r}\left(\eta_i - \eta_{r-1}\right) + h\sum_{i=0}^{r} \frac{b_i}{a_r}\zeta_i + \frac{h}{a_r}\epsilon_r \,.$$

Notice, that (2.13) implies (2.7) if only

$$\eta_0^N = y_0 + \epsilon_0^N$$

is uniformly bounded for all $N \in \mathbf{N}'$, this being a very natural condition, since generally even

$$\lim_{\substack{N\to\infty\\ N\in\mathbf{N}'}} \eta_0^N = y_0$$

should hold for the approximations of the exact initial value y_0. In any case, due to (2.13), (2.9), and (2.11), there exists a constant C_8 with

$$\|\eta_r^N - \eta_{r-1}^N\|_\infty \le C_8 h \tag{2.15}$$

uniformly for all $N \in \mathbf{N}'$.

Combining (2.12), (2.13), and (2.15), we proved the existence of a constant L with

$$\|\eta_{j+1}^N - \eta_j^N\|_\infty \le Lh \qquad (j = 0, \ldots, N-1) \tag{2.16}$$

uniformly for all $N \in \mathbf{N}'$.

Consequently, the piecewise linear continuous interpolant $\eta^N(\cdot)$ is Lipschitz continuous with Lipschitz constant L uniformly for all $N \in \mathbf{N}'$.

Step 3. The estimates (2.6) and (2.16) show that the sequence $\left(\eta^N(\cdot)\right)_{N\in\mathbf{N}'}$ is uniformly bounded and equicontinuous. By Arzela's theorem, there exists a subsequence

$$\left(\eta^N(\cdot)\right)_{N\in\mathbf{N}''}, \; \mathbf{N}'' \subset \mathbf{N}',$$

converging uniformly to a continuous function

$$y(\cdot) : I \longrightarrow \mathbf{R}^n,$$

which is Lipschitz on I with the same Lipschitz constant L. Therefore, all approximations $\eta^N(\cdot)$ together with their limit function $y(\cdot)$ are absolutely continuous and almost everywhere differentiable with derivatives $\dot\eta^N(\cdot)$, $\dot y(\cdot)$ bounded by L. By Alaoglu's theorem there exists a subsequence

$$\left(\dot\eta^N(\cdot)\right)_{N\in\mathbf{N}'''}, \; \mathbf{N}''' \subset \mathbf{N}'',$$

weakly* convergent in $L_\infty(I)^n$ to a function $z(\cdot) \in L_\infty(I)^n$.

Without loss of generality we may assume $\mathbf{N}''' = \mathbf{N}''$, and from

$$\eta^N(t) = \eta_0^N + \int_{t_0}^t \dot\eta^N(\tau)\, d\tau$$

for all $t \in I$ and all $N \in \mathbf{N}'$ we conclude

$$\lim_{\substack{N\to\infty\\ N\in\mathbf{N}''}} \eta_0^N = y(t_0)$$

and

$$z(t) = \dot y(t)$$

for almost all $t \in I$.

Assuming now

$$\lim_{\substack{N\to\infty\\ N\in\mathbf{N}'}} \eta_0^N = y_0,$$

we have shown that the subsequence $\left(\eta^N(\cdot)\right)_{N\in\mathbf{N}''}$ converges uniformly to an absolutely continuous function $y(\cdot)$ with correct initial value $y(t_0) = y_0$, and the sequence of derivatives $\left(\dot{\eta}^N(\cdot)\right)_{N\in\mathbf{N}''}$ is weakly* convergent in $L_\infty(I)^n$ to the derivative $\dot{y}(\cdot)$ of $y(\cdot)$.

Step 4. Until now, we did not exploit fully any consistency condition. Imitating any classical consistency proof with the aid of Taylor expansions is impossible, on the one hand because Taylor expansions for set-valued mappings are not yet fully developed, on the other hand because F anyway does not possess enough smoothness properties, e.g. if F is modelling discontinuous right-hand sides. Hence we have to analyse the difference inclusions directly to show that $y(\cdot)$ solves the differential inclusion.

For this purpose, let

$$\left(\eta^N(\cdot)\right)_{N\in\mathbf{N}''}, \ \mathbf{N}''\subset\mathbf{N}',$$

be any subsequence solving (2.1) and (2.2) for $N \in \mathbf{N}''$ and converging uniformly to a continuous function $y(\cdot) : I \to \mathbf{R}^n$. At any rate, the left-hand side of difference equation (2.1) should approximate in some sense the exact relative increment

$$\frac{1}{s}\left(y(t+s) - y(t)\right)$$

of $y(\cdot)$ for almost all $t \in I$, whereas the right-hand side should stay near $F(t, y(t))$.

Choose $t \in (t_0, T)$ and let $s > 0$ with $t + s \in (t_0, T)$. Then points t and $t + s$ can be approximated by grid points

$$\lim_{\substack{N\to\infty\\ N\in\mathbf{N}''}} t_{j(N,t)} = t, \tag{2.17}$$

$$\lim_{\substack{N\to\infty\\ N\in\mathbf{N}''}} t_{j(N,t+s)} = t+s \tag{2.18}$$

with

$$t_{j(N,t+s)} = t_{j(N,t)} + k(N,s)rh .$$

Adding differences for fixed, sufficiently large $N \in \mathbf{N}''$, exploiting equation (2.1) and condition (2.14), yields

$$\frac{1}{rk(N,s)}\sum_{\mu=0}^{r-1}\sum_{k=0}^{k(N,s)-1}\sum_{i=0}^{r} b_i\zeta^N_{j(N,t)-r+i+kr+\mu} \tag{2.19}$$

$$+\frac{1}{rk(N,s)}\sum_{\mu=0}^{r-1}\sum_{k=0}^{k(N,s)-1} \epsilon^N_{j(N,t)+kr+\mu} \tag{2.20}$$

$$= \frac{1}{rk(N,s)h}\sum_{\mu=0}^{r-1}\sum_{k=0}^{k(N,s)-1}\sum_{i=0}^{r} a_i\left(\eta^N_{j(N,t)-r+i+kr+\mu} - \eta^N_{j(N,t)-r+kr+\mu}\right)$$

$$= \frac{1}{rk(N,s)h}\sum_{i=1}^{r} a_i\sum_{\nu=0}^{i-1}\sum_{\mu=0}^{r-1}\sum_{k=0}^{k(N,s)-1}\left(\eta^N_{j(N,t)-r+\nu+1+kr+\mu} - \eta^N_{j(N,t)-r+\nu+kr+\mu}\right)$$

$$= \frac{1}{rk(N,s)h}\sum_{i=1}^{r} a_i\sum_{\nu=0}^{i-1}\left(\eta^N_{j(N,t)-r+\nu+k(N,s)r} - \eta^N_{j(N,t)-r+\nu}\right) \tag{2.21}$$

where

$$\zeta^N_{j(N,t)-r+i+kr+\mu} \in F\left(t_{j(N,t)-r+i+kr+\mu},\ \eta_{j(N,t)-r+i+kr+\mu}\right) \tag{2.22}$$

for all $i = 0, \ldots, r$, $k = 0, \ldots, k(N,s) - 1$, $\mu = 0, \ldots, r-1$, $N \in \mathbb{N}''$.

For $N \to \infty$ the expression (2.21) converges to

$$\left(\sum_{i=1}^{r} i a_i\right) \frac{1}{s} \left(y(t+s) - y(t)\right) . \tag{2.23}$$

Strengthening (2.11), we assume that for $N \to \infty$ the maximum norm

$$\max_{j=r,\ldots,N} \|\epsilon_j^N\|_\infty$$

of the errors converges to 0, then the convex combination (2.20) converges for $N \to \infty$ to $0_{\mathbb{R}^n}$ as well.

Now we relate expression (2.19) to $F(t, y(t))$. According to (2.22), all selections involved in (2.19) are selected on the time interval

$$\left[t_{j(N,t)-r}, t_{j(N,t+s)}\right] . \tag{2.24}$$

Choose $\delta > 0$ arbitrary. Then, due to (2.16), (2.17), (2.18), and the uniform convergence of $\left(\eta^N(\cdot)\right)_{N \in \mathbb{N}''}$ to $y(\cdot)$, there exists an index $N(s) \in \mathbb{N}''$ with

$$\begin{aligned}
\|\eta^N(\tau) - y(t)\|_\infty &\leq \|\eta^N(\tau) - \eta^N(t)\|_\infty + \|\eta^N(t) - y(t)\|_\infty \\
&\leq L|\tau - t| + \frac{\delta}{2} \\
&\leq L2s + \frac{\delta}{2}
\end{aligned}$$

for all $N \in \mathbb{N}''$ with $N \geq N(s)$ and all τ in the interval (2.24). Decrease s if necessary, such that $2s \leq \delta$ and $L2s \leq \frac{\delta}{2}$ holds, then we have

$$|\tau - t| \leq \delta , \tag{2.25}$$

$$\|\eta^N(\tau) - y(t)\|_\infty \leq \delta \tag{2.26}$$

for all $N \in \mathbb{N}''$ with $N \geq N(s)$ and all τ in the interval (2.24).

This result can be interpreted best using the notation

$$B(z, \rho) = \{x \in X : d(x, z) \leq \rho\}$$

for the ball with radius $\rho \geq 0$ around some point z in a metric space X with distance d, and

$$B(K, \rho) = \{x \in X : d(x, K) \leq \rho\}$$

for a subset K of X, where

$$d(x, K) = \inf_{k \in K} d(x, k)$$

denotes the distance between a point $x \in X$ and the set K. In $\mathbf{R}^{1+n}$ and $\mathbf{R}^n$ we always use the distance induced by the maximum norm $\|\cdot\|_\infty$.

With these notations in mind, (2.25) and (2.26) show that expression (2.19) is an element of the set

$$\frac{1}{rk(N,s)} \sum_{\mu=0}^{r-1} \sum_{k=0}^{k(N,s)-1} \sum_{i=0}^{r} b_i F\left(B((t,y(t)),\delta)\right) \tag{2.27}$$

for all $N \in \mathbf{N}''$ with $N \geq N(s)$, and all s with $2s \leq \delta$, $L2s \leq \frac{\delta}{2}$, where we formally define $F(\tau,\xi) = \emptyset$ for all $\tau \notin I$.

Assuming

$$b_i \geq 0 \qquad (i = 0, \ldots, r)\ ,$$

the set (2.27) is contained in

$$\left(\sum_{i=0}^{r} b_i\right) \operatorname{conv}\left(F\left(B((t,y(t)),\delta)\right)\right)\ .$$

Hence, from (2.23) we get

$$\left(\sum_{i=1}^{r} i a_i\right) \frac{1}{s}\left(y(t+s) - y(t)\right) \in \left(\sum_{i=0}^{r} b_i\right) \operatorname{cl}\left(\operatorname{conv}\left(F\left(B((t,y(t)),\delta)\right)\right)\right)$$

for all s with $2s \leq \delta$, $L2s \leq \frac{\delta}{2}$.

Introducing now the full consistency condition

$$\sum_{i=0}^{r} a_i = 0, \quad \sum_{i=0}^{r} i a_i = \sum_{i=0}^{r} b_i\ , \tag{2.28}$$

we conclude $\sum_{i=0}^{r} b_i > 0$, since $\lambda = 1$ is a simple zero of the polynomial (2.4), and hence

$$\frac{1}{s}\left(y(t+s) - y(t)\right) \in \operatorname{cl}\left(\operatorname{conv}\left(F(B((t,y(t)),\delta))\right)\right)$$

for all $t \in (t_0, T)$ and all $s > 0$ with $t + s \in (t_0, T)$ and $2s \leq \delta$, $L2s \leq \frac{\delta}{2}$.

Since $y(\cdot)$ is absolutely continuous and $\delta > 0$ arbitrary we finally arrive at the fundamental result

$$\dot{y}(t) \in \bigcap_{\delta>0} \operatorname{cl}\left(\operatorname{conv}\left(F\left(B((t,y(t)),\delta)\right)\right)\right) \tag{2.29}$$

for almost all $t \in I$.

Hence, on all the assumptions introduced by now, we have proved that the limit function $y(\cdot)$ is a "generalized solution" of Initial Value Problem 1.6 in an appropriate sense, cp. in this connection e.g. L. Cesari's book [5].

Step 5. Finally, nothing remains but to impose additional conditions guaranteeing

$$F(t,y(t)) = \bigcap_{\delta>0} \operatorname{cl}\left(\operatorname{conv}\left(F\left(B((t,y(t)),\delta)\right)\right)\right)\ , \tag{2.30}$$

i.e. L. Cesari's property (Q) along the graph of $y(\cdot)$, for almost all $t \in I$.

Since the right-hand side of (2.30) is closed and convex, we assume that the sets $F(t,x)$ are closed and convex for all $(t,x) \in I \times \mathbf{R}^n$. Due to the growth condition (2.3), all these sets are compact and convex, hence $F(\cdot,\cdot)$ maps $I \times \mathbf{R}^n$ into the set of all non-empty, compact and convex subsets of $\mathbf{R}^n$. Then, as is well-known, the right-hand side of (2.30) is upper semicontinuous as a set-valued mapping depending on $t \in I$ and $x = y(t) \in \mathbf{R}^n$. Hence, mandatorily we have to require, that F is upper semicontinuous at (t,x) in the following sense:

For every $\rho > 0$ there exists $\delta > 0$ with

$$F\left(B\left((t,x),\delta\right)\right) \subset B\left(F(t,x),\rho\right) . \tag{2.31}$$

This property ensures that (2.30) holds. Indeed, choose $\rho > 0$ and $\delta > 0$ according to (2.31). Since $B(F(t,x),\rho)$ is closed and convex, we have

$$\mathrm{cl}\left(\mathrm{conv}\left(F\left(B\left((t,x),\delta\right)\right)\right)\right) \subset B\left(F(t,x),\rho\right) ,$$

hence

$$\bigcap_{\delta>0} \mathrm{cl}\left(\mathrm{conv}\left(F\left(B\left((t,x),\delta\right)\right)\right)\right) \subset B\left(F(t,x),\rho\right) .$$

Because $\rho > 0$ was chosen arbitrarily and $F(t,x)$ is closed, we get

$$\bigcap_{\delta>0} \mathrm{cl}\left(\mathrm{conv}\left(F\left(B\left((t,x),\delta\right)\right)\right)\right) \subset F(t,x) ,$$

the reverse inclusion being trivially satisfied.

In view of the above results, some condition like the Scorza-Dragoni property, i.e. for every $\epsilon > 0$ there exists a closed subset $I_\epsilon \subset I$ with $\mu(I \setminus I_\epsilon) < \epsilon$ such that the restriction $F|_{I_\epsilon \times \mathbf{R}^n}$ is upper semicontinuous, would be sufficient to ensure that the limit function $y(\cdot)$ solves Initial Value Problem 1.6. For simplicity, we assume that the set-valued mapping F is upper semicontinuous on the whole set $I \times \mathbf{R}^n$. By the way, in case $b_r \neq 0$ upper semicontinuity of $F(t_j,\cdot)$ is also the essential tool to prove the existence of a solution η_j of the difference inclusion (2.1), (2.2), e.g. by Kakutani's fixed point theorem.

Gathering all the assumptions introduced in Step 1 to Step 5 we arrive at the following

2.2. Convergence Theorem. *Let the following assumptions be satisfied for Initial Value Problem 1.6 and Linear Multistep Method 2.1:*
i) F maps $I \times \mathbf{R}^n$ into the set of all non-empty closed and convex subsets of $\mathbf{R}^n$.
ii) F satisfies the growth condition

$$\|\zeta\|_\infty \leq a\|x\|_\infty + b$$

for all $\zeta \in F(t,x)$ and all $t \in I$, $x \in \mathbf{R}^n$ with constants $a \geq 0$, $b \geq 0$.
iii) F is upper semicontinuous.
iv) The strong root condition is satisfied, i.e. all zeros λ of the polynomial

$$\sum_{i=0}^{r} a_i \lambda^i$$

have absolute value $|\lambda| < 1$ *except the simple zero* $\lambda = 1$.
v) The method is consistent, i.e.

$$\sum_{i=0}^{r} a_i = 0, \quad \sum_{i=0}^{r} i a_i = \sum_{i=0}^{r} b_i \ .$$

vi) The coefficients b_i *are nonnegative* $(i = 0, \ldots, r)$.
vii) The starting values satisfy

$$\|\eta_{j+1}^N - \eta_j^N\|_\infty \le hM \qquad (j = 0, \ldots, r-2)$$

for all $N \in \mathbf{N}'$ *with a constant* M *which is independent of the stepsize* $h = \frac{T-t_0}{N}$.
viii) The approximations of the initial value y_0 *satisfy*

$$\lim_{\substack{N\to\infty \\ N\in\mathbf{N}'}} \eta_0^N = y_0 \ .$$

ix) For $N \to \infty$ *the maximum norm*

$$\max_{j=r,\ldots,N} \|\epsilon_j^N\|_\infty$$

of the errors converges to 0.

Then the sequence

$$\left(\eta^N(\cdot)\right)_{N\in\mathbf{N}'}$$

of piecewise linear continuous interpolants of the grid functions

$$\left(\eta_0^N, \ldots, \eta_N^N\right)$$

contains a subsequence

$$\left(\eta^N(\cdot)\right)_{N\in\mathbf{N}''}, \qquad \mathbf{N}'' \subset \mathbf{N}',$$

which converges uniformly to a solution $y(\cdot)$ *of Initial Value Problem 1.6.*

The sequence of derivatives

$$\left(\dot\eta^N(\cdot)\right)_{N\in\mathbf{N}''}$$

is weakly convergent in* $L_\infty(I)^n$ *to* $\dot y(\cdot)$. ■

Step 4 of the proof shows that every uniformly convergent subsequence of the sequence $\left(\eta^N(\cdot)\right)_{N\in\mathbf{N}'}$ converges to a solution of Initial Value Problem 1.6. Assume, that this problem has a unique solution $y(\cdot)$. If the whole sequence $\left(\eta^N(\cdot)\right)_{N\in\mathbf{N}'}$ would not converge to $y(\cdot)$, then there would exist $\epsilon > 0$ such that for all $\delta \in \mathbf{N}'$ there is at least one $N_\delta \in \mathbf{N}'$ with $N_\delta \ge \delta$ and $\|\eta^{N_\delta}(\cdot) - y(\cdot)\|_\infty > \epsilon$. By Convergence Theorem 2.2 the subsequence

$$\left(\eta^{N_\delta}(\cdot)\right)_{\delta\in\mathbf{N}'}$$

would contain a further subsequence, converging uniformly to a solution $\hat y(\cdot)$ satisfying $\|\hat y(\cdot) - y(\cdot)\|_\infty \ge \epsilon$. This contradiction proves

2.3. Corollary. *On the assumptions of Convergence Theorem 2.2 every uniformly convergent subsequence of the sequence* $\left(\eta^N(\cdot)\right)_{N \in \mathbf{N}'}$ *converges to a solution of Initial Value Problem 1.6. If the solution of this problem is unique, then the whole sequence converges uniformly to this solution.* ■

The general steps of the proof of Convergence Theorem 2.2 were based mainly on K. Taubert's papers. Another approach more in the spririt of F. Stummel's general discretization theory is present in H.-D. Niepage's report [24] and in [26]. Selection strategies, performing the selection in (2.2) not randomly but in a special way to ensure convergence to solutions with better qualitative behaviour, can be found for Euler's method e.g. in [2]. Strategies for selection with respect to reference trajectories are treated in [7], for explicit linear multistep methods cp. [20].

Concluding, we apply Euler's method and a special 3-step method, proposed in [17], to Example 1.2. The 3-step method is defined by the set of coefficients

$$\begin{aligned} a_0 &= -0.81 & b_0 &= \tfrac{725}{1200} \\ a_1 &= -0.99 & b_1 &= \tfrac{488}{1200} \\ a_2 &= 0.8 & b_2 &= \tfrac{3119}{1200} \\ a_3 &= 1 & b_3 &= 0 \end{aligned}$$

and would be consistent and therefore convergent of order 3 for single-valued smooth right-hand sides. The plot of the approximation to the solution $y(\cdot)$ in Fig. 5 for stepsize $h = 0.001$ shows heavy oscillations due to "almost instability", when the solution hits the manifold of discontinuity of the right-hand side $\{x \in \mathbf{R}^2 : x_2 = 0\}$.

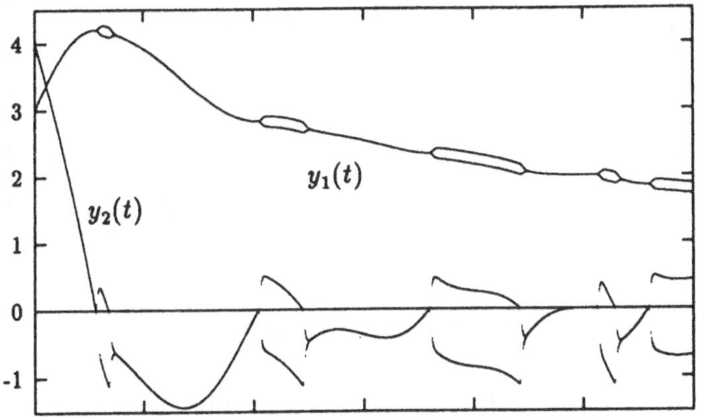

Figure 5: 3-step method with stepsize $h = 0.001$

The performance of Euler's method is much better, although the oscillations never disappear totally, compare the plots of the approximations to the second solution component $y_2(\cdot)$ in Fig. 6 and Fig. 7 for stepsizes $h = 0.001$ and $h = 0.0025$ respectively.

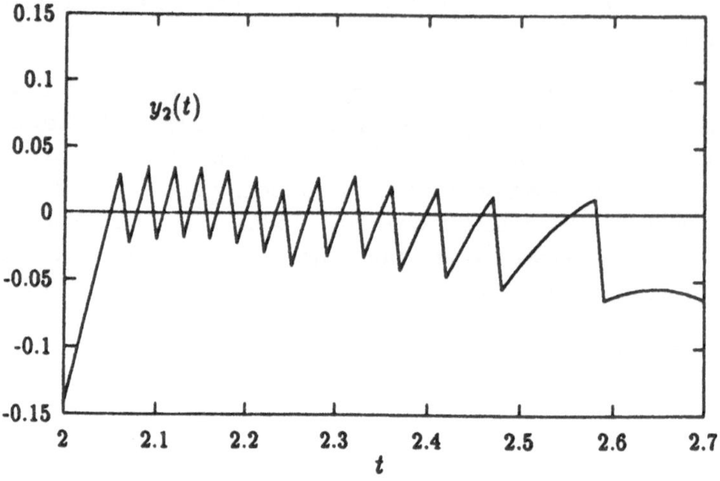

Figure 6: Euler method with stepsize $h = 0.01$

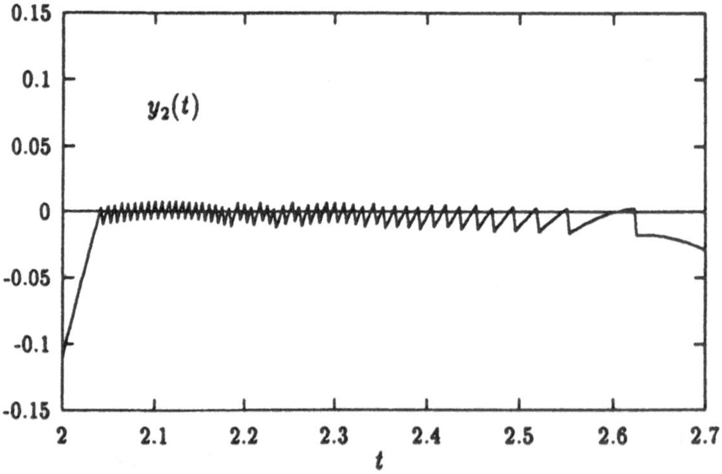

Figure 7: Euler method with stepsize $h = 0.0025$

Higher order convergence cannot be expected without imposing stronger assumptions on the right-hand side or changing the whole type of the algorithm, cp. in this connection D. Stewart's transformation method [34], [33] or V. Veliov's results on higher order approximations of the whole solution set of Initial Value Problem 1.6 by the whole solution set of an approximating difference inclusion [38], [39], [40].

3 Higher Order Convergence

Having proved the basic convergence theorem for difference approximations to Initial Value Problem 1.6 we now outline the ideas underlying the proof of higher order convergence results. First of all, to get convergence of the whole sequence of approximations, one should introduce some conditions guaranteeing uniqueness of the solution of Initial Value Problem 1.6. In view of differential inclusions, describing differential equations with right-hand side discontinuous with respect to the state variables, the classical Lipschitz condition is not adequate. But there is a weaker condition, well-known in the field of stiff differential equations, cp. the books of K. Dekker and J. G. Verwer [6] and J. C. Butcher [4], which is applicable to differential inclusions as well.

3.1. Definition. *Let* $\mathbf{R}^n$ *be endowed with the scalar product* $(\cdot|\cdot)$ *and the corresponding induced norm* $\|\cdot\|$. *A set-valued mapping*

$$F : I \times \mathbf{R}^n \Longrightarrow \mathbf{R}^n$$

satisfies a one-sided Lipschitz condition with one-sided Lipschitz constant λ *if*

$$(l_1 - l_2 | x_1 - x_2) \leq \lambda \|x_1 - x_2\|^2 \tag{3.1}$$

holds for all $x_1, x_2 \in \mathbf{R}^n$ *and all* $l_1 \in F(t, x_1)$, $l_2 \in F(t, x_2)$ *uniformly for all* $t \in I$. ∎

Notice that the one-sided Lipschitz constant λ can be negative, zero, or positive, cp. the discussion of Examples 1.3 and 1.4. Naturally, it depends on F, but on the choice of the inner product $(\cdot|\cdot)$ as well, cp. [6] for more details.

3.2. Lemma. *Let* F *satisfy a one-sided Lipschitz condition with one-sided Lipschitz constant* λ *according to Definition 3.1, and let*

$$y(\cdot) : I \longrightarrow \mathbf{R}^n, \qquad \tilde{y}(\cdot) : I \longrightarrow \mathbf{R}^n$$

be absolutely continuous and satisfy

$$y'(t) \in F(t, y(t)), \qquad \tilde{y}'(t) \in F(t, \tilde{y}(t))$$

for almost all $t \in I$.

Then the following estimate holds

$$\|\tilde{y}(t) - y(t)\| \leq \exp(\lambda(t - t_0)) \, \|\tilde{y}(t_0) - y(t_0)\|$$

for all $t \in I$.

Proof. The absolutely continuous function $\|\tilde{y}(\cdot) - y(\cdot)\|^2$ satisfies

$$\begin{aligned} & \frac{d}{dt}\|\tilde{y}(t) - y(t)\|^2 \\ = \ & 2\,(\tilde{y}'(t) - y'(t) | \tilde{y}(t) - y(t)) \\ \leq \ & 2\lambda \|\tilde{y}(t) - y(t)\|^2 \end{aligned}$$

for almost all $t \in I$.

Hence, the absolutely continuous function

$$\Phi(t) = \exp\left(-2\lambda(t-t_0)\right) \|\tilde{y}(t) - y(t)\|^2$$

has derivative

$$\begin{aligned}\Phi'(t) &= -2\lambda \exp\left(-2\lambda(t-t_0)\right) \|\tilde{y}(t) - y(t)\|^2 \\ &\quad + \exp\left(-2\lambda(t-t_0)\right) \frac{d}{dt}\|\tilde{y}(t) - y(t)\|^2 \\ &\leq 0\end{aligned}$$

for almost all $t \in I$. Therefore

$$\begin{aligned}\Phi(t) &= \Phi(t_0) + \int_{t_0}^{t} \Phi'(\tau)\, d\tau \\ &\leq \Phi(t_0)\end{aligned}$$

for all $t \in I$. ■

Especially, Lemma 3.2 implies that the solution of Initial Value Problem 1.6 is unique if F satisfies a one-sided Lipschitz condition. Moreover, it clarifies the eminent rôle played by antitone ($\lambda = 0$) or strictly antitone ($\lambda < 0$) right-hand sides. In the following we demonstrate the importance of the one-sided Lipschitz condition for order of convergence proofs by means of the classical explicit Euler method for space dimension 1. Having illustrated this way the general proof structure and the significance of stability and consistency properties, in general shared only by certain classes of implicit Runge-Kutta methods, we summarize the relevant results for these methods afterwards.

Naturally, the classical Euler method is a special case of Linear Multistep Method 2.1 and of s-Stage Runge-Kutta Method 3.5. For easy reference we define this method explicitly for differential inclusions.

3.3. Euler Method. *Choose a stepsize*

$$h = \frac{T - t_0}{N} \qquad (N \in \mathbf{N}' \subset \mathbf{N})$$

and a starting approximation

$$\eta_0 = y_0 + \epsilon_0 \in \mathbf{R}^n .$$

For $j = 0, \ldots, N-1$ solve the difference inclusions

$$\zeta_j \in F(t_j, \eta_j)\,, \tag{3.2}$$

$$\eta_{j+1} = \eta_j + h\zeta_j + h\epsilon_{j+1}\,. \tag{3.3}$$

■

Here again $\epsilon_0 = \epsilon_0^N$ denotes the error of the initial approximation, and $\epsilon_{j+1} = \epsilon_{j+1}^N$ reflects all errors made in the selection procedure (3.2) and in the subsequent calculation (3.3) of the next approximation.

Following current approaches to difference methods for stiff differential equations and for differential inclusions satisfying a one-sided Lipschitz condition we analyse the global discretization error directly using piecewise smoothness of the exact solution instead of smoothness properties of the right-hand side F.

The structure of the different parts of the global discretization error

$$\eta_{j+1} - y_{j+1}$$

at some grid point t_{j+1} is illustrated in Fig. 8.

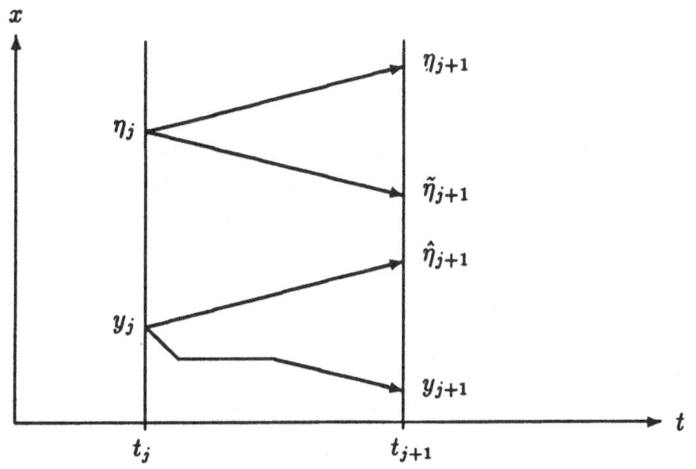

Figure 8: Global discretization error

Here, $y(\cdot)$ denotes the exact solution of Initial Value Problem 1.6, and $\hat{\eta}_{j+1}$ is the result of a single exact Euler step starting from the value y_j of the exact solution at the grid point t_j using a special selection $\hat{\zeta}_j$. Similarily, $\tilde{\eta}_{j+1}$ is the result of a single exact Euler step starting from the last computed approximation η_j using just the selection ζ_j chosen during the calculation of the next approximation η_{j+1}. According to the estimate

$$\begin{aligned} \|\eta_{j+1} - y_{j+1}\| &= \| (\eta_{j+1} - \tilde{\eta}_{j+1}) + (\tilde{\eta}_{j+1} - \hat{\eta}_{j+1}) + (\hat{\eta}_{j+1} - y_{j+1}) \| \\ &\leq \|\eta_{j+1} - \tilde{\eta}_{j+1}\| + \|\tilde{\eta}_{j+1} - \hat{\eta}_{j+1}\| + \|\hat{\eta}_{j+1} - y_{j+1}\| , \end{aligned} \tag{3.4}$$

in the following steps the different parts of the global discretization error are analysed separately.

Step 1. Since η_{j+1} satisfies the equation (3.3) with perturbation $h\epsilon_{j+1}$ and $\tilde{\eta}_{j+1}$ satisfies the same equation without perturbation, we get the estimate

$$\|\eta_{j+1} - \tilde{\eta}_{j+1}\| \leq h\|\epsilon_{j+1}\| \,. \tag{3.5}$$

The corresponding estimate for more general difference methods is a consequence of appropriate stability properties. E.g., for s-stage Runge-Kutta methods the notion of BS-stability introduced by R. Frank, J. Schneid, and W. Ueberhuber, cp. [13], [12], is very well suited.

Step 2. The same stability considerations apply to $\|\hat{\eta}_{j+1} - y_{j+1}\|$, if y_{j+1} could be regarded as solution of the inclusion

$$\hat{\zeta}_j \in F(t_j, y_j) \,, \tag{3.6}$$

$$y_{j+1} = y_j + h\hat{\zeta}_j + h\hat{\epsilon}_{j+1} \tag{3.7}$$

with a certain perturbation $\hat{\epsilon}_{j+1}$, whereas $\hat{\eta}_{j+1}$ solves the unperturbed equation

$$\hat{\eta}_{j+1} = y_j + h\hat{\zeta}_j \tag{3.8}$$

with the same selection $\hat{\zeta}_j$.

We now derive a representation of $\hat{\epsilon}_{j+1}$ and $\hat{\zeta}_j$ exploiting only smoothness properties of the exact solution $y(\cdot)$. In order to include discontinuities of the derivative of $y(\cdot)$ we assume $y(\cdot)$ to be piecewise Lipschitz continuously differentiable in the following sense:

The whole interval I can be subdivided into finitely many subintervals $[\tau_\nu, \tau_{\nu+1}]$ with

$$t_0 = \tau_0 < \tau_1 < \ldots < \tau_m = T$$

such that $y(\cdot)$ is continuously differentiable and $y'(\cdot)$ Lipschitz continuous with Lipschitz constant L_1 on each interval $[\tau_\nu, \tau_{\nu+1}]$, where at the end points $y'(\cdot)$ has to be understood as one-sided derivative, e.g.

$$y'(\tau_\nu) = y'_+(\tau_\nu) = \lim_{\sigma \to 0+} \frac{1}{\sigma}\left(y(\tau_\nu + \sigma) - y(\tau_\nu)\right) \,.$$

If we assume the graph of F to be closed, then it follows

$$y'_+(t) \in F(t, y(t))$$

for all $t \in [t_0, T)$.

This result suggests to choose the selection

$$\hat{\zeta}_j = y'_+(t_j) \,, \tag{3.9}$$

and to estimate $\hat{\epsilon}_{j+1}$ by a Taylor expansion. Since $y(\cdot)$ is absolutely continuous, we get

$$y_{j+1} = y_j + hy'_+(t_j) + \int_{t_j}^{t_{j+1}} y'(\tau) - y'_+(t_j)\, d\tau \,,$$

hence

$$h\hat{\epsilon}_{j+1} = \int_{t_j}^{t_{j+1}} y'(\tau) - y'_+(t_j)\, d\tau \ .$$

If the interval (t_j, t_{j+1}) does not contain any of the points τ_ν, then we can exploit Lipschitz continuity of $y'(\cdot)$ on $[t_j, t_{j+1}]$ and get

$$\|\hat{\eta}_{j+1} - y_{j+1}\| \leq \frac{1}{2} L_1 h^2 \ . \tag{3.10}$$

If the interval (t_j, t_{j+1}) contains at least one of the points τ_ν, then we can merely exploit boundedness of $y'(\cdot)$, say by a constant L_0, and get

$$\|\hat{\eta}_{j+1} - y_{j+1}\| \leq 2L_0 h \ , \tag{3.11}$$

notice, that the latter situation can only occur at most $m-1$ times independently of $N \in \mathbb{N}'$.

To get higher order estimates for $\|\hat{\eta}_{j+1} - y_{j+1}\|$ at least on subintervals where $y(\cdot)$ is smooth one has to select more elaborate difference methods and exploit higher order consistency conditions, e.g. the simplifying conditions of s-stage Runge-Kutta methods, and again appropriate stability concepts like BS-stability.

Step 3. At last, we have to estimate $\|\tilde{\eta}_{j+1} - \hat{\eta}_{j+1}\|$, where $\tilde{\eta}_{j+1}$ solves the inclusion

$$\zeta_j \in F(t_j, \eta_j) \ , \tag{3.12}$$

$$\tilde{\eta}_{j+1} = \eta_j + h\zeta_j \ , \tag{3.13}$$

and $\hat{\eta}_{j+1}$ solves the inclusion

$$\hat{\zeta}_j \in F(t_j, y_j) \ , \tag{3.14}$$

$$\hat{\eta}_{j+1} = y_j + h\hat{\zeta}_j \ . \tag{3.15}$$

If F is single-valued and satisfies the classical Lipschitz condition with Lipschitz constant C, then we would get

$$\begin{aligned} \|\tilde{\eta}_{j+1} - \hat{\eta}_{j+1}\| &\leq \|\eta_j - y_j\| + h\|\zeta_j - \hat{\zeta}_j\| \\ &\leq (1 + Ch)\|\eta_j - y_j\| \ , \end{aligned}$$

i.e. the so-called C-stability of the difference method. It is well-known, that this approach to C-stability is not adequate for stiff ordinary differential equations. The same holds true for differential equations with discontinuous right-hand sides and for differential inclusions. Hence, we assume that F satisfies a one-sided Lipschitz condition in the sense of Definition 3.1 with one-sided Lipschitz constant λ. This results in the estimate

$$\begin{aligned} (\tilde{\eta}_{j+1} - \hat{\eta}_{j+1} | \eta_j - y_j) &= \left(\eta_j + h\zeta_j - y_j - h\hat{\zeta}_j | \eta_j - y_j\right) \\ &= (\eta_j - y_j | \eta_j - y_j) + h\left(\zeta_j - \hat{\zeta}_j | \eta_j - y_j\right) \\ &\leq (1 + \lambda h)\|\eta_j - y_j\|^2 \ . \end{aligned} \tag{3.16}$$

In general, this estimate cannot be converted into a C-stability inequality, instead one has to switch over to other methods, e.g. implicit s-stage Runge-Kutta methods with the necessary consistency and stability properties. But before summarizing the relevant results we show how the classical Euler method is related to C-stability in a very special situation.

Assume e.g. that space dimension n is equal to 1, $(\cdot|\cdot)$ the usual product and $\|\cdot\|$ the absolute value. Then the following two cases can occur:

Either

$$\operatorname{sgn}(\tilde{\eta}_{j+1} - \hat{\eta}_{j+1}) = \operatorname{sgn}(\eta_j - y_j) \neq 0 ,$$

then in fact (3.16) implies $1 + \lambda h > 0$ and

$$\|\tilde{\eta}_{j+1} - \hat{\eta}_{j+1}\| \le (1 + \lambda h)\|\eta_j - y_j\| ; \tag{3.17}$$

or else there exists $\mu \in [0,1]$ with

$$\begin{aligned} & (\eta_j - y_j) + \mu\left((\tilde{\eta}_{j+1} - \hat{\eta}_{j+1}) - (\eta_j - y_j)\right) \\ = \ & (\eta_j - y_j) + \mu h\left(\zeta_j - \hat{\zeta}_j\right) \\ = \ & 0 , \end{aligned}$$

since due to (3.13), (3.15)

$$\tilde{\eta}_{j+1} - \hat{\eta}_{j+1} = \eta_j - y_j + h\left(\zeta_j - \hat{\zeta}_j\right) ,$$

whence we conclude finally

$$\tilde{\eta}_{j+1} - \hat{\eta}_{j+1} = (1-\mu)h\left(\zeta_j - \hat{\zeta}_j\right) . \tag{3.18}$$

From (3.9) we know already

$$\|\hat{\zeta}_j\| = \|y'_+(t_j)\| \le L_0 ,$$

remembering Step 1 of the proof of Convergence Theorem 2.2 we infer in addition

$$\|\zeta_j\| \le \tilde{L}_0 , \tag{3.19}$$

if a growth condition is satisfied, if all starting approximations η_0^N are bounded uniformly for $N \in \mathbf{N}'$, and if the mean error

$$\frac{T - t_0}{N} \sum_{\mu=1}^{N} \|\epsilon_\mu\|$$

is uniformly bounded for $N \in \mathbf{N}'$ as well.

Assuming these conditions we get from (3.18) the estimate

$$\|\tilde{\eta}_{j+1} - \hat{\eta}_{j+1}\| \le h\left(\tilde{L}_0 + L_0\right) \tag{3.20}$$

with a constant $\tilde{L}_0 + L_0$ independent of $N \in \mathbf{N}'$.

Hence, for space dimension $n = 1$, either the C-stability inequality (3.17) holds for the transition from grid point t_j to t_{j+1}, or a reset (3.20) of errors occurs. For space dimension $n > 1$ such behaviour cannot be expected any longer, and one has to choose methods satisfying the C-stability inequality on the whole grid.

Step 4. Finally, we solve the difference inequality, implicitly defined by (3.4), exploiting all the estimates (3.5), (3.10) or alternatively (3.11), (3.17) or alternatively (3.20).

Assume that case (3.17) occurs for subsequent indices

$$j = j_0,\ j = j_0 + 1, \ldots, j = j_1 - 1 ,$$

where $j_0 = 0$ or case (3.20) applies to index $j = j_0 - 1$, and $j_1 = N$ or case (3.20) applies to index $j = j_1$. Then we get from (3.4)

$$\|\eta_{j_1} - y_{j_1}\| \leq (1 + \lambda h)^{j_1 - j_0} \|\eta_{j_0} - y_{j_0}\| \tag{3.21}$$

$$+ \sum_{j=j_0+1}^{j_1} (1 + \lambda h)^{j_1 - j} \left[\|\eta_j - \tilde{\eta}_j\| + \|\hat{\eta}_j - y_j\|\right] . \tag{3.22}$$

In case $j_0 = 0$ the initial error in (3.21) is

$$\|\eta_{j_0} - y_{j_0}\| = \|\epsilon_0\| ,$$

else (3.4), (3.5), (3.20), and (3.10) or alternatively (3.11) imply

$$\|\eta_{j_0} - y_{j_0}\| \leq h\|\epsilon_{j_0}\| + h\left(\tilde{L}_0 + L\right) + \begin{cases} \frac{1}{2} L_1 h^2 \\ 2 L_0 h \end{cases} ,$$

and similarily in (3.22) for $j = j_0 + 1, \ldots, j_1$

$$\|\eta_j - \tilde{\eta}_j\| + \|\hat{\eta}_j - y_j\| \leq h\|\epsilon_j\| + \begin{cases} \frac{1}{2} L_1 h^2 \\ 2 L_0 h \end{cases} .$$

Assume now the existence of constants c_0, c_1 with

$$\|\epsilon_0^N\| \leq c_0 h ,$$

$$\frac{T - t_0}{N} \sum_{j=1}^{N} \|\epsilon_j^N\| \leq c_1 h$$

for all $N \in \mathbf{N}'$, exploit the fact that the alternative (3.11) occurs at most $m - 1$ times independently of $N \in \mathbf{N}'$, and notice that all the factors $(1 + \lambda h)^j$ for $j = 0, \ldots, N$ are uniformly bounded by 1, if $\lambda \leq 0$, and by $\exp(\lambda(T - t_0))$, if $\lambda \geq 0$. Then we conclude

$$\|\eta_j^N - y_j\| \leq c_2 h$$

for all $j = 0, \ldots, N$ with a constant c_2, which is independent of $N \in \mathbf{N}'$.

Collecting all assumptions and results of Step 1 to Step 4 we arrive at the following

3.4. Theorem. *Let $F : I \times \mathbf{R} \Longrightarrow \mathbf{R}$ be a set-valued mapping with closed graph, satisfying a one-sided Lipschitz condition and a growth condition, and let the exact solution $y(\cdot)$ of Initial Value Problem 1.6 exist and be piecewise Lipschitz continuously differentiable on I.*

Let the initial error $|\epsilon_0^N|$ and the mean error

$$\frac{T-t_0}{N}\sum_{j=1}^{N}|\epsilon_j^N|$$

be of order 1 as functions of the stepsize $h = \frac{T-t_0}{N}$.

Then the order of convergence of Euler Method 3.3 is equal to 1. ∎

This result explains the fact, that numerical experiments with Euler's method suggested first order convergence for this special situation.

We mentioned in Step 1 to Step 4 above possible extensions to other difference methods, e.g. implicit s-stage Runge-Kutta methods. For these methods the analogue of estimate (3.5) requires BS-stability, higher order consistency proofs along the lines of Step 2 require in addition that an appropriate set of so-called simplifying conditions holds, a C-stability inequality like (3.17) in Step 3 has to be satisfied on the whole grid, and the discussion leading to (3.20) and hence the growth condition is not needed any longer. The reasoning in Step 4 remains essentially unchanged, showing that globally first order convergence can be expected and higher order convergence on subintervals where the solution is sufficiently smooth.

We summarize the relevant results which were proved by A. Kastner-Maresch in [18], [19], who succeeded to adapt proof techniques for stiff ordinary differential equations to differential inclusions, starting from a paper of C. M. Elliott [9] on first order convergence for special implicit methods.

3.5. s-Stage Runge-Kutta Method. *Fix $s \in \mathbf{N}$ and the scheme of coefficients*

$$\begin{array}{c|ccc} c_1 & a_{11} & \dots & a_{1s} \\ \vdots & \vdots & & \vdots \\ c_s & a_{s1} & \dots & a_{ss} \\ \hline & b_1 & \dots & b_s \end{array}$$

with $0 \le c_\mu \le 1$ for $\mu = 1, \dots, s$. Choose a stepsize

$$h = \frac{T-t_0}{N} \qquad (N \in \mathbf{N}' \subset \mathbf{N})$$

and a starting approximation

$$\eta_0 = y_0 + \epsilon_0 \in \mathbf{R}^n . \tag{3.23}$$

For $j = 0, \dots, N-1$ solve the difference inclusions

$$Y^\mu = \eta_j + h\sum_{\nu=1}^{s} a_{\mu\nu} Z^\nu + h\mathcal{E}_{j+1}^\mu , \tag{3.24}$$

$$Z^\mu \in F(t_j + c_\mu h, Y^\mu) \qquad (\mu = 1, \dots, s) , \tag{3.25}$$

and compute the next approximation

$$\eta_{j+1} = \eta_j + h \sum_{\nu=1}^{s} b_\nu Z^\nu + h\epsilon_{j+1} . \qquad (3.26)$$

∎

Here, Y^μ resp. Z^μ are approximations resp. selections at some auxiliary points $t_j + c_\mu h$ $(\mu = 1, \ldots, s)$, and in general (3.24), (3.25) constitute an implicit system of inclusions. As for the classical Runge-Kutta method the next approximation is obtained in (3.26) by equating the difference quotient $(\eta_{j+1} - \eta_j)/h$ to a weighted mean of the selections Z^μ. Naturally, the auxiliary vectors Y^μ, Z^μ do not only depend on the step number $j = 0, \ldots, N-1$ but also on the discretization parameter $N \in \mathbb{N}'$, just as the input error ϵ_0 in (3.23), the perturbations $\mathcal{E}^\mu_{j+1}$ in (3.24), and the rounding errors ϵ_{j+1} in (3.26).

As motivated by (3.4) and Step 1 of the proof of Theorem 3.4 we compare the solution $\tilde{\eta}_{j+1}$ of the unperturbed inclusion (3.24), (3.25), (3.26) with $\mathcal{E}^\mu_{j+1} = 0_{\mathbb{R}^n}$ $(\mu = 1, \ldots, s)$, $\epsilon_{j+1} = 0_{\mathbb{R}^n}$ with η_{j+1} and call the difference method BS-stable, if there exists a constant C_1, independent of the vector η_j and the chosen selections, with

$$\|\eta_{j+1} - \tilde{\eta}_{j+1}\| \leq hC_1 \max\left\{\|\mathcal{E}^1_{j+1}\|, \ldots, \|\mathcal{E}^s_{j+1}\|, \|\epsilon_{j+1}\|\right\}$$

for all $j = 0, \ldots, N-1$ and all sufficiently large $N \in \mathbb{N}'$.

Motivated by Step 2 of the proof of Theorem 3.4 we introduce the simplifying conditions (3.27), (3.28), which guarantee that the exact value y_{j+1} can be interpreted as the solution of the inclusions (3.24), (3.25), (3.26) with η_j replaced by y_j and suitably chosen perturbations of appropriate order in h. Then, together with BS-stability, we get the analogues of (3.10) and (3.11), namely

$$\|\hat{\eta}_{j+1} - y_{j+1}\| \leq C_2 h^{p+1}$$

if the exact solution $y(\cdot)$ is p-times Lipschitz continuously differentiable on $[t_j, t_{j+1}]$, and

$$\|\hat{\eta}_{j+1} - y_{j+1}\| \leq C_3 h ,$$

if $y(\cdot)$ is piecewise Lipschitz continuously differentiable, where the constants C_2 and C_3 are independent of $j = 0, \ldots, N-1$ and $N \in \mathbb{N}'$ and where $\hat{\eta}_{j+1}$ is the solution of inclusions (3.24), (3.25), (3.26) likewise with η_j replaced by y_j, but without any perturbations.

Motivated by Step 3 of the proof of Theorem 3.4 we call the difference method C-stable, if there exists a constant C, independent of the vectors η_j and y_j and all chosen selections, with

$$\|\tilde{\eta}_{j+1} - \hat{\eta}_{j+1}\| \leq (1 + Ch)\|\eta_j - y_j\|$$

for all $j = 0, \ldots, N-1$ and all sufficiently large $N \in \mathbb{N}'$.

Finally, a global error estimate like in Step 4 of the proof of Theorem 3.4 yields the following general result, compare [18], [19].

3.6. Theorem. *Let the following assumptions be satisfied for Initial Value Problem 1.6 and s-Stage Runge-Kutta Method 3.5:*
i) The set-valued mapping $F : I \times \mathbf{R}^n \Longrightarrow \mathbf{R}^n$ *has closed graph.*
ii) The exact solution $y(\cdot)$ *exists and is piecewise p-times Lipschitz continuously differentiable in* I *for some* $p \in \mathbf{N}$.
iii) The simplifying conditions $B(p)$, *i.e.*

$$\sum_{\nu=1}^{s} b_\nu c_\nu^{k-1} = \frac{1}{k} \tag{3.27}$$

for $k = 1, \ldots, p$, *and* $C(p)$, *i.e.*

$$\sum_{\nu=1}^{s} a_{\mu\nu} c_\nu^{k-1} = \frac{1}{k} c_\mu^k \tag{3.28}$$

for $\mu = 1, \ldots, s$ *and* $k = 1, \ldots, p$ *hold.*
iv) The method is BS-stable and C-stable.
v) The initial error $\|\epsilon_0\|$ *and the mean error*

$$\frac{T - t_0}{N} \sum_{j=1}^{N} \max \left\{ \|\mathcal{E}_j^1\|, \ldots, \|\mathcal{E}_j^s\|, \|\epsilon_j\| \right\}$$

are of order 1 as functions of the stepsize $h = \frac{T - t_0}{N}$.

Then s-Stage Runge-Kutta Method 3.5 is convergent of order 1. *Moreover, the order of convergence is equal to p on subintervals where the solution is p-times Lipschitz continuously differentiable and where the corresponding initial error and mean error are of order p.* ∎

Naturally, to guarantee a priori existence of a solution of Initial Value Problem 1.6, assumption i) has to be sharpened, cp. Section 2 where we assumed F to be non-empty-, convex- and compact-valued upper semicontinuous and to satisfy a growth condition. The proof of BS-stability and C-stability for concrete methods, e.g. all Gauss methods, requires F to satisfy a one-sided Lipschitz condition, which in turn yields uniqueness of the exact solution. We conclude with an application of the simplest Gauss method, namely the implicit midpoint rule, to Example 1.4.

3.7. Example. The implicit midpoint rule is defined by the difference inclusion

$$\eta_{j+1} \in \eta_j + hF\left(t_j + \frac{1}{2}h, \frac{1}{2}(\eta_j + \eta_{j+1})\right) ,$$

or in more formal notation as perturbed implicit one-stage Runge-Kutta method

$$\begin{aligned}
\eta_0 &= y_0 + \epsilon_0 , \\
Y^1 &= \eta_j + \frac{1}{2} h Z^1 + h \mathcal{E}_{j+1}^1 , \\
Z^1 &\in F\left(t_j + \frac{1}{2}h, Y^1\right) , \\
\eta_{j+1} &= \eta_j + h Z^1 + h \epsilon_{j+1}
\end{aligned}$$

for $j = 0, \ldots, N-1$ with Butcher array

$$\begin{array}{c|c} c_1 & a_{11} \\ \hline & b_1 \end{array} = \begin{array}{c|c} \frac{1}{2} & \frac{1}{2} \\ \hline & 1 \end{array}.$$

This method satisfies simplifying conditions $B(1)$, $B(2)$, and $C(1)$. Moreover, it is BS-stable and C-stable for set-valued mappings F satisfying a one-sided Lipschitz condition. We want to apply this method to Example 1.4 with F given by (1.9), initial value $y_0 = 1$, and varying stiffness parameter $L > 0$. F satisfies a one-sided Lipschitz condition with one-sided Lipschitz constant $-L$ and has closed graph. Moreover, F is non-empty, convex- and compact-valued upper semicontinuous and satisfies a growth condition. Therefore, we know from Convergence Theorem 2.2 that a global solution $y(\cdot)$ exists on the whole interval $I = [0, 2]$, uniquely determined by its initial value $y(0) = 1$. In fact, this solution is piecewise smooth. Hence, we infer from Theorem 3.6 that the implicit midpoint rule is convergent of order 1, if the initial error $\|\epsilon_0\|$ and the mean error

$$\frac{T - t_0}{N} \sum_{j=1}^{N} \max \left\{ \|\mathcal{E}_j^1\|, \|\epsilon_j\| \right\}$$

are of order 1.

Moreover, the order of convergence is equal to 2 on subintervals where the solution is twice Lipschitz continuously differentiable and where the corresponding initial error and mean error are of order 2, although the condition $C(2)$ does not hold. To circumvent this difficulty one can use the same reasoning as in [21].

These results are illustrated by the following Tables 1 to 3 partly from [18]. Table 1 lists the error at the point $t = 1$ as a function of stiffness parameter L and discretization parameter N. On the interval $[0, 1]$ the solution is smooth, which explains second order convergence of implicit midpoint rule on this interval even for large L.

	L		
N	$2 \cdot 10^0$	$2 \cdot 10^3$	$2 \cdot 10^6$
20	$8.685 \cdot 10^{-5}$	$8.908 \cdot 10^{-5}$	$1.988 \cdot 10^{-4}$
40	$2.167 \cdot 10^{-5}$	$1.957 \cdot 10^{-6}$	$4.958 \cdot 10^{-5}$
80	$5.416 \cdot 10^{-6}$	$1.655 \cdot 10^{-8}$	$1.228 \cdot 10^{-5}$
160	$1.354 \cdot 10^{-6}$	$4.145 \cdot 10^{-9}$	$2.953 \cdot 10^{-6}$
320	$3.384 \cdot 10^{-7}$	$1.036 \cdot 10^{-9}$	$6.332 \cdot 10^{-7}$

Table 1: Error at $t = 1$ starting at $t = 0$

Table 2 lists the error at the point $t = 2$. Now, order of convergence is reduced to 1 on the interval $[0, 2]$, since $y'(\cdot)$ is discontinuous at the point $t = 1$. To avoid that this point is always a grid point, we chose N to be odd.

N	L		
	$2 \cdot 10^0$	$2 \cdot 10^3$	$2 \cdot 10^6$
41	$3.098 \cdot 10^{-2}$	$3.098 \cdot 10^{-2}$	$3.087 \cdot 10^{-2}$
81	$1.569 \cdot 10^{-2}$	$1.570 \cdot 10^{-2}$	$1.567 \cdot 10^{-2}$
161	$7.900 \cdot 10^{-3}$	$7.900 \cdot 10^{-3}$	$7.900 \cdot 10^{-3}$
321	$3.970 \cdot 10^{-3}$	$3.970 \cdot 10^{-3}$	$3.960 \cdot 10^{-3}$
641	$1.986 \cdot 10^{-3}$	$1.984 \cdot 10^{-3}$	$1.986 \cdot 10^{-3}$
1281	$9.939 \cdot 10^{-4}$	$9.939 \cdot 10^{-4}$	$9.938 \cdot 10^{-4}$

Table 2: Error at $t = 2$ starting at $t = 0$

These results suggest a combination of the implicit midpoint rule with a suitable localization procedure for detecting possible discontinuities of $y'(\cdot)$ to ensure second order convergence globally on the whole interval $[0, 2]$. This was done with a heuristic approach in [18] yielding the results in Table 3.

N	L		
	$2 \cdot 10^0$	$2 \cdot 10^3$	$2 \cdot 10^6$
41	$8.150 \cdot 10^{-4}$	$8.492 \cdot 10^{-4}$	$9.759 \cdot 10^{-4}$
81	$2.149 \cdot 10^{-4}$	$2.303 \cdot 10^{-4}$	$2.053 \cdot 10^{-4}$
161	$5.521 \cdot 10^{-5}$	$6.082 \cdot 10^{-5}$	$6.007 \cdot 10^{-5}$
321	$1.399 \cdot 10^{-5}$	$1.537 \cdot 10^{-5}$	$1.525 \cdot 10^{-5}$
641	$3.520 \cdot 10^{-6}$	$3.864 \cdot 10^{-6}$	$3.849 \cdot 10^{-6}$
1281	$8.834 \cdot 10^{-7}$	$9.687 \cdot 10^{-7}$	$9.667 \cdot 10^{-7}$

Table 3: Error at $t = 2$ starting at $t = 0$ with localization

Bibliography

[1] R. Ansorge and K. Taubert. Set-valued collectively compact operators and applications to set-valued differential equations. *Computing*, 23:333–343, 1979.

[2] J.-P. Aubin and A. Cellina. *Differential Inclusions*. Springer-Verlag, Berlin, 1984.

[3] J.-P. Aubin and H. Frankowska. *Set-Valued Analysis*. Volume 2 of *Systems and Control: Foundations and Applications*, Birkhäuser, Boston–Basel–Berlin, 1990.

[4] J. C. Butcher. *The Numerical Analysis of Ordinary Differential Equations*. John Wiley and Sons, Chichester–New York–Brisbane–Toronto–Singapore, 1987.

[5] L. Cesari. *Optimization—Theory and Applications*. Volume 17 of *Applications of Mathematics*, Springer-Verlag, New York–Heidelberg–Berlin, 1983.

[6] K. Dekker and J. G. Verwer. *Stability of Runge-Kutta Methods for Stiff Nonlinear Differential Equations*. Volume 2 of *CWI Monographs*, North-Holland, Amsterdam–New York–Oxford, 1984.

[7] A. L. Dontchev and E. M. Farkhi. Error estimates for discretized differential inclusions. *Computing*, 41:349–358, 1989.

[8] A. L. Dontchev and F. Lempio. Difference methods for differential inclusions: A survey. 1991. Preprint.

[9] C. M. Elliott. On the convergence of a one-step method for the numerical solution of an ordinary differential inclusion. *IMA Journal of Numerical Analysis*, 5:3–21, 1985.

[10] A. F. Filippov. Differential equations with discontinuous right-hand side. *AMS Transl.*, 42:199–231, 1964.

[11] A. F. Filippov. *Differential Equations with Discontinuous Righthand Side. Mathematics and Its Applications*, Kluwer Academic Publishers, Dordrecht–Boston–London, 1988.

[12] R. Frank, J. Schneid, and W. Ueberhuber. Order results for implicit Runge-Kutta methods applied to stiff systems. *SIAM J. Numer. Anal.*, 22(3):515–534, 1985.

[13] R. Frank, J. Schneid, and W. Ueberhuber. Stability properties of implicit Runge-Kutta methods. *SIAM J. Numer. Anal.*, 22(3):497–514, 1985.

[14] E. Hairer, S. P. Norsett, and G. Wanner. *Solving Ordinary Differential Equations I: Nonstiff Problems*. Springer-Verlag, Berlin–Heidelberg–New York–London–Paris–Tokyo, 1987.

[15] M. R. Hestenes. *Calculus of Variations and Optimal control Theory. Applied Mathematics Series*, John Wiley and Sons, Inc., New York–London–Sydney, 1966.

[16] A. D. Ioffe and V. M. Tichomirov. *Theorie der Extremalaufgaben.* VEB Deutscher Verlag der Wissenschaften, Berlin, 1979.

[17] H. Kahl. *Direkte Methoden für optimale Steuerungsprobleme.* Diplomarbeit, Universität Bayreuth, Bayreuth, 1986.

[18] A. Kastner-Maresch. Differenzenverfahren höherer Ordnung für Differentialinklusionen. Dissertation, Universität Bayreuth, 1990.

[19] A. Kastner-Maresch. Implicit Runge-Kutta methods for differential inclusions. Numer. Func. Anal. and Optimiz. To appear.

[20] A. Kastner-Maresch and F. Lempio. Difference methods with selection strategies for differential inclusions. 1991. Preprint.

[21] J. F. B. M. Kraaijevanger. B-convergence of the implicit midpoint rule and the trapezoidal rule. *BIT*, 25:652–666, 1985.

[22] F. Lempio. Modified Euler methods for monotone differential inclusions. Lecture, delivered at the Workshop on Set-Valued Analysis and Differential Inclusions, Pamporovo, Bulgaria, 1990.

[23] F. Lempio. The numerical treatment of differential inclusions. In *Proceedings CDE'IV, Rousse, Bulgaria*, 1989.

[24] H.-D. Niepage. *Convergence of Multistep Methods for Differential Equations with Multivalued Right Hand Side.* Seminarbericht der Humboldt-Universität zu Berlin, Humboldt-Universität, Berlin, 1980. No. 32.

[25] H.-D. Niepage. Inverse stability and convergence of difference approximations for boundary value problems for differential inclusions. *Numer. Funct. Anal. and Optimiz.*, 9(7+8):761–778, 1987.

[26] H.-D. Niepage and W. Wendt. On the discrete convergence of multistep methods for differential inclusions. *Numer. Funct. Anal. and Optimiz.*, 9(5+6):591–617, 1987.

[27] B. T. Polyak. *Introduction to Optimizaton. Translations Series in Mathematics and Engineering*, Optimization Software, Inc., Publications Division, New York, 1987.

[28] L. S. Pontryagin, V. G. Boltyanskii, R. V. Gamkrelidze, and E. F. Mishchenko. *The Mathematical Theory of Optimal Processes.* Interscience Publ., New York, 1962.

[29] K. Schilling. An algorithm to solve boundary value problems for differential inclusions and applications in optimal control. *Numer. Funct. Anal. and Optimiz.*, 10(7 & 8):733–764, 1989.

[30] K. Schilling. Simpliziale Algorithmen zur Berechnung von Fixpunkten mengenwertiger Operatoren. Dissertation, Universität Bayreuth, 1985.

[31] K. Schittkowski. NLPQL: A FORTRAN subroutine solving constrained nonlinear programming problems. *Annals of Operations Research*, 5:485–500, 1985.

[32] H. J. Stetter. *Analysis of Discretization Methods for Ordinary Differential Equations*. Volume 23 of *Springer Tracts in Natural Philosophy*, Springer-Verlag, New York, 1973.

[33] D. Stewart. A high accuracy method for solving ODEs with discontinuous right-hand side. *Numer. Math.*, 58:299–328, 1990.

[34] D. Stewart. *High Accuracy Numerical Methods for Ordinary Differential Equations with Discontinuous Right-Hand Side*. PhD thesis, University of Queensland, St. Lucia (Australia), 1989.

[35] K. Taubert. Converging multistep methods for initial value problems involving multivalued maps. *Computing*, 27:123–136, 1981.

[36] K. Taubert. Differenzenverfahren für Schwingungen mit trockener und zäher Reibung und für Regelungssysteme. *Num. Math.*, 26:379–395, 1976.

[37] K. Taubert. Über die Approximation nicht-klassischer Lösungen von Anfangswertaufgaben. Dissertation, Universität Hamburg, 1973.

[38] V. M. Veliov. Approximations to differential inclusions by discrete inclusions. IIASA Working Paper WP-89-017, 1989.

[39] V. M. Veliov. Second order discrete approximation to linear differential inclusions. SIAM J. Numer. Anal. To appear.

[40] V. M. Veliov. Second order discrete approximations to strongly convex differential inclusions. *Systems and Control Letters*, 13:263 - 269, 1989.

Ekeland's Variational Principle, Convex Functions and Asplund–Spaces

D. Pallaschke

Abstract

In this paper we discuss the following topics in infinite dimensional optimization theory: The Ekeland variational principle and some of its equivalent formulations, differential properties of convex functions in Banach Spaces and the theory of Asplund Spaces.

0 Introduction

In this paper we present two topics from infinite dimensional optimization theory, namely the Ekeland Variational Principle and the differential properties of continuous convex functions defined on a non–empty open convex subset of a real Banach–space. This presentation is taken from the paper of J. P. Penot [1] concerning the Ekeland Variational Property and from the introductory part of the Lecture Notes of R. R. Phelps [3] on convex Functions and Asplund Spaces, concerning the part on convex functions.

This paper is a part of the lecture on "Infinite dimensional Optimization Theory" which was given by the author during the summer–semester 1990 at the Humboldt–University of Berlin.
Specially the author thanks Doz. Dr. Bernd Kummer from Berlin for many discussions on this topic during the stay at the Humboldt–University.

Moreover the author thanks Prof. Dr. J. Jahn and Dr. T. Staib from the University of Erlangen and Prof. Dr. J. Zowe from the University of Bayreuth for many valuable remarks during the presentation of this topic at the Summer–School in Thurnau.

1 The Equivalence of the Ekeland Variational Principle to the Flower–Petal and the Drop–Theorem

1.1 The Ekeland Variational Principle

We start with the "Basic Ekeland Variational Principle":

Theorem 1.1.1
Let (M,d) be a complete metric space and

$$f: M \to \overline{\mathbb{R}} := \mathbb{R} \cup \{+\infty, -\infty\}$$

a lower–semicontinuous function which is bounded from below and not improper, i.e. not everywhere $+\infty$.
Then there exists a point $a \in M$, such that for all $x \in M \setminus \{a\}$

$$f(a) < f(x) + d(x,a) .$$

Proof: Let us define a binary relation

$$R \subseteq M \times M$$

by

$$R := \{(x,y) \in M \times M \mid y \in R(x)\}$$

where

$$R(x) := \{z \in M \mid f(z) + d(x,z) \leq f(x)\} .$$

Obviously, for every $x \in M$ we have $(x,x) \in R$ and from the triangle inequality follows the transitivity of R.
Since f is lower–semicontinuous, for every $x \in M$ the set $R(x)$ is closed. Now choose a sequence $(x_n)_{n\in\mathbb{N}}$ with:

$$\alpha) \qquad x_{n+1} \in R(x_n)$$

and

$$\beta) \qquad f(x_{n+1}) \leq \inf_{z\in R(x_n)} f(z) + \frac{1}{n}.$$

From α) follows, that $(x_k)_{k\in\mathbb{N}}$ is a Cauchy–sequence which converges to a point $a \in \bigcap_{k\in\mathbb{N}} R(x_k)$.
Namely

$$f(x_{n+1}) + d(x_{n+1},x_n) \leq f(x_n)$$

implies for every $k \in \mathbb{N}$

$$\sum_{n=1}^{k} d(x_{n+1},x_n) \leq \sum_{n=1}^{k} (f(x_n)-f(x_{n+1})) \leq f(x_1)-f(x_{k+1}).$$

Since $f(x_1) - f(x_{k+1})$ is bounded, it follows that $(x_k)_{k\in\mathbb{N}}$ is a Cauchy–sequence which converges to a point $a \in M$.

From the transitivity of $R \subseteq M \times M$ follows, that for every $k \in \mathbb{N}$

$$R(x_{k+1}) \subseteq R(x_k)$$

and hence

$$a \in \bigcap_{k \in \mathbb{N}} R(x_k) .$$

Now β) implies, that

$$R(a) = \bigcap_{k \in \mathbb{N}} R(x_k) = \{a\} .$$

This follows from the Baire–Category Theorem together with the fact that $\operatorname{diam}(R(x_{k+1})) \le \frac{2}{k}$.

Namely, let $x \in R(x_{k+1})$ then

$$f(x) + d(x,x_{k+1}) \le f(x_{k+1})$$

and hence

$$d(x,x_{k+1}) \le f(x_{k+1}) - f(x) \le \inf_{z \in R(x_k)} f(z) + \frac{1}{k} - f(x) \le \frac{1}{k} ,$$

since $R(x_{k+1}) \subseteq R(x_k)$ implies

$$\inf_{z \in R(x_k)} f(z) - f(x) \le 0 .$$

From the transitivity of $R \subseteq M \times M$ follows that

$$R(a) \subseteq R(x_k)$$

for every $k \in \mathbb{N}$ and hence $R(a) = \{a\}$.

Now let $x \in M \setminus \{a\}$, then $x \notin R(a)$ and therefore $f(x) + d(x,a) > f(a)$.

□

From this we deduce the "Altered Ekeland Variational Principle":

Theorem 1.1.2
Let (M,d) be a complete metric space and

$$f : M \to \overline{\mathbb{R}}$$

a lower–semicontinuous function which is bounded from below and not improper. Then for every $\gamma > 0$ and every $x_0 \in M$ there exists an element $a \in M$ such that

i) $f(a) < f(x) + \gamma \cdot d(x,a)$, for all $x \in M \setminus \{a\}$
ii) $f(a) \leq f(x_0) - \gamma \cdot d(a,x_0)$.

Remark: From ii) follows, that $d(a,x_0) \leq \frac{1}{\gamma} \cdot (f(x_0) - f(a))$.

Now let $\epsilon > 0$ be given and let $x_0 \in M$
be such that

$$f(x_0) \leq \inf_{x \in M} f(x) + \epsilon .$$

There we have

$$d(a,x_0) \leq \frac{1}{\gamma} \cdot (f(x_0) - f(a)) \leq \frac{1}{\gamma} \cdot (\inf_{x \in M} f(x) + \epsilon - f(a)) \leq \frac{\epsilon}{\gamma} .$$

Proof: This theorem is a simple consequence of the "Basic Ekeland–Variational Principle" which is to apply to the closed subset

$$S := \{x \in M \mid f(x) + \gamma \cdot d(x,x_0) \le f(x_0)\} \subseteq M$$

and the function

$$\frac{1}{\gamma} \cdot (f \mid S) . \qquad \square$$

1.2 Geometrical Aspects of the Ekeland Variational Principle

In this part we will state three geometric variants of the Ekeland Variational Principle.

A: Conical support points with respect to Bishop–Phelps cones – The Icecream Theorem:

Let $(E, \|\ \|)$ be a real normed space, $K \subseteq E$ a closed convex cone with vertex 0 and $S \subseteq E$ a closed subset. A point $x_0 \in S$ is called a "conical support point" of S with respect to K if

$$S \cap (K + x_0) = \{x_0\} .$$

Not every convex cone will yield such points, a class of cones with this property is of the form

$$K := \mathbb{R}_+ \cdot B = \{\lambda y \mid \lambda \ge 0, y \in B\}$$

where B is a non–empty closed convex bounded set with $0 \notin B$. A special subclass of this cones are the "Bishop–Phelps cones" given by

$$K(\alpha,x^*) := \{y \in E \mid \alpha \cdot \|y\| \le \langle x^*,y \rangle\} ,$$

where $x^* \in E'$ (i.e. the topological goal of E), $\|x^*\| = 1$ and $0 < \alpha < 1$.
In this case, a basis $B \subsetneq E$ of $K(\alpha,x^*)$ is given by

$$B := \{y \in E \mid \|y\| \leq 1, < x^*,y > \geq \alpha\} ;$$

i.e.

$$K(\alpha,x^*) = \mathbb{R}_+ \cdot B .$$

With the identical proof of the Basic Ekeland Variational Principle, the existence of conical support points with respect to Bishop–Phelps cones can be shown: (see [2]).

Theorem A (Icecream–Theorem)
Let $(E,\| \|)$ be a real Banach–space, $S \subsetneq E$ a closed subset, $0 < \alpha < 1$ and $x^* \in E'$ with $\|x^*\| = 1$ such that $\sup_{z \in S} < x^*,z > < \infty$.

Then for any $z \in S$ there exists an element $x_0 \in S$ with

$$x_0 \in K(\alpha,x^*) + z$$

and

$$S \cap (K(\alpha,x^*) + x_0) = \{x_0\} .$$

Remark: For illustration we draw the following picture:

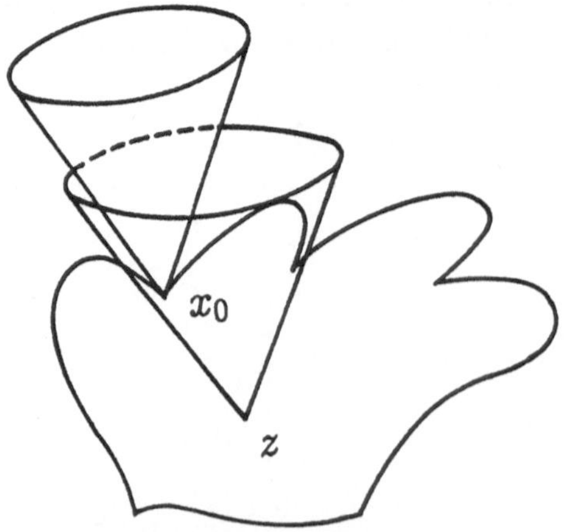

B: The Flower–Petal Theorem

Let (E,d) be a complete metric space, $a,b \in E$ with $a \neq b$ and $\gamma \in \mathbb{R}_+ \setminus \{0\}$ be given.
The "Flower–Petal" associated with $a,b \in E$ and the parameter $\gamma > 0$ is then defined by

$$P_\gamma(a,b) := \{x \in E \mid \gamma \cdot d(a,x) + d(x,b) \leq d(a,b)\}$$

The point $a \in E$ is called the "vertex" of the petal and $b \in E$ the leaf bud.
In the case, where $E = \mathbb{R}^2$, endowed with the Euclidean metric, a petal looks like

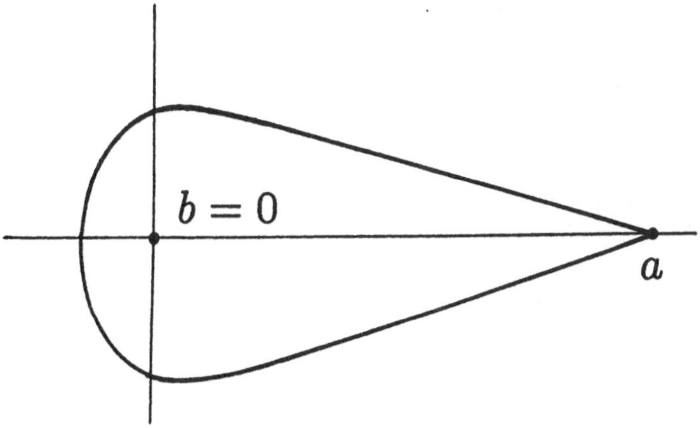

Observe, that for $0 < \gamma_1 < \gamma_2$ the following inclusion holds:

$$P_{\gamma_1}(a,b) \supseteq P_{\gamma_2}(a,b) .$$

A geometric variant of the Ekeland Variational Principle is the

Theorem B (The Flower–Petal Theorem)
Let $X \subsetneq E$ be a complete subset of a metric space (E,d), $x_0 \in X$ and $b \in E \setminus X$. Then for every $\gamma > 0$ there exists an element $a \in X \cap P_\gamma(x_0,b)$ such that

$$P_\gamma(a,b) \cap X = \{a\} .$$

Remark: For illustration we draw the following picture:

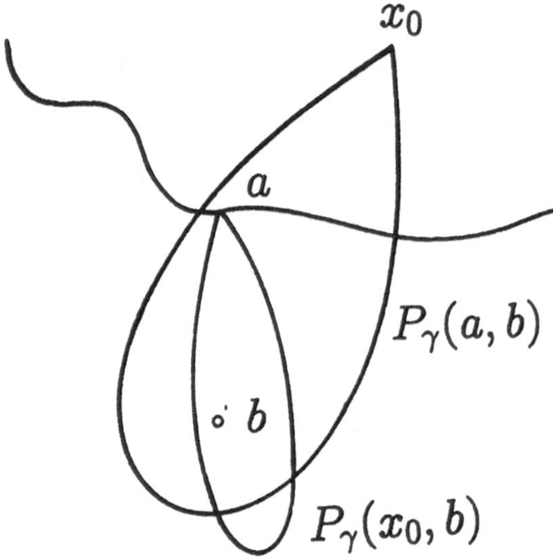

C: The Drop–Theorem

Let $(E, \| \ \|)$ be a real Banach–space $a,b \in E$ and $r \in \mathbb{R}_+ \setminus \{0\}$.
By

$$B := B(b,x) := \{x \in E \mid \|x-b\| \leq r\}$$

we denote the ball with center b and radius r.
For $a \notin B$ we denote by

$$D(a,B) := \text{conv}(a,B)$$

the "Drop generated by a and B", i.e. the closed convex hull of a with the ball B.

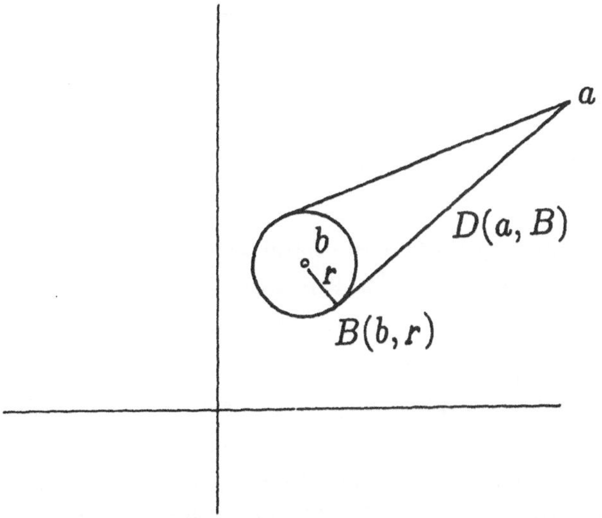

As a further geometrical variant of the Ekeland Variational Principle we state:

Theorem C (Drop–Theorem)
Let $(E, \| \ \|)$ be a real Banach–Space $C \subsetneq E$ a closed subset and $B := B(b,r)$ a ball with center b and radius r.
If $r < \text{dist}(b,C)$ then there exists an element $a \in C$ such that

$$D(a,B) \cap C = \{a\} .$$

Remark:
We illustrate this by the following picture:

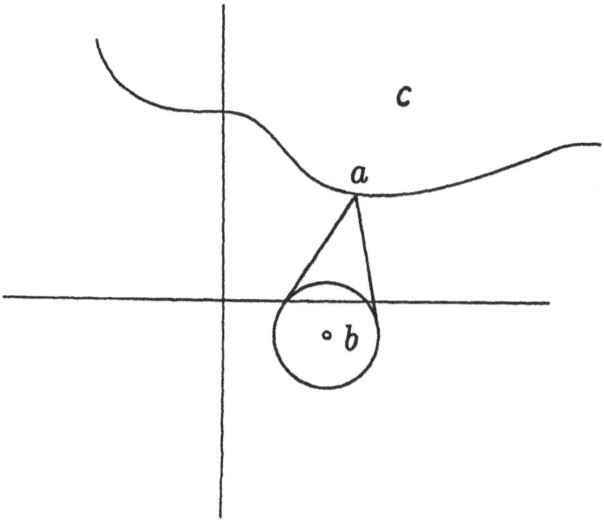

1.3 The Proof of Equivalences

In this part we prove the following chain of implications:

Altered Ekeland Variational Principle ⟹ Flower–Petal Theorem ⟹ Drop–Theorem ⟹ Altered Ekeland Variational Principle.

Theorem 1.3.1
The altered Ekeland Variational Principle implies the Flower–Petal Theorem.

Proof: In the Flower–Petal Theorem we assume that (E,d) is a metric space, $X \subsetneq E$ a complete subset and that $x_0 \in X$, $b \in E \setminus X$ and $\gamma > 0$ are given. In the notation of the altered Ekeland Variational Principle we put $M := X$, endowed with the induced metric and define

$$f\colon M \to \mathbb{R}$$

by

$$f(x) := d(x,b).$$

Since f is continuous and $\operatorname{dist}(b,X) \geq r > 0$, the altered Ekeland Variational Principle implies the existence of an element $a \in M = X$, such that

i) $f(a) < f(x) + \gamma \cdot d(x,a)$, for all $x \in M \setminus \{a\}$

and

ii) $f(a) \leq f(x_0) - \gamma \cdot d(a,x_0)$.

Using the definition of the Flower–Petal i) implies

$$P_\gamma(a,b) \cap X = \{a\}$$

and ii) implies

$$a \in X \cap P_\gamma(x_0,b) \ . \qquad \square$$

To show the next implication, we state the following Lemma which can be proved by a straightforward calculation:

Lemma 1.3.2
Let $(E, \|\ \|)$ be a real normed vector space, $a,b \in E$ and $r > 0$ such that

$$a \notin B := B(a,r) \ .$$

Put $t := \|a-b\|$.
Then for every $0 < \gamma < \frac{t-r}{t+r}$ we have

$$D(a,B) \subseteq P_\gamma(a,b) \ .$$

Theorem 1.3.3
The Flower–Petal Theorem implies the Drop Theorem.

Proof: In the Drop Theorem the following assumptions are made: Let $(E, \|\ \|)$ be a Banach–Space, $C \subseteq E$ a closed subset, $b \in E \setminus C$ and $0 < r < \text{dist}(b,C)$. To apply the Flower–Petal Theorem we choose an arbitrary element $x_0 \in C$ and put

$$X := C \cap D(x_0,B)$$

with

$$B := B(b,r) \ .$$

For

$$d := \text{dist}(b,C)$$

and

$$\gamma_0 := \frac{d-r}{d+r} > 0 ,$$

the Flower–Petal Theorem implies the existence of a point $a \in X$ with

$$a \in P_{\gamma_0}(x_0,b) \cap X$$

and

$$\{a\} = P_{\gamma_0}(a,b) \cap X .$$

Since $a \in D(x_0,B)$ we have $D(a,B) \subseteq D(x_0,B)$ and by the above Lemma $D(a,B) \subseteq P_{\gamma_0}(a,b)$.
Hence

$$a \in D(a,B) \cap C = D(a,B) \cap (D(x_0,B) \cap C) \subseteq P_{\gamma_0}(a,b) \cap X = \{a\}$$

gives

$$D(a,B) \cap C = \{a\} .$$

□

To prove the last implication the metric space (M,d) has to be embedded into a Banach–Space.
Using the variant of the proof of the Basic Ekeland Variation Principle, which is needed for the Icecream–Theorem, the proof of this last implication is done. It is relatively technical and we refer to [1] .

2 Convex Functions and Monotone Operators

2.1 Basic Properties of Convex Functions

In the following we assume, that $(E,\| \|)$ is a real Banach–space and $D \subsetneq E$ a non–empty open convex set.

A function

$$f : D \to \mathbb{R}$$

is said to be "**convex**" if for any two elements $x,y \in D$ and any $t \in [0,1]$ the inequality

$$f(tx + (1-t)y) \leq t \cdot f(x) + (1-t \cdot f(y)$$

holds.

A convex function

$$p: E \to \mathbb{R}$$

which is positively homogeneous, i.e. $p(tx) = t \cdot p(x)$ for all $x \in E$ and all $t \geq 0$, is said to be "**sublinear**".

Without prove we state the following:

Lemma 2.1.1

Let

$$f: D \to \mathbb{R}$$

be a convex function which is continuous at $x_0 \in D$.

Then for any direction $u \in E$, the directional derivative

$$d^+f\big|_{x_0}(u) \quad := \lim_{\substack{t \to 0 \\ t > 0}} \frac{f(x_0+tu)-f(x_0)}{t}$$

$$= \inf_{t>0} \frac{f(x_0+tu)-f(x_0)}{t}$$

exists and the mapping

$$d^+f\big|_{x_0}: \quad E \to \mathbb{R}$$

$$u \to d^+f\big|_{x}(u)$$

is sublinear.

A function

$$f: D \to \mathbb{R}$$

is said to be **"Gateau–differentiable"** at $x_0 \in D$ if for every $u \in E$

$$df\big|_{x_0}(u) := \lim_{t \to 0} \frac{f(x_0+tu)-f(x_0)}{t}$$

exists. The positively homogeneous mapping

$$df\big|_{x_0}: \quad E \to \mathbb{R}$$

$$u \to df\big|_{x_0}(u)$$

is called the **"Gateau–differential"**.

Thus a convex function is Gateau–differentiable at $x_0 \in D$ if and only if $d^+f\big|_{x_0}$ is linear.

Another well known fact is stated in the following

Lemma 2.1.2

i) Every continuous convex function is locally Lipschitz

ii) if E is finite dimensional then every convex function

$$f : D \to \mathbb{R}$$

defined on a non–empty open convex subset $D \subseteq E$, is continuous

iii) for a convex function f of one variable, $f'(x)$ exists for all but countable many points of D.

An important tool in convex analysis is the notion of the subdifferential.
Let $C \subseteq E$ be an arbitrary convex subset and

$$f: C \to \mathbb{R}$$

a convex function. Then for a point $x_0 \in C$, the **"subdifferential of f at x_0"** is defined by

$$\partial f\big|_{x_0} := \{x^* \in E' \mid < x^*, x - x_0 > \leq f(x) - f(x_0) \text{ for all } x \in C\} .$$

A standard fact on the subdifferential is the following

Theorem 2.1.3

Let $(E, \|\ \|)$ be a real Banach–space and $D \subseteq E$ a non–empty open convex subset.
Let $f : D \to \mathbb{R}$ be a convex function which is continuous at $x_0 \in D$.

Then the subdifferential $\partial f\big|_{x_0} \subsetneq E'$ is a non–empty weak–*–compact convex subset of E'. Moreover the mapping $x \to \partial f\big|_x$ is locally bounded at x_0, i.e. there exists a neighborhood U of x_0 and a $M > 0$ such that for $x \in U$ and all $x^* \in \partial f\big|_x$ we have $\|x^*\| \leq M$.

2.2 Monotone Operators

The set–valued mapping defined by the subdifferential of a continuous convex function is a special monotone operator:

More precisely:
Let $(E,\| \ \|)$ be a real Banach–space. A mapping

$$T: E \to \mathbb{P}(E')$$

is said to be "**monotone**" if for every $x,y \in E$ and every $x^* \in T(x)$, $y^* \in T(y)$ the inequality

$$< x^* - y^*, x - y > \geq 0$$

holds.

With

$$D(T) := \{x \in E \,|\, T(x) \neq \phi\}$$

we denote the "**effective domain**" of T.

Now let $D \subseteq E$ be a non–empty open convex subset and

$$f: D \to \mathbb{R}$$

a continuous convex function.
Define

$$T: E \to \mathbb{P}(E')$$

by

$$T(x) := \begin{cases} \partial f|_x & , \ x \in D \\ \phi & , \ x \in E \setminus D \end{cases}$$

Then $D(T) = D$ and for every $x,y \in D$ and for every $x,y \in D$ and $x^* \in \partial f|_y$, $y^* \in \partial f|_y$ we have $< x^*, y - x > \leq f(y) - f(x)$ and $< y^*, x - y > \leq f(x) - f(y)$ which implies $< x^* - y^*, x - y > \geq 0$.

This is the natural generalization of the fact, that the derivative of a differentiable convex function is monotone.

Let us now recall the notation of upper–semi–continuity. Therefore let X,Y be real topological vector spaces and

$$T: X \to \mathbb{P}(Y)$$

a mapping. For a subset $A \subseteq X$ we put $T(A) := \bigcup_{x \in A} T(x)$. Then T is said to be "**upper–semi–continuous**" at $x_0 \in X$ if for every open set $V \subseteq Y$ with $V \supseteq T(x_0)$ there exists an open set $U \subseteq X$ with $x_0 \in U$ such that $T(u) \subseteq V$.

In this presentation we are mostly interested in the case where $X := E$ and $Y = E'$, where E' is either endowed with the weak $-*-$ topology or the norm–topology; where E is always endowed with the norm–topology. We will use the expression "norm–weak– * –upper–semi–continuous" respectively "norm–norm–upper–semi continuous".

The mapping

$$T: E \to \mathbb{P}(E')$$

is

i) norm–weak– * –upper–semi continuous at $x_0 \in E$ if and only if for every weak– * –open set $V \supseteq T(x_0)$ and every sequence $(x_n)_{n\in\mathbb{N}}$ which is norm–convergent to x_0, there exists an index n_0, such that for all $n \geq n_0$

$$T(x_n) \subseteq V.$$

ii) norm–norm–upper–semi continuous at $x_0 \in E$ if and only if for every norm–open set $V \supseteq T(x_0)$ there exists a $\delta > 0$ such that

$$T(B(x_0,\delta)) \subseteq V .$$

If $T(x_0)$ consists of a single point, this is equivalent to

$$\lim_{\substack{\delta \to 0 \\ \delta > 0}} \operatorname{diam}(T(B(x_0,\delta))) = 0 .$$

The following result is well known:

Theorem 2.2.1

Let $(E,\| \ \|)$ be a real Banach–space, $D \subseteq E$ a non–empty open convex set and

$$f: D \to \mathbb{R}$$

a continuous convex function.

Then the mapping

$$\partial f\colon D \to \mathbb{P}(E')$$
$$x \to \partial f|_x$$

is norm–weak–*–upper semi continuous.

Proof: Let $W \supseteq \partial f|_x$ be a weak–*–open subset of E' and $(x_n)_{n\in\mathbb{N}}$ a sequence of elements of D with $\lim x_n = x_0$. Then we have to show the existence of an index n_0, such that for $n \geq n_0$ the inclusion $\partial f|_{x_n} \subseteq W$ holds.

Let us assume that this is not true. Then there exists a weak–*–open subset $W \supseteq \partial f|_{x_0}$ of E' and a sequence $(x_n)_{n\in\mathbb{N}}$ of elements of D with $\lim x_n = x_0$ which contains a subsequence $(x_{n_k})_{k\in\mathbb{N}}$ such that for all $k \geq k_0$ there exist elements $x^*_{n_k} \in \partial f\big|_{x_{n_k}} \setminus W$. Since the subdifferential is locally bounded, the sequence $(x^*_{n_k})_{k\geq k_0}$ has a weak–*–cluster point $x^* \in E' \setminus W$.
Since for every $y \in D$ we have

$$< x^*, y - x_0 > = \lim_k < x^*_{n_k}, y - x_{n_k} > \leq \lim(f(y) - f(x_{n_k})) = f(y) - f(x_0)\,,$$

it follows that $x^* \in \partial f\big|_{x_0}$.
Therefore

$$x^* \in \partial f\big|_{x_0} \setminus W = \phi\,,$$

which is a contradiction. □

To study the problem on which subsets of D the subdifferential is single valued and norm–norm upper–semi continuous, we introduce the following notation;

Let $0 < \alpha < 1$. A subset $M \subseteq E$ is called "**α–meager**" if for every $x \in M$ and every $\epsilon > 0$ there exists a $z \in B(x,\epsilon)$ and an element $x^* \in E'$, $\|x^*\| = 1$, such that for the Bishop–Phelps cone $K(\alpha,x^*)$ the following holds:

$$M \cap (z + \text{int } K(\alpha,x^*)) = \phi .$$

A subset $M \subsetneq E$ is said to be **"angle–small"** if for every $0 < \alpha < 1$ it can be expressed as a countable union of α–meager sets.

Every angle–small set is of first category, i.e. the countable union of nowhere dense sets.

Now we give a result of D. Preiss and L. Zajicek [3] which implies that the points of nondifferentiability of a continuous convex function are smaller than a meager set.

Theorem 2.2.2
Let $(E, \| \ \|)$ be a real Banach–space with separable dual E' and let

$$T: E \to \mathbb{P}(E')$$

be a monotone operator with effective domain $D(T)$.
Then there exists an angle small subset $A \subsetneq D(T)$ such that

$$T|(D(T) \setminus A): \ D(T) \setminus A \to \mathbb{P}(E')$$

is single valued and norm–norm upper semicontinuous.

Proof: Let us consider the following set

$$A := \{x \in D(T) \,|\, \lim_{\substack{\delta \to 0 \\ \delta > 0}} \text{diam}[T(B(x,\delta))] > 0\} .$$

Obviously the operator T is on $D(T)\setminus A$ single valued and norm–norm upper–semi continuous. Thus it remains to show that A is angle–small.
Therefore we write

$$A = \bigcup_{n=1}^{\infty} A_n$$

with

$$A_n := \{x \in D(T) \mid \lim_{\substack{\delta \to 0 \\ \delta > 0}} \operatorname{diam} [T(B(x,\delta))] > \tfrac{1}{n}\}$$

and show that every A_n is an angle–small set
Thus, for $0 < \alpha < 1$ we have to represent

$$A_n = \bigcup_{k=1}^{\infty} A_{n,k}$$

where $A_{n,k}$ is α–meager. This set $A_{n,k}$ can be constructed as follows. Since E′ is separable, we choose a dense sequence $(x_k^*)_{n \in \mathbb{N}}$ of E′ such that $x_k^* \in E' \setminus \{0\}$ and put

$$A_{n,k} := \{x \in A_n \mid \operatorname{dist}(x_k^*, T(x)) < \tfrac{\alpha}{4n}\} .$$

Obviously we have

$$A_n = \bigcup_{k=1}^{\infty} A_{n,k} .$$

Now we show, that every $A_{n,k}$ is α–meager.
Let $x \in A_{n,k}$ and $\epsilon > 0$ be given. Since $x \in A_n$, there exists a $\delta \in (0,\epsilon)$ and points $z_1, z_2 \in B(x,\delta)$ and $z_i^* \in T(z_i)$, $i \in \{1,2\}$ such that $\|z_1^* - z_2^*\| > \frac{1}{n}$.
Hence for every $x^* \in T(x)$ we have

$$\max\{\|z_1^* - x^*\|, \|z_2^* - x^*\|\} > \frac{1}{2n} .$$

Since by definition of $A_{n,k}$ we have $\operatorname{dist}(x_k^*, T(x)) < \frac{\alpha}{4n}$, we choose $x^* \in T(x)$ specially so, that

$$\|x_k^* - x^*\| < \frac{\alpha}{4n}.$$

From this follows, that there exists an element $z \in B(x,\epsilon)$ and an element $z^* \in T(z)$ such that

$$\|z^* - x_k^*\| \geq \|z^* - x^*\| - \|x_k^* - x^*\| > \frac{1}{2n} - \frac{\alpha}{4n} > \frac{\alpha}{2n}.$$

Now we show that

$$A_{n,k} \cap \left[z + \text{int } K\left(\alpha, \frac{z^*-x_k^*}{\|z^*-x_k^*\|}\right) \right] = \phi .$$

First observe that

$$z + \text{int } K\left(\alpha, \frac{z^*-x_k^*}{\|z^*-x_k^*\|}\right) = \{y \in E \,|\, \alpha\cdot \|z^*-x_k^*\| \cdot \|y-z\| < <z^*-x_k^*, y-z>\} .$$

Now let us take an element $y \in D(T)$
with

$$\alpha\cdot \|z^* - x_k^*\| \cdot \|y - z\| < <z^* - x_k^*, y - z> .$$

Then for every $y^* \in T(y)$ we have

$$\begin{aligned} < y^* - x_k^*, y - z > &= < y^* - z^*, y - z > + < z^* - x_k^*, y - z > \\ &\geq <z^* - x_k^*, y - z> \\ &> \alpha\cdot \|z^*-x_k^*\| \cdot \|y-z\| \\ &> \frac{\alpha}{4n}\cdot \|y-z\| \end{aligned}$$

since the operator T is monotone.

Hence we have

$$\|y^* - x_k^*\| > \frac{\alpha}{4n}$$

and this implies that $y \notin A_{n,k}$. □

3 Differentiability of Convex Functions and Asplund–Spaces

3.1 Frechet–Differentiability of Convex Functions

Let $(E,\|\ \|)$ be a real Banach–space and $U \subseteq E$ an open subset.
A function

$$f : U \to \mathbb{R}$$

is said to be "**Frechet–differentiable**" at $x_0 \in U$ if there exists a continuous linear functional $df|_{x_0} \in E'$ such that for every $\epsilon > 0$ there is a $\delta > 0$ with

$$|f(x_0+h) - f(x_0) - df|_{x_0}(h)| \leq \epsilon \cdot \|h\|$$

for all $h \in E$ with $\|h\| < \delta$ and $x_0 + h \in U$.

First we have:

Lemma 3.1.1
Let$(E,\|\ \|)$ be a real Banach–space, $D \subseteq E$ a non–empty open convex set and

$$f: D \to \mathbb{R}$$

a continuous convex function.
Then f is Frechet–differentiable at $x_0 \in D$ if and only if for every $\epsilon > 0$ there is a $\delta > 0$ such that for all $y \in E$ with $\|y\| = 1$ and $0 < t < \delta$

$$f(x_0 + t\cdot y) + f(x_0 - t\cdot y) - 2\cdot f(x_0) \leq t\cdot\epsilon .$$

Proof: " $\Longrightarrow$ ": Let us assume, that $df\big|_{x_0} \in E'$ exists. Then for every $\epsilon > 0$ there exists a $\delta > 0$ such that

$$f(x_0 + t\cdot y) - f(x_0) - df\big|_{x_0}(y) \leq \tfrac{\epsilon}{2}\cdot t\cdot \|y\|$$

for $0 < t < \delta$ and $\|y\| = 1$.
Substitution y by $-y$ and adding the corresponding inequalities, we get

$$f(x_0 + t\cdot y) + f(x_0 - t\cdot y) - 2\cdot f(x_0) \leq \epsilon\cdot\|y\| = \epsilon\cdot t .$$

" $\Longleftarrow$ ": Now let us assume that for every $\epsilon > 0$ there exists a $\delta > 0$ such that for all $y \in E$ with $\|y\| = 1$ and all $0 < t < \delta$

$$(*) \qquad f(x_0+t\cdot y) + f(x_0-t\cdot y) - 2\cdot f(x_0) \leq \epsilon\cdot t .$$

Since f is continuous, we choose an element $x^* \in \partial f\big|_{x_0}$.
From the definition of the subdifferential follows:

$$\begin{aligned} <x^*,t\cdot y> &= <x^*,(x_0+t\cdot y) - x_0> \leq f(x_0+t\cdot y) - f(x_0) \\ -<x^*,ty> &= <x^*,(x_0-t\cdot y) - x_0> \leq f(x_0-t\cdot y) - f(x_0) . \end{aligned}$$

Together with (*) we get

$$0 \leq f(x_0+t\cdot y) - f(x_0) - <x^*,t\cdot y> \;\leq\; t\epsilon + f(x_0) - f(x_0-t\cdot y) - <x^*,t\cdot y>$$

and hence

$$|f(x_0+ty) - f(x_0) - < x^*,ty >| \leq t\cdot\epsilon$$

since $f(x_0)-t\cdot y) - < x^*,t\cdot y > \leq 0$ by definition of x^* .
Therefore

$$df\Big|_{x_0}(y) = < x^*,y > .$$ □

Using the above Lemma, we prove:

Theorem 3.1.2
Let $(E,\|\ \|)$ be a real Banach–space, $D \subseteq E$ a non–empty open convex set and

$$f: D \to \mathbb{R}$$

a continuous convex function.
Then the set

$$G := \{x \in D|\ df|_x \in E' \text{ exists}\}$$

is a G_δ–set.
Remark: We do not claim that $G \neq \phi$.

Proof: For $n \in \mathbb{N}$ we define

$$G_n:=\left\{x \in D \;\middle|\; \text{there exists a } \delta > 0 \text{ with } \sup_{\|y\|=1}\left[\frac{f(x+\delta y)+f(x-\delta y)-2f(x)}{\delta}\right] < \frac{1}{n}\right\}.$$

Since for fixed $x,y \in E$ the functions $t \to \frac{1}{t}\cdot(f(x \pm t\cdot y) - f(x))$ are non–increasing for $t \to 0$ and $t > 0$, it follows form Lemma 3.1.1 that $G = \bigcap_{n=1}^{\infty} G_n$. Thus it remains to show that for every $n \in \mathbb{N}$ the set $G_n \subseteq D$ is open.
Let $x_0 \in G_n$, since f is locally Lipschitz, there exists a $\delta_1 > 0$ and a constant $M>0$ such that for all $u,v \in B(x_0,\delta_1)$

$$|f(u) - f(v)| \le M \cdot \|u-v\| .$$

From the definition of G_n follows, that there exists a $\delta > 0$ and a $0 < r < \frac{1}{n}$ such that for all $y \in E$ with $\|y\| = 1$

$$x_0 \pm \delta y \in D$$

and

$$\frac{f(x_0+\delta y)+f(x_0-\delta y)-2f(x_0)}{\delta} \le r .$$

Now choose δ_2 such that $0 < \delta_2 < \min(\delta,\delta_1)$ and $B(x_0,\delta_2) \subseteq D$; and such that for all $y \in E$ with $\|y\| = 1$ and $z \in B(x_0,\delta_2)$

$$z \pm \delta_2 \cdot y \in D .$$

Then we have

$$\begin{aligned}
&\tfrac{1}{\delta}(f(z+\delta y) + f(z-\delta y) - 2f(z)) \\
&= \tfrac{1}{\delta}[(f(z+\delta y) - f(x_0+\delta y)) + f(x_0+\delta y) + (f(z-\delta y) - f(x_0-\delta y)) + f(x_0-\delta y) \\
&\qquad - 2(f(z) - f(x_0) - 2f(x_0)] \\
&\le \tfrac{1}{\delta}(f(x_0+\delta y) + f(x_0-\delta y) - 2f(x_0)) + \tfrac{1}{\delta}|f(z+\delta y) - f(x_0+\delta y)| + \tfrac{1}{\delta}|f(z-\delta y) - \\
&\qquad - f(x_0-\delta y)| + \tfrac{2}{\delta}|f(z) - f(x_0)| \\
&\le r + \frac{4M}{\delta} \|z-x_0\| \\
&\le r + \frac{4M}{\delta} \cdot \delta_2 \\
&< \frac{1}{n},
\end{aligned}$$

for a suitable small $\delta_2 > 0$. Hence G_n is open. □

Next we show:

Theorem 3.1.3
Let $(E,\|\ \|)$ be a real Banach–space, $D \subseteq E$ a non–empty open convex set and

$$f\colon D \to \mathbb{R}$$

a continuous convex function, which is Frechet–differentiable at $x_0 \in D$.
Then the mapping

$$\begin{aligned} \partial f\colon\ & D \to \mathbb{P}(E') \\ & x \to \partial f|_x \end{aligned}$$

is norm–norm upper–semi continuous at x_0.

Proof: Let us write $df\big|_{x_0}(h) = \langle x_0^*, h\rangle$ with $x_0^* \in E'$ and let $V \subseteq E'$ be a norm–open neighborhood of x_0^*. Then we have to show that there exists a norm–open neighborhood $U \subseteq D$ of x_0, such that $\partial f(U) \subseteq V$.
Let us assume that this is not true. Then there exists an $\epsilon > 0$ and a sequence $(x_n)_{n\in\mathbb{N}}$ with $x_n \in D$, $\lim x_n = x_0$ and elements $x_n^* \in \partial f\big|_{x_n}$, such that
$\|x_n^* - x_0^*\| \geq 2\epsilon$.
Hence there exists elements $z_n \in E$ with $\|z_n\| = 1$,
such that

(*) $\langle x_n^* - x_0^*, z_n\rangle \geq 2\epsilon$. Since f is Frechet–differentiable at $x_0 \in D$,
there exists for the above chosen $\epsilon > 0$ a $\delta > 0$
such that for all $y \in E$ with $\|y\| \leq \delta$ and $x_0 + y \in D$
(**) $f(x_0+y) - f(x_0) - \langle x_0^*, y\rangle \leq \epsilon\cdot\|y\|$.
From $x_n^* \in \partial f\big|_{x_n}$ follows that for all $y \in E$ with $x_0 + y \in D$

$$\langle x_n^*, (x_0+y) - x_n\rangle \leq f(x_0+y) - f(x_n)$$

holds; i.e.

(***) $\quad < x_n^*, y > \leq f(x_0+y) - f(x_0) + < x_n^*, x_n - x_0 > + f(x_0) - f(x_n)$.
Now put $y_n := \delta \cdot z_n$, then we get from (*) – (***):

$$2 \cdot \epsilon \cdot \delta \leq < x_n^* - x_0^*, y_n > = < x_n^*, y_n > - < x_0^*, y_n >$$
$$\leq f(x_0+y_n) - f(x_0) - < x_0^*, y_n > + < x_n^*, x_n - x_0 > + f(x_0) - f(x_n)$$
$$\leq \epsilon \cdot \delta + < x_n^*, x_n - x_0 > + f(x_0) - f(x_n) .$$

Since the subdifferential is locally bounded, we have

$$|< x_n^*, x_n - x_0 >| \leq \|x_n^*\| \cdot \|x_n - x_0\| \leq M \cdot \|x_n - x_0\| ,$$

and since f is continuous there exists a $n_0 \in \mathbb{N}$ such that for all $n \geq n_0$:

$$|< x_n^*, x_n - x_0 > + f(x_0) - f(x_n)| \leq \frac{\epsilon\delta}{10} .$$

Hence for $n \geq n_0$ this implies

$$2\epsilon\delta \leq \epsilon\delta + \frac{\epsilon\delta}{10}$$

which is a contradiction. □

Let T be a set–valued mapping with effective domain D(T). A **"selection"** φ for T is a single valued mapping satisfying $\varphi(x) \in T(x)$ for each $x \in D(T)$.

Corollary 3.1.4
Let $(E, \| \ \|)$ be a real Banach–space, $D \subseteq E$ a non–empty convex set, $x_0 \in D$ and

$$f: D \to \mathbb{R}$$

a continuous convex function.

Then f is Frechet–differentiable at x_0 if and only if there exists a selection φ of ∂f which is norm–norm continuous at x_0.

Proof: "$\Leftarrow$": Since for every $u \in E$ with $x_0 + u \in D$ we have:

$$\begin{aligned} 0 &\le f(x_0+u) - f(x_0) - < \varphi(x_0),u > \\ &\le < \varphi(x_0+u) - \varphi(x_0),u > \\ &\le \|\varphi(x_0+u) - \varphi(x_0)\| \cdot \|u\|, \end{aligned}$$

it follows from the continuity of φ that f is Frechet–differentiable at x_0.

"$\Rightarrow$": If f is Frechet–differentiable at x_0, then $\partial f\big|_{x_0}$ consists only of a single element. Since ∂f is norm–norm upper semi–continuous at x_0, also every selection is norm–norm continuous at x_0.

Corollary 3.1.5

Every Frechet–differentiable convex function, defined on a non–empty open convex subset of a real Banach–space is a C^1–function.

3.2 Mazur's Theorem and Asplund–Spaces

The Theorem of S. Mazur motivates the following part of this chapter:

Theorem 3.2.1 (S. Mazur)

Let $(E,\| \ \|)$ be a real separable Banach–space, $D \subseteq E$ an open convex set and $f: D \to \mathbb{R}$ a continuous convex function.

Then there exists a dense G_δ–set $A \subseteq D$ such that for each $x \in A$ the Gateau differential $df|_x$ exists.

Proof: First we show, that $df|_x$ does not exist on an F_σ–set. Namely let $(x_k)_{k\in\mathbb{N}}$ be a dense subset of the unit–sphere.

For $m,n \in \mathbb{N}$ we define

$$A_{n,m} := \{x \in D \mid < x^*-y^*, x_n > \geq \frac{1}{m}, \text{ for all } x^*, y^* \in \partial f|_x\}.$$

So, if for $x_0 \in D$ the Gateau–differential does not exist, then this is equivalent that $x_0 \in A_{n,m}$ for some $n,m \in \mathbb{N}$.
Now we show that $A_{n,m}$ is a closed subset of E in the norm–topology. Let $(z_k)_{k\in\mathbb{N}}$, $z_k \in A_{n,m}$ be a sequence which converges to $z \in D$.
Then there exist $x_k^*, y_k^* \in \partial f\big|_{z_k}$ such that $< x_k^*-y_k^*, x_n > \geq \frac{1}{m}$. Since $\partial f|_x$ is weak– * –compact and E is separable we can assume that x_k^* converges to x^*, and y_k^* converges to y^*.
For an arbitrary element $y \in D$ we have

$$< x^*, y-z > \; = \lim_k < x_k^*, y-z_k > \; \leq \lim_k (f(y)-f(z_k)) = f(y) - f(z) ;$$

and hence $x^* \in \partial f|_z$, resp. $y^* \in \partial f|_z$.
Moreover

$$< x^*-y^*, x_n > = \lim_k < x_k^*-y_k^*, x_n > \geq \frac{1}{m},$$

which implies that $z \in A_{n,m}$.
Now we show, that

$$A := \bigcap_{n,m\in\mathbb{N}} (D \setminus A_{n,m})$$

is a dense subset of D. By construction, A is a G_δ–set, since $A_{n,m}$ is closed.
Let $x_0 \in D$. For every $n \in \mathbb{N}$ we define

$$\tilde{f}_n : I_n \to \mathbb{R}$$

by

$$\tilde{f}_n(r) := f(x_0 + r \cdot x_n)$$

and

$$I_n := \{r \in \mathbb{R} | \ x_0 + r \cdot x_n \in D\} .$$

This convex functions are differentiable up to a countable subset. Now we approximate $x_0 \in D$ by a point $\tilde{x} := x_0 + \tilde{r}\, x_n$, such that $\tilde{f}_n'(\tilde{r})$ exists.
If $x^*,y^* \in \partial f|_{\tilde{x}}$, then for $\tilde{z} \in x_0 + \mathbb{R} \cdot x_n$ we have $< x^*,\tilde{z} > = < y^*,\tilde{z} >$, hence $\tilde{x} \in \bigcap_{m=1}^{\infty} (D \setminus A_{n,m})$. Therefore $A \subseteq D$ is a dense G_δ–subset of D. □

This famous theorem of Mazur from 1933 leaded to a direction of research which was started in 1968 by E. Asplund.

Definition 3.2.2
A real Banach–space $(E,\| \ \|)$ is called an **"Asplund–Space"** (resp. **"weak Asplund–Space"**) if for every continuous convex function

$$f: D \to \mathbb{R}$$

which is defined on a non–empty open convex set $D \subseteq E$ there exists a dense G_δ–subset $A \subseteq D$, such that f is Frechet– (resp. Gateau–) differentiable on A.

In this notation the theorem of S. Mazur can be stated as:

"Every real separable Banach–Space is a weak Asplund–Space".

The results of part 2.2 und 3.1 imply the following

Theorem 3.2.3
Let $(E,\| \ \|)$ be a real Banach–space with separable dual space E'.
Then E is an Asplund–space
Proof: Let $D \subseteq E$ be a non–empty open convex subset and

$$f: D \to \mathbb{R}$$

a continuous convex function. Then

$$\partial f: D \to \mathbb{P}(E')$$

a monotone operator. From the theorem of Preiss–Zajicek follows the existence of a dense G_δ–subset $A \subsetneq D$ such that $\partial f|A$ is a single–valued and norm–norm–upper–semi–continuous. Hence any selection of ∂f on A is norm–norm continuous and therefore f is Frechet–differentiable in the points of A.

□

Without proof we state the following characterization [3]

Theorem 3.2.4
Let $(E,\| \ \|)$ be a real separable Banach–space. Then E is an Asplund–space if and only if E' is separable.

We finish this part with the following two examples:

Example 1:
The norm

$$\| \ \| : \ell^1 \to \mathbb{R}_+$$

with

$$\|x\| := \sum_n |x_n|$$

is

i) Gateau–differentiable in any point

$x := (x_n)_{n\in\mathbb{N}} \in \ell^1$ with $x_n \neq 0$ for every $n \in \mathbb{N}$

and

ii) in no point of ℓ^1 Frechet–differentiable.

We start with

i) Let $x = (x_n)_{n\in\mathbb{N}} \in \ell^1$ be such that for some $n \in \mathbb{N}$ $x_n = 0$.
Let $\delta_n = (0,0,...,1,0 ...) \in \ell^1$ the sequence with the value 1 at position n and zero else.
Then

$$\|x+t\delta_n\| - \|x\| = |t| \text{ , and hence}$$

$$\frac{\|x+t\delta_n\| - \|x\|}{t} = \frac{|t|}{t} \text{ has no limit for } t \to 0 .$$

Now let $x := (x_n)_{n\in\mathbb{N}} \in \ell^1$ be such that for all $n \in \mathbb{N}$ we have $x_n \neq 0$ and let $y := (y_n)_{n\in\mathbb{N}} \in \ell^1$ be with $\|y\| = 1$.
For a given $\epsilon > 0$ there exists an integer N, such that

$$\sum_{n>N} |y_n| \le \frac{\epsilon}{2} .$$

Furthermore there exists a $\delta > 0$, such that for all $n \in \{1,...,N\}$ and all $t \in \mathbb{R}$ with $|t| \le \delta$

$$\operatorname{sign}(x_n + ty_n) = \operatorname{sign} x_n .$$

Hence we have for $t \neq 0$ and $|t| \le \delta$

$$\left| \frac{1}{t} (\|x+ty\| - \|x\| - \sum_{n=1}^{\infty} y_n \cdot \operatorname{sign} x_n \right|$$

$$\leq \left| \frac{1}{t} \cdot \sum_{n=1}^{N} (|x_n+ty_n| - |x_n| - \sum_{n=1}^{N} y_n \operatorname{sign} x_n \right|$$

$$+ \left| \frac{1}{t} \sum_{n>N} (|x_n+ty_n| - |x_n|) - \sum_{n>N} y_n \operatorname{sign} x_n \right|$$

$$\leq \left| \frac{1}{t} \sum_{n=1}^{N} (|x_n+ty_n| - |x_n|) - \sum_{n=1}^{N} y_n \operatorname{sign} x_n \right|$$

$$+ \left| \frac{|t|}{t} \sum_{n>N} |x_n| \right| + \sum_{n>N} |y_n| \leq \epsilon ,$$

because of $|x_n+t\cdot y_n| - |x_n| \leq |t\cdot y_n|$
and of the fact that the first summand vanishes. This follows from the fact that $|t| \leq \delta$ since:

$$\begin{aligned} &\tfrac{1}{t}(|x_n+ty_n| - |x_n|) - y_n \operatorname{sign} x_n \\ &\qquad = \tfrac{1}{t}(\operatorname{sign} x_n \cdot (x_n+ty_n) - \operatorname{sign} x_n \cdot x_n) - y_n \operatorname{sign} x_n \\ &\qquad = \operatorname{sign} x_n (\tfrac{1}{t}(x_n+ty_n-x_n) - y_n) = 0 . \end{aligned}$$

Now we show part
ii). To prove this part of the example it suffices to show that for every $x :=$ $(x_n)_{n\in\mathbb{N}} \in \ell^1$ with $x_n \neq 0$ for each $n \in \mathbb{N}$, the Frechet–differential does not exist. For such an element $x := (x_n)_{n\in\mathbb{N}} \in \ell^1$, we construct a sequence $(y^{(m)})_{m\in\mathbb{N}}$ of elements of ℓ^1 as follows:

$$y^{(m)} := (0,\dots,0,-2x_m,-2x_{m+1},-2x_{m+2},\dots) .$$

Then

$$\lim_{m\to\infty} \|y^{(m)}\| = 0 .$$

Since by part i) the Frechet–differential is given by:

$$(\operatorname{sign} x_n)_{n\in\mathbb{N}} \in \ell^\infty ,$$

we have:

$$\left| \|x+y^{(m)}\| - \|x\| - \sum_{n=1}^{\infty} (y^{(m)})_n \cdot \operatorname{sign} x_n \right|$$

$$= \left| \sum_{i=1}^{m-1} (|x_i+0| - |x_i| + \sum_{i=m}^{\infty} |x_i-2x_i| - |x_i| + \sum_{n=m}^{\infty} 2x_n \operatorname{sign} x_n \right|$$

$$= 2 \sum_{n=m}^{\infty} |x_n| = \|y^{(m)}\| .$$

Hence we have

$$\left| \|x+y^{(m)}\| - \|x\| - \sum_{n=1}^{\infty} (y^{(m)})_n \operatorname{sign} x_n \right| = \|y^{(m)}\| ,$$

which means that the Frechet–differential does not exist at this point. From this example follows that ℓ^1 is no Asplund–Space.

Example 2:

The sublinear mapping

$$p : \ell^\infty \to \mathbb{R}_+$$

given by

$$p(x) := \limsup_{n\to\infty} |x_n|$$

is in no point of ℓ^∞ Gateau–differentiable.

First we show, that p is a continuous sublinear function.

Let $x,y \in \ell^\infty$ and $\lambda > 0$.

Obviously we have

$$p(\lambda \cdot x) = \lambda \cdot p(x) ,$$

and

$$\begin{aligned} p(x+y) &= \limsup_{n\to\infty} |x_n+y_n| \le \limsup_{n\to\infty} (|x_n| + |y_n|) \\ &\le \limsup_{n\to\infty} |x_n| + \limsup_{n\to\infty} |y_n| = p(x) + p(y) . \end{aligned}$$

Since

$$0 \le p(x) = \limsup_{n\to\infty} |x| \le \sup_{n\in\mathbb{N}} | = \|x\|$$

we have proved the continuity of p.

Now we show that p is in no point of ℓ^∞ Gateau–differentiable:

If $x_0 \in \ell^\infty$ such that $p(x_0) = 0$ then for $x_0 = (x_1^0, x_2^0, \ldots, x_n^0 \ldots)$

we have of course

$$\lim_{n\to\infty} x_n^0 = 0 .$$

Now let

$$y := (1,1,1,\ldots,1,\ldots) ,$$

then for $t > 0$ we have,

$$\tfrac{1}{t}(p(x_0+ty) - p(x_0)) = \tfrac{|t|}{t} ,$$

and this limit does not exist for $t \to 0$.

Now let $p(x_0) > 0$; then because of the positive homogeneity we may assume that $p(x_0) = 1$.
Since $p(x_0) = p(-x_0)$ we may assume that there exists a positive subsequence $(x_{n_i}^0)_{i\in\mathbb{N}}$, $x_{n_i}^0 > 0$, such that $\lim_i x_{n_i}^0 = 1.$
This means that $p(x_0) = \lim_i x_{n_i}^0$.
Now we define an element $y = (y_n)_{n\in\mathbb{N}} \in \ell^\infty$ in the following way:

$$y_n = \begin{cases} 0 & \text{if either } n \neq n_i \text{, with } i \text{ odd} \\ 1 & \text{if } n = n_i \text{, with } i \text{ even.} \end{cases}$$

Then obviously

$$\frac{p(x+ty)-(p(x)}{t} = \begin{cases} 1, & \text{if } t > 0 \\ 0, & \text{if } t < 0 \end{cases} ,$$

i.e. the Gateau–differential of p does not exist in x_0 .

From this example follows, that ℓ^∞ is no weak Asplund–Space.

References

[1] PENOT, J.–P.: The Drop Theorem, the Petal Theorem and Ekeland's Variational Principle, Nonlinear Analysis–Theory, Methods Applications, Vol. 10 (1986) pp.813–822

[2] PHELPS, R. R.: Support Cones in Banach–Spaces and their Applications, Advances in Mathematics 13 (1974) pp. 1–19

[3] PHELPS, R. R.: Convex Functions, Monotone Operators and Differentiability, Lecture Notes in Mathematics, Vol. 1364, Springer–Verlag, Berlin – Heidelberg – New York (1989)

TOPICS IN MODERN COMPUTATIONAL METHODS FOR OPTIMAL CONTROL PROBLEMS

EKKEHARD W. SACHS
UNIVERSITÄT TRIER
FB IV - MATHEMATIK
POSTFACH 3825
5500 TRIER
GERMANY

Optimal control problems represent an important and interesting family of optimization problems. They differ from other optimization problems in so far as the description of the system to be controlled always involves some dynamics. This part is usually described by a system of differential equations or integral equations or difference equations. Depending on the model the type of differential equations is either ordinary or partial or a combination of both. The special feature about optimal control problems is that they are formulated in a function space, hence in an infinite dimensional space, and that they involve some kind of dynamics.

Starting in the 60ies researchers in engineering and mathematics have worked on the numerical solution of optimal control problems. There are mainly two approaches: In the first one looks at the necessary optimality conditions (e.g. Pontryagin's maximum principle) and solves these. This approach often involves differential equations solver especially designed for the requirements of optimal control problems such as discontinuities. The other approach consists in discretizing the dynamics such that a finite dimensional optimization problem appears which is solved by standard optimization software.

In this review article we want to restrict ourselves to a few aspects of modern developments in this field. Our intention is to show that the two approaches outlined above should not be viewed totally separated. If both viewpoints take each other into account, this synthesis can lead to very efficient methods and explain various numerical effects.

We want to focus in the second chapter on the gradient projection method. This is a very robust but also locally rather slow method. This disadvantage can be overcome by Newton type versions of it. We want to show how for example the convergence rate in the finite dimensional case can be predicted and explained by looking at the underlying infinite dimensional problem. Also, new developments like finite identification of active indices are considered from the point of view of optimal control.

The third chapter deals with quasi-Newton methods which have been used very successfully in finite dimensional optimization during the past two decades. Their application to optimal control problems can be interpreted much better by the use of recent results in the convergence analysis for infinite dimensional problems.

The fourth chapter shows how the description of the dynamics of the system can be used in the design of quasi Newton methods. This leads to a new method with substantial improvement over existing methods in its convergence features.

We emphasize that this paper reviews only a few aspects of new developments in this area. Among those methods which will not be mentioned are SQP methods, globalization by trust region methods, homotopy methods, new variants of the shooting method, multigrid methods, hierarchical preconditioning techniques, etc. The paper is organized in such a way, that we try to transfer the main ideas of the methods. In order to keep the notation at a reasonable level we have decided to refer the reader to the original papers with regard to a rigorous derivation of statements and e.g. the precise choice of function spaces or Fréchet-differentiability. For similar reasons, literature is not cited within the text, but at the end of each chapter one finds a list of references. Since this paper is a review article its emphasis is not on originality and therefore it contains material cited in the lists of references. Questions and requests for references by the reader to the author are encouraged and more than welcome.

1. Overview of Control Problems.

1.1. Classes of Optimal Control Problems.

We introduce briefly the notation for the control problems which we consider in the sequel.

1.1.1. Optimal Control Problems with Ordinary Differential Equations.

Let

$$f : \mathbb{R}_n \times \mathbb{R}_m \times \mathbb{R} \to \mathbb{R}_n$$

$$L : \mathbb{R}_n \times \mathbb{R}_m \times \mathbb{R} \to \mathbb{R}$$

be smooth functions. Let U denote the set of *controls* and X the set of *states*.
Usual choice of function spaces

$$\text{Control Space } L_m^\infty[0,T]; \qquad \text{State Space } W_n^{1,\infty}[0,T]. \tag{1.1}$$

We use the notation

$$L_m^p[0,T] = L^p([0,T];\mathbb{R}_m)$$

and similarly $C_n[0,T]$ and the Sobolev space $W_n^{1,r}[0,T]$.

This leads to the optimal control problem:
Minimize

$$\int_0^T L(x(t),u(t),t)\,dt \tag{1.2}$$

such that $(x,u) \in W_n^{1,\infty}[0,T] \times L_m^\infty[0,T]$ satisfies

$$\dot{x}(t) = f(x(t),u(t),t) \text{ a.e. } [0,T], \quad r(x(0),x(T)) = 0 \tag{1.3}$$

where

$$r : \mathbb{R}_{2n} \to \mathbb{R}$$

describes the boundary condition. In the sequel often

$$r(x(0), x(T)) = x(0).$$

1.1.2. Optimal Control Problems with Partial Differential Equations.

In various applications in engineering and economics the process to be controlled is described by a partial differential equation instead of an ordinary differential equation. These systems are sometimes called distributed parameter systems.
Let $c, l \in C^1(\mathbb{R}_3), k \in C^1(\mathbb{R}_2), g \in C^1(\mathbb{R})$ and an initial value $y_0 \in C(\Omega)$ be given. Ω is a bounded domain with boundary Γ.
Semilinear Parabolic Differential Equation

$$(1.4) \qquad \begin{array}{rcll} y_t(t,x) & = & (\Delta y)(t,x) + c(t,x,y(t,x)) & t \in (0,T), x \in \Omega, \\ y(0,x) & = & y_0(x) & x \in \Omega, \\ \frac{\partial y}{\partial n}(t,\xi) & = & g(\xi)u(t) & t \in (0,T), \xi \in \Gamma. \end{array}$$

Here we have chosen *control* u, *state* y.
The problems consists of finding u, y which minimize

$$(1.5) \qquad \int_0^T \int_\Omega l(t,x,y(t,x))\, dx\, dt + \int_0^T k(t,u(t))\, dt.$$

Other classes of problems include:

- nonlinear differential operator
- nonlinear boundary condition
- cost function depending on $y(\cdot, T)$.

1.1.3. Constraints for Optimal Control Problems.

Two important classes of constraints:

Control Constraints:
Pointwise $\quad h(t, u(t)) \leq 0$
Integral type $\quad h(u(\cdot)) \leq 0$ or $= 0$.

State Constraints:
Pointwise $\quad h(t, x(t)) \leq 0$
Integral type $\quad h(x(\cdot)) \leq 0$ or $= 0$.

1.2. First Order Methods.

1.2.1. Elimination of State Variable.

We assume in this section that there exists a *solution operator*

$$S : L_m^\infty[0,T] \to W_n^{1,\infty}[0,T],$$

where $x = S(u)$ satisfies the boundary value problem (1.3) and S is defined for each $u \in L_m^\infty[0,T]$. For this purpose let $r(x_1, x_2) = x_1$.

Hence the optimal control problem in (1.3) and (1.2) can be written as an *unconstrained minimization problem*:

$$\text{Minimize } \phi(u) = \int_0^T L(S(u)(t), u(t), t)\, dt. \tag{1.6}$$

Similarly, we suppose that in the partial differential equation (1.4) for each $u \in U$ there exists $y = S(u) \in Y$ which solves (1.4). The choice of the function spaces U, Y is difficult and depends on the nonlinearities. The objective (1.5) can be written as

$$\int_0^T \int_\Omega l(t, x, S(u)(t,x))\, dx\, dt + \int_0^T k(t, u(t))\, dt. \tag{1.7}$$

Both examples lead to *infinite dimensional* optimization problems.

1.2.2. Computation of the Gradient.

We consider the problem:

$$\text{Minimize } \phi(u) \text{ on } U. \tag{1.8}$$

For the unconstrained case let $U = L_m^\infty[0,T]$. Then the Fréchet-derivative of ϕ at the solution u^* is given by an element in $(L_m^\infty[0,T])^*$, the dual space. However, it is well known, that for $f, L \in C^1$ the gradient at u^* belongs to $L_m^2[0,T]$ although the function ϕ does not need to be Fréchet-differentiable in $L_m^2[0,T]$:

$$\nabla \phi(u)(t) = p^T(t) f_u(S(u)(t), u(t), t) + L_u(S(u)(t), u(t), t), \tag{1.9}$$

where p solves the adjoint equation

$$\begin{aligned} -\dot{p}(t) &= p^T(t) f_x(x(t), u(t), t) + L_x(x(t), u(t), t), \\ p(T) &= 0. \end{aligned} \tag{1.10}$$

Define the *Hamiltonian*

$$H(x, u, t) = L(x, u, t) + p(t)^T f(x, u, t). \tag{1.11}$$

Short notation for (1.9) and (1.10) is

$$\nabla \phi = H_u, \quad -\dot{p} = H_x, \quad p(T) = 0.$$

1.2.3. Gradient Method.

In summary, the gradient method for optimal control problems involves the following steps:

- Given u_k
- Compute solution p_k of adjoint equation (1.10)
- Compute gradient $s_k = - \nabla \phi(u_k)$
- Determine step length α_k
- Set $u_{k+1} = u_k + \alpha_k s_k$
- Return

Termination criterion: Stop if

$$\| \nabla \phi(u_k)\| \leq \epsilon$$

for small ϵ.

1.2.4. Optimal Controls with Simple Bounds.

We consider two examples where the optimal control can be characterized in such an explicit way that is renders the use of certain projection methods from optimization. In a constrained minimization problem the smallest function value of ϕ has to be found on a closed convex set U. The necessary optimality condition states that if $u_* \in U$ is optimal, then for all $u \in U$

$$\phi'(u_*)(u - u_*) \geq 0.$$

This can be rewritten for an optimal control problem in the case $m = n = 1$ as follows: Let $s_* = \nabla\phi(u_*)$, then for all $u \in U$

$$\int_0^T s_*(t)(u(t) - u_*(t))dt \geq 0. \tag{1.12}$$

If

$$U = \{u \in L^\infty[0,T] : -1 \leq u(t) \leq 1\}$$

then (1.12) can be written as

$$u_*(t) = -sgn(s_*(t)) \text{ a.e. on } [0,T]. \tag{1.13}$$

Here *sgn* is defined as

$$sgn(s) = \begin{cases} 1 & \text{if } s > 0 \\ -1 & \text{if } s < 0. \end{cases}$$

If a control u_* is determined by (1.13) it is called a *bang-bang control.* Intervals where $s_*(t) = 0$ are called *singular.*

If the objective function (1.2) contains an energy term

$$L(x,u,t) = \bar{L}(x,u,t) + \frac{1}{2}u^2$$

then (1.13) yields

$$u_*(t) = -sgn(s_*(t) + u_*(t)) \text{ a.e. on } [0,T].$$

Then

$$u_*(t) = -sat(s_*(t)) \text{ a.e. on } [0,T] \tag{1.14}$$

where

$$sat(s) = \begin{cases} 1 & \text{if } s > 1 \\ s & \text{if } s \in [-1,1] \\ -1 & \text{if } s < -1. \end{cases}$$

It should be noted that in both cases the *sat*– and *sgn*–functions are inexpensive to evaluate.

1.2.5. Gradient Projection Method.

Consider the problem to minimize $\phi(u)$ on a convex closed feasible set U. The direction of the negative gradient has to considered with care because

$$u_k - \alpha_k \nabla \phi(u_k) \in U$$

is not true in general. One way to solve this is to project the step into U.

- Given u_k
- Compute solution p_k of adjoint eqn (1.10)
- Compute gradient $s_k = -\nabla \phi(u_k)$
- Determine step length α_k
- Set $u_{k+1} = sat(u_k + \alpha_k s_k)$
- Return

Termination criterion: Stop if

$$\|u_{k+1} - u_k\| \le \epsilon$$

for small ϵ.
At each iterate u_k we have

$$\phi(sat(u_k - \alpha_k \nabla \phi(u_k))) < \phi(u_k), \tag{1.15}$$

if u_k is not a stationary point, i.e. $P(u_k - \nabla\phi(u_k)) = P(u_k)$.

1.2.6. Convergence Rates.

The convergence behavior of the gradient projection method is quite different from e.g. a Newton type method. One way to classify how fast the iterates converge to a solution of the problem is to distinguish between convergence rates.

DEFINITION 1.1. *A sequence $\{u_k\}_{\mathbb{N}}$ converges to u_* at a* linear *rate if there exists $c \in (0,1), k_0 \in \mathbb{N}$ with*

$$\|u_{k+1} - u_*\| \leq c\|u_k - u_*\| \quad k \geq k_0.$$

It converges at a superlinear *rate if*

$$\lim_{k \to \infty} \frac{\|u_{k+1} - u_*\|}{\|u_k - u_*\|} = 0.$$

It converges at a quadratic *rate if for some $d > 0$*

$$\|u_{k+1} - u_*\| \leq d\|u_k - u_*\|^2 \quad k \geq 0.$$

First order methods, in general, are at best linearly convergent. Improvement can be obtained by using higher order methods.

1.3. Second Order Methods.

1.3.1. Newton's Method.

If we take second order terms in an expansion of ϕ at u_k into account, we obtain Newton's method.

- Given u_k
- Solve $\nabla^2\phi(u_k)s_k = -\nabla\phi(u_k)$
- Set $u_{k+1} = u_k + s_k$
- Return

We terminate the algorithm on a small gradient. The advantage of this method is that one can show that Newton's method has a quadratic rate of convergence. This statement holds under certain assumptions among which one has to check that the starting point is sufficiently close to the solution. This requirement can be relaxed in theory and practice through a globalization strategy such as a step size rule or trust region method. In this article we will not focus on these aspects.

1.3.2. Computation of Second Derivative.

As it is in the case for any other optimization problem the disadvantage of Newton's method is the computation of the second derivative. We simplify to $m = 1, n = 1$. Let

$$< u, v >= \int_0^T u(t)v(t)dt.$$

The second derivative of ϕ is given by

$$(1.16)\quad \begin{aligned} < w, \nabla^2\phi(u)v >= & < \xi(w), H_{xx}(x,u,\cdot)\xi(v) > + < w, H_{ux}(x,u,\cdot)\xi(v) > \\ & + < \xi(w), H_{xu}(x,u,\cdot)v > + < w, H_{uu}(x,u,\cdot)v > \end{aligned}$$

where x solves (1.3), p solves (1.10) and $\xi(w)$ solves

$$(1.17)\quad \begin{aligned} \dot{\xi}(t) &= \xi(t)f_x(x(t),u(t),t) + f_u(x(t),u(t),t)w(t), \\ \xi(0) &= 0. \end{aligned}$$

The rigorous mathematical justification for the previous lines is a rather tedious task and requires a careful choice of the function spaces.

1.3.3. Quasi Newton Methods.

In the context of quasi-Newton methods we replace the Hessian $\nabla^2\phi(u)$ by an approximation B.

- Given u_k and B_k
- Solve $B_k s_k = - \nabla\, \phi(u_k)$
- Set $u_{k+1} = u_k + s_k$
- Update B_k to B_{k+1}
- Return

The advantage is that under proper choice of B_k one can show and observe numerically a superlinear convergence rate if u_0 close to u_* and B_0 close to $\nabla\phi(u_*)$. The update considered to be the most successful is the BFGS update (Broyden, Fletcher, Goldfarb, Shanno)

Set $y_k = \nabla\phi(u_{k+1}) - \nabla\phi(u_k)$ and

$$(1.18)\quad B_{k+1} = B_k + \frac{1}{< s_k, y_k >} < y_k, \cdot > y_k - \frac{1}{< s_k, B_k s_k >} < B_k s_k, \cdot > B_k s_k.$$

REFERENCES

[1] V. F. Demyanov and A. M. Rubinov. *Approximate Methods in Optimization Theory*. Elsevier, New York, 1970.

[2] J. E. Dennis and R. B. Schnabel. *Numerical Methods for Unconstrained Optimization and Nonlinear Equations*. Prentice-Hall, Englewood Cliffs, N.J, 1983.

[3] W. A. Gruver and E. Sachs. *Algorithmic Methods In Optimal Control.* Pitman, London, 1980.

[4] A. Kirsch, W. Warth, and J. Werner. *Notwendige Optimalitätsbedingungen und ihre Anwendungen*. Lect. Notes in Economics and Mathematical Systems. Springer, 1978.

[5] J. L. Lions. *Optimal Control of Systems Governed by Partial Differential Equations*. Springer, Berlin, 1971.

[6] K. C. P. Machielsen. *Numerical solution of optimal control problems with state constraints by sequential quadratic programming in function space.* CWI Tract. Centrum voor Wiskunde en Informatica, Amsterdam, 1988.

[7] E. Polak. *Computational Methods in Optimization.* Academic Press, New York, 1971.

[8] F. Tröltzsch. *Optimality Conditions for Parabolic Control Problems and Applications*. Teubner, Leipzig, 1984.

2. Gradient Projection Methods.

2.1. Introduction.

2.1.1. Optimization Problem.

Let H be a Hilbert space and $U \subset H$ a bounded, closed, convex subset of H. Furthermore let

$$\phi : H \to \mathbb{R}.$$

The goal is to find $u_* \in U$, such that for all $u \in U$

$$\phi(u_*) \leq \phi(u).$$

ASSUMPTION 2.1. *U has a simple structure, i.e. the calculation of $P(h)$ must be inexpensive.*

The projection $P : H \to U$ is defined by

$$\|P(h) - h\| \leq \|u - h\| \text{ for all } u \in U.$$

2.1.2. Algorithm.

Given $u_k \in U$
- Compute $\nabla\phi(u_k)$
- Determine a step length α_k
- If $\|u_k - P(u_k - \alpha_k \nabla \phi(u_k))\| \leq \epsilon$, stop
- Set $u_{k+1} = P(u_k - \alpha_k \nabla \phi(u_k))$

For termination criterion see Theorem 2.1.

2.1.3. Step Length Rules.

We present two step length rules one of which is implementable and widely used. Recent developments allow to replace the step length rule by a trust region method also for the projection method. This approach, however has not yet been tested in the context with optimal control problems.

Exact step length rule

$$\phi(u_{k+1}) = \min_{\alpha>0} \phi(P(u_k - \alpha \nabla \phi(u_k)))$$

The disadvantage is that it requires an iterative method to determine α_k and the computation of the projection at each iteration for the determination of α_k.

Armijo step length rule
Select $\gamma > 0, \quad \beta, \delta \in (0,1)$. Set $\alpha_k = \gamma$, if

$$\phi(u_k) - \phi(P(u_k - \gamma \nabla \phi(u_k))) \geq \delta < \nabla\phi(u_k), u_k - P(u_k - \gamma \nabla \phi(u_k)) > .$$

Otherwise determine α_k such that $\alpha_k = \beta^{m_k}\gamma$ where m_k is the smallest number $m \in \mathbb{N}$ with

$$\phi(u_k) - \phi(P(u_k - \beta^m\gamma \nabla \phi(u_k))) \geq \delta < \nabla\phi(u_k), u_k - P(u_k - \beta^m\gamma \nabla \phi(u_k)) > .$$

2.2. Convergence Analysis.

2.2.1. Convergence Theorems.

The following theorems are two examples from standard results in the convergence analysis of gradient projection methods.

THEOREM 2.1. *Let ϕ have a Lipschitz continuous Fréchet-derivative. If $u_k \in U$ with*

$$u_k - P(u_k - \alpha \nabla \phi(u_k)) = 0$$

for some $\alpha > 0$, u_k is called a stationary point, i.e. u_k is a solution for ϕ convex.
If u_k is not a stationary point, then there are step sizes α_k, such that

$$\phi(u_{k+1}) < \phi(u_k).$$

THEOREM 2.2. *Let ϕ be Fréchet-differentiable with Lipschitz continuous derivative. Then we have*

$$\lim_{k\to\infty} \phi(u_k) = \rho, \quad \lim_{k\to\infty} < \nabla\phi(u_k), u_k - u_{k+1} >= 0$$

if the Armijo step size rule is used.

2.2.2. Example.

To illustrate further results we consider a simple example. The feasible set of controls U is defined as

$$U = \{u \in L^\infty[0,T] : a^-(t) \leq u(t) \leq a^+(t) \quad \text{a.e. on} \quad [0,T]\}$$

where $a^-, a^+ \in L^\infty[0,T]$ are fixed functions such that $a^-(t) < a^+(t)$ a.e. $[0,T]$.
We choose

$$L(x,u,t) = \frac{1}{2}x^2, \quad f(x,u,t) = bu,$$

$$x(0) = 1, \quad T = 1, \quad a^+(\cdot) = -a^-(\cdot) = 0.5,$$

and three choices for b

$$\begin{array}{ll} \text{Example 1} & b(t) = (t - 0.5)^3, \\ \text{Example 2} & b(t) = (t - 0.5)^3 + 0.03, \\ \text{Example 3} & b(t) = \cos(3t). \end{array}$$

We discretize this optimal control problem with

$$\min \sum_{i=0}^{N-1} L^i(x^i, u^i), \tag{2.1}$$

where x depends on u through a discretization of (1.3) such as

$$x^{i+1} = x^i + f^i(x^i, u^i), \quad i = 0, 1, ..., N-1, \tag{2.2}$$

and

$$L^i, f^i \in C^1(\mathbb{R}_2), \quad i = 0, 1, ..., N-1, \tag{2.3}$$

are functions arising from the discretization process. In particular, we choose

$$N = D_N, \quad L^i(x, u) = \frac{1}{2D_N} x^2, \quad f^i(x, u) = \frac{1}{D_N} b(\frac{1}{D_N}) u.$$

The set of active indices is defined as

$$I(u) = \{i : u^i = a_i^- = -0.5 \text{ or } u^i = a_i^+ = 0.5\}.$$

The number of steps needed to compute the solution is identical with the number of steps needed to identify $D_N - 1$ active indices

$$k_{act} = \min \{k \in \mathbb{N} \ : \ |I(u_k)| = D_N - 1\}.$$

The results are shown in Table 2.1.

TABLE 2.1
Finite Identification of Active Indices

D_N	k_{act} for ex.1	k_{act} for ex.2	k_{act} for ex.3
4	163	56	6
8	779	53	7
16	5032	76	14
32	36472	127	24
64	277582	231	56

The results indicate that there is large discrepancy in the results of the three examples. Our goal is to explain why e.g. Example 1 requires such a large number of steps whereas only a slight perturbation such as Example 2 yields quite different results. We want to emphasize that our observations can only be explained from the underlying infinite dimensional problem.

There are two approaches for an analysis of the convergence of the gradient projection method:

- Convergence Rate
- Identification of Active Indices.

2.3. Convergence Rates.

2.3.1. Growth Function.

DEFINITION 2.1. *The function* ϕ *satisfies the growth condition of order* ν *at* u_* *if there are* $c, \sigma_1 > 0$ *and* $\nu > 0$ *with*

$$\gamma(\sigma) \geq c\sigma^\nu \text{ for all } \sigma \in (0, \sigma_1)$$

where

$$\gamma(\sigma) = \inf_{u \in U, \|u - u_*\| \geq \sigma} \phi'(u_*)(u - u_*).$$

Sometimes

$$\gamma(\sigma) = \inf_{u \in U, \|u - u_*\| \geq \sigma} \phi(u) - \phi(u_*).$$

is considered instead. The following theorem gives an example when the growth condition is satisfied.

THEOREM 2.3. *Let* $\phi : U \to \mathbb{R}$ *be Fréchet-differentiable in a* L^1*-neighborhood of the solution* $u_* \in U$ *with* $s \in C[0,T]$

$$\phi'(u_*)(u - u_*) = \int_0^T s(t)(u(t) - u_*(t))dt.$$

If s *has no zero in* $[0,T]$, *then the growth condition of order 1 holds. If* s *has finitely many zeroes on* $[0,T]$ *and all of them are simple, then the growth condition of order 2 holds.*

Further extensions include:

- s in $L^\infty[0,T]$
- other orders of growth.

It is easy to check (see Section 2.4.4) that the growth condition of order 2 is true for Examples 2 and 3 and false for Example 1.

2.3.2. Convergence Rates.

The next theorem tells us how the growth condition is related to the rate of convergence.

THEOREM 2.4. *If*

$$\gamma(\sigma) > 0 \quad \text{for} \quad \sigma > 0$$

holds, then

$$\phi(u_k) - \phi(u_*) = O(\frac{1}{k}).$$

If the growth condition is true, then for $\lambda \in (0,1)$, *k sufficiently large*

$$\begin{array}{rcll} \phi(u_k) - \phi(u_*) & = & O(k^{\nu/(2-\nu)}) & \text{if} \quad \nu > 2 \\ \phi(u_k) - \phi(u_*) & = & O(\lambda^k) & \text{if} \quad 2 \geq \nu > 1 \\ \phi(u_k) - \phi(u_*) & = & 0 & \text{if} \quad \nu = 1. \end{array}$$

This gives us an indication why Example 1 exhibits slow convergence (no growth condition in this case). The function s, as we shall see in Section 2.4.4 is for these three examples given by bp and has the same zeroes as b (for Example 1 a zero with multiplicity three). This theorem, however, holds only in Hilbert space. All of our estimates for the growth condition are in $L^1[0,T]$. The inequality

$$\|u - u_*\|_1 \leq c\|u - u_*\|_2$$

holds in general but is of no use here. Since all u lie in U, a pointwise bounded set, it is easy to see that

$$\|u - u_*\|_2^2 \leq c\|u - u_*\|_1.$$

Therefore, in an L^2-setting we have a growth condition of order 4 and the theorem yields a sublinear rate of convergence.

Next we consider the problem how Theorem 2.4 can be applied to a finite dimensional discretized control problem. For finite dimensional problems it is not possible to distinguish in such a refined way between various growth conditions. Also, in the case of our example, the growth condition does not hold for the discretized problem (see next Section).

2.3.3. Convergence Rates under Perturbation of Data.

Consider a control problem of the type (1.3) and (1.2) with a discretization (2.1) and (2.2). Then the objective function is a nonlinear function ϕ_N. In the finite dimensional case, however, we have

LEMMA 2.5. *In the finite dimensional case the growth condition*

$$\inf_{u \in U_N, \|u - u_*^N\| \geq \sigma} \phi_N'(u_*^N)^T(u - u_*^N) \geq c\sigma^\nu$$

is satisfied for all $\nu > 0$ *if and only if no component of the vector* $\phi_N'(u_*^N)$ *vanishes.*

This implies that even if the growth condition holds for the infinite dimensional problem it might *not be true* for any of the finite dimensional discretized problems.

We can utilize the results for (P) if we interprete (P_N) as a perturbation of (P). We have

$$\gamma_N(\sigma) \geq c_1\sigma^2 - \varepsilon_N\,, \quad \sigma \in (0, \sigma_1)$$

$$\gamma_N(\sigma) \geq c_2\sigma^2, \qquad \sigma \in (\sigma_2, \sigma_1)$$

which implies linear convergence outside a σ_2-ball, i.e.

$$\phi(u_{k+1}^N) - \phi(u_*^N) \leq \lambda(\phi(u_k^N) - \phi(u_*^N))$$

for all k with

$$\|u_k^N - u_*^N\| > \sigma_2.$$

This shows that the estimates for the rate of convergence in the infinite dimensional case carry over to the finite dimensional discrete version of the optimal control problem up to a small ball around the solution. Therefore, the difference in magnitude of the iteration numbers in Table 2.1 can explained with this perturbation result. Next we focus on finding a reason why the number of iterations increases when the discretization is refined.

2.4. Identification of Active Indices.

Since in the case of bang-bang controls all indices of the constraints are active, one might ask the following question: When are all the active indices known ? This is an important question even for nonlinearly constrained optimization problems because it allows to switch to an equality constrained optimization problem which does not need any active set strategy.

2.4.1. Finite Identification.

ASSUMPTION 2.2 (NONDEGENERACY CONDITION). *Let u_* be a solution of*

(2.4) $$\textit{Minimize } \phi(u) \textit{ s. t. } u \in U$$

$$U = \{u \in \mathbb{R}_n : c_i(u) \leq 0, i = 1, \ldots t\}.$$

The set of active indices is given by

$$I(u) = \{i \in \{1, \ldots, t\} : c_i(u) = 0\}.$$

The nondegeneracy condition is satisfied at u_*, if

$$\nabla c_i(u_*), i \in I(u_*)$$

are linearly independent and

$$\nabla\phi(u_*) = \sum_{i\in I(u_*)} \lambda_i \nabla c_i(u_*)$$

implies

$$\lambda_i > 0 \text{ for all } i \in I(u_*).$$

THEOREM 2.6. *If*

$$u_k \in U \text{ with } \lim_{k\to\infty} u_k = u_*$$

is an infinite sequence of iteration points, which is produced by the gradient projection method such that the step sizes stay bounded away from 0, then there is an index k_0 with

$$I(u_k) = I(u_*) \text{ for all } k \geq k_0.$$

As a consequence of this theorem (2.4) can be replaced after finitely many steps by

$$\min \phi(u) \text{ with } c_i(u) = 0, \; i \in I(u_k).$$

2.4.2. Identification for Infinite Dimensional Problem.

Does the finite identification of active indices also hold for infinite dimensional problems, e.g. when U is as follows

$$U = \{u \in L^\infty[0,T] : |u(t)| \leq 1 \text{ a.e. } [0,T]\} \quad ?$$

Obviously, there is *no* finite identification of the set of active indices. Let us consider Table 2.1 again: The fact that the magnitude of the numbers of iterations needed to find the solution differs from one example to another can be explained by results on the convergence rate. What remains unsatisfactory is the fact that the number of iterations needed until the active indices are identified is *dependent on the dimension* of the problem.

We reformulate the notion of an *active* index for problems under perturbation of data.

LEMMA 2.7. *Let $s_k, s_*, u_k, u_* \in L^\infty[0,T]$ and $\alpha_k \in \mathbb{R}$ be given such that*

$$\alpha_k \geq \alpha > 0 \text{ for all } k \in \mathbb{N} \tag{2.5}$$

and

$$u_{k+1}(t) = sat(u_k(t) - \alpha_k s_k(t)) \text{ a.e. on } [0,T]. \tag{2.6}$$

Suppose that

$$\lim_{k\to\infty} \|s_k - s_*\|_{L^\infty[0,T]} = 0,$$

and for all $\varepsilon > 0$

$$|u_*(t)| = 1 \ a.e. \ on \ J_\varepsilon = \{t \in [0,T] : |s_*(t)| \geq \varepsilon\}. \tag{2.7}$$

Then for all $\varepsilon > 0$ *there exists* $k_\varepsilon \in \mathbb{N}$ *with*

$$u_k(t) = u_*(t) \ for \ all \ k \geq k_\varepsilon, \ a.e. \ t \in J_\varepsilon. \tag{2.8}$$

The size of the interval J_ε where finite identification occurs depends on the shape of the graph of s_*. If s_* has only finitely many zeroes in $[0,T]$, then the length $m(J_\varepsilon)$ of J_ε converges to T

$$\lim_{\varepsilon \to 0} m(J_\varepsilon) = T.$$

2.4.3. Mesh Independence of Finite Identification .

We choose as a termination criterion the usual test for stationarity:

$$k_N(\varepsilon) = \min\{k \ : \ \|u_k^N - u_{k-1}^N\|_H < \varepsilon\}$$

$$k(\varepsilon) = \min\{k \ : \ \|u_k - u_{k-1}\|_H < \varepsilon\}.$$

THEOREM 2.8. *Let* $u_k^N \in H$ *and assume that* $u_k^N \to u_k$ *in* H *for all* k. *Then for all* $\varepsilon, \delta > 0$ *there is* $\hat{N}_{\varepsilon,\delta}$ *such that if* $N \geq \hat{N}_{\varepsilon,\delta}$ *then*

$$k(\varepsilon + \delta) \leq k_N(\varepsilon) \leq k(\varepsilon).$$

The result of this theorem can be interpreted in such a way that the number of steps needed to terminate for the discretized problem can be estimated from above and below by the corresponding termination index for the infinite dimensional problem. This result claims that the termination criterion proposed is independent on the dimension of the problem if the discretization is fine enough. We check this statement for an example in the next section.

2.4.4. Numerical Results, ODE.

Let us reconsider our previously discussed example. Minimize

$$\frac{1}{2}\int_0^1 x(t)^2 dt$$

subject to

$$\dot{x}(t) = b(t)u(t), \quad x(0) = 1, \quad t \in [0,1]$$

and

$$u \in U = \{u \in L^2[0,1] \ : |u(t)| \leq 0.5 \text{ a.e. in } [0,1]\}.$$

The optimal control satisfies the following bang-bang-principle

$$u_*(t) = -0.5 \text{ sgn } p_*(t)b(t) \text{ a.e. in } [0,1]$$

where p_* solves the adjoint equation

$$-\dot{p}(t) = x_*(t), \quad p(1) = 0, \quad t \in [0,1]. \tag{2.9}$$

Since p_* is positive on $[0,1)$ for all our choices of b, the function b determines the type of growth for the switching function $s_* = bp_*/2$. For the discrete case

$$U_N = \{u \in L^2[0,1] \ : u = \sum_{i=0}^{D_N-1} u^i \chi_i, \quad |u^i| \leq 0.5 \ i = 0,..,D_N - 1\},$$

where χ_i denotes the characteristic function on $(\frac{i}{D_N}, \frac{i+1}{D_N}]$. Then

$$\phi_N(u_N) = \frac{1}{2}\int_0^1 (\sum_{i=0}^{D_N-1} x^i\chi_i)^2 dt = \sum_{i=0}^{D_N-1} \frac{1}{2D_N}(x^i)^2$$

where x^i is determined by Euler's scheme

$$x^{i+1} = x^i + \frac{1}{D_N} b_i u^i, \quad i = 0,...,D_N - 1, \quad x_0 = 1,$$

$$\text{with } b_i = b(\frac{i}{D_N}), \quad i = 0,...,D_N - 1.$$

The gradient of ϕ_N can be computed by

$$s_N = \sum_{i=0}^{D_N-1} s_i \chi_i,$$

where

$$s_i = 0.5 b_i p_i, \quad i = 0,..,D_N - 1$$

and p_i is the solution of difference equation (discretization of (2.9)).

Example 1	$b(t) = (t - 0.5)^3$
Example 2	$b(t) = (t - 0.5)^3 + 0.03$
Example 3	$b(t) = \cos(3t)$
Example 4	$b(t) = (t - 0.5)\|t - 0.5\|$

Terminate the algorithm when a certain tolerance for the discrete L^2-norm of the steps or the projected gradient is reached:

$$k_N(10^{-4}) = \min \{k \in \mathbb{N} : \|u_k^N - u_{k-1}^N\| \leq 10^{-4}\}.$$

TABLE 2.2
Termination Based on Small Steps

D_N	k_N for ex.1	k_N for ex.2	k_N for ex.3
4	6	8	24
8	13	30	8
16	12	31	14
32	12	31	12
64	12	31	11
128	12	31	11
256	12	31	11

Another termination criterion which we used is based on the length of the interval corresponding to all active indices. Termination occurs when a certain fixed percentage of the length of the total interval is reached:

$$k^p = \min \{k \in \mathbb{N} : |I(u_k)|/D_N \geq p/100\}.$$

TABLE 2.3
Termination Based on Relative Length of Interval of Active Indices

D_N	k^{90} for ex.1	k^{90} for ex.2	k^{90} for ex.3	k^{80} for ex.1
4	164	59	24	164
8	780	56	21	779
16	5032	76	14	734
32	4879	74	13	1065
64	9104	85	16	1048
128	9032	85	16	1040
256	8617	84	16	1069

The results show that the termination of the algorithm is no longer dependent on the dimension if the discretization is fine enough.

2.4.5. Numerical Results, PDE.

Let $y(t,x)$ denote the temperature at time $t \in [0,T]$ and at location $x \in [0,1]$. The following differential equation describes a heat conduction problem with memory. The boundary control problem for a pseudoparabolic differential equation is given by

$$\begin{aligned}
y_t(t,x) &= y_{xx}(t,x) + \varepsilon y_{xtx}(t,x), && x \in (0,1), t \in (0,T), \\
y(0,x) &= 0, && x \in (0,1), \\
y_x(t,1) &= 0, && t \in (0,T), \\
-y_x(t,0) &= \varepsilon y_{xt}(t,0) + u(t), && t \in (0,T].
\end{aligned}$$

$$\phi(u) = \int_0^1 \bar{\phi}(y(T,x) - z(x))dx + \frac{\alpha}{2}\int_0^T u(t)^2 dt$$

where

$$\bar{\phi} : \mathbb{R} \to \mathbb{R}$$

is a given continuously differentiable function and $\alpha > 0, \varepsilon > 0, z$ are given constants. The set of feasible controls is given by

$$U = \{u \in L^2[0,T] : |u(t)| \leq 1 \quad \text{a.e. in } [0,T]\}.$$

We define for $u \in L^2[0,T]$ the solution $y(T,\cdot)$

$$S(u)\ (x) := y(T,x) = \sum_{j=0}^{\infty} a_j^2 b_j c_j(x) \int_0^T e^{-\lambda_j(T-s)} u(s)ds$$

where for $j = 0, 1, 2, \ldots$

$$\begin{array}{rcl} c_j(x) & = & \cos j\pi x, \qquad x \in [0,1] \\ \mu_j & = & j^2\pi^2, \\ b_j & = & 1/(1+\varepsilon\mu_j), \\ \lambda_j & = & \mu_j/(1+\varepsilon\mu_j), \\ a_j & = & \sqrt{2}, \quad j > 0, \quad a_0 = 1. \end{array}$$

If we use the adjoint operator S^* of $S : L^2[0,T] \to C[0,1]$, the optimal control u_* satisfies

$$u_*(t) = -sgn\ (S^*(\bar{\phi}'(S(u_*)(\cdot) - z(\cdot)))(t) + \alpha u_*(t)), t \in [0,T],$$

or equivalently

$$u_*(t) = -sat\ (\frac{1}{\alpha} S^*(\bar{\phi}'(S(u_*)(\cdot) - z(\cdot)))(t)).$$

Approximate S by

$$S_N(u)\ (x) = \sum_{j=0}^{j_N} a_j^2 b_j c_j(x) \int_0^T e^{-\lambda_j(T-s)} u(s)ds$$

and replace U by U_N as in control problem above. Integration along the space variable x from 0 to 1 is replaced by a quadrature rule. Hence ϕ_N is

$$\phi_N(v) = \frac{1}{l_N+1}\sum_{l=0}^{l_N} \bar{\phi}(S_N(\sum_{i=0}^{D_N-1} v_i\chi_i)\ (x_l) - z(x_l)) + \frac{\alpha}{2D_N}\sum_{i=0}^{D_N-1} v_i^2.$$

We set

$$T = 0.5,\ \varepsilon = 10^{-4},\ \alpha = 0.1,\ j_N = 50,$$

$$l_N = 100,\ \bar{\phi}(y) = y^4,\ z(x) = 3(x - 0.5).$$

The termination criterion is set to be

$$iter = \min\ \{k : \|u_{k+1}^N - u_k^N\| \leq 10^{-6}\}.$$

This termination criterion is reasonable because u_* is not a bang-bang control and has a singular arc. The interval $[l_{act}, r_{act}]$ represents the singular part, i.e. the control has values inside the interval $(-1,1)$.

TABLE 2.4
Pseudoparabolic Control Problem

D_N	$\lvert I(u_{iter})\rvert$	l_{act}	r_{act}	iter
10	9 (90%)	0.2000	0.2500	32
25	21 (84%)	0.2000	0.2800	31
50	44 (88%)	0.2100	0.2700	28
75	65 (87%)	0.2067	0.2733	29
100	86 (86%)	0.2050	0.2750	30
150	130 (87%)	0.2067	0.2733	29
200	173 (87%)	0.2050	0.2725	29

REFERENCES

[1] D. B. Bertsekas. On the Goldstein-Levitin-Polyak gradient projection method. *IEEE Trans. Autom. Control*, (2):174–184, 1976.

[2] D. B. Bertsekas. Projected Newton methods for optimization problems with simple constraints. *SIAM J. Control and Optimization*, 20:221–246, 1982.

[3] J. V. Burke and J. Moré. On the identification of active constraints. Technical Report ANL/MCS-TM-82, Argonne National Laboratory, Math. and Comp. Science Div. Report, 1986.

[4] P. H. Calamai and J. Moré. Projected gradient methods for linearly constrained problems. *Math. Programming*, 39:93–116, 1987.

[5] J. C. Dunn. Global and asymptotic convergence rate estimates for a class of projected gradient processes. *SIAM J. Control and Optimization*, 19:368–400, 1981.

[6] J. C. Dunn. On the convergence of projected gradient processes to singular critical points. *J. Optim. Th. Appl.*, 55(2):203–215, 1987.

[7] J. C. Dunn and E. W. Sachs. The effect of perturbations on the convergence rates of optimization algorithms. *Applied Math. and Optimization*, 10:143–147, 1983.

[8] A. A. Goldstein. On Newton's method. *Numer. Math.*, 7:391–393, 1965.

[9] C. T. Kelley and E. W. Sachs. Mesh independence of the gradient projection method for optimal control problems. *SIAM J. Control and Optimization*, to appear.

[10] E. S. Levitin and B. T. Polyak. Constrained optimization methods. *USSR Comput. Math. Phys.*, 6:1–50, 1966.

[11] E. W. Sachs. Convergence of algorithms for perturbed optimization problems. *Annals of Operations Research*, 27:311–342, 1990.

3. Quasi-Newton Methods and Optimal Control Problems.

3.1. Broyden's Method.

3.1.1. Broyden's Update.

Consider $F : I\!R_N \to I\!R_N$ and find $u_* \in I\!R_N$ such that

$$F(u_*) = 0.$$

For minimization problems set for example $F = \nabla\phi$. We review some of the results for quasi-Newton methods which avoid the calculation of the Jacobian of F. No symmetry of F' needed is at this point.
Given u_k, B_k, solve

$$B_k s_k = -F(u_k)$$

and set

$$u_{k+1} = u_k + s_k, \quad y_k = F(u_{k+1}) - F(u_k).$$

Question: What is a good choice for B_{k+1} ?

- Secant equation $B_{k+1}s_k = y_k$ should hold.
- B_{k+1} should retain information from B_k.

Let B_{k+1} be a solution of

(3.1) $$\min_{B \in I\!R_{N\times N}, y_k = Bs_k} \|B - B_k\|$$

LEMMA 3.1. *Problem (3.1) is solved by the Broyden update*

$$B_{k+1} = B_k + \frac{(y_k - B_k s_k)s_k^T}{s_k^T s_k}$$

if we choose the l^2- or Frobenius-norm.

The proof is by verification. If we require in (3.1) also that B_k and B are symmetric we obtain the PSB-update (Powell-Symmetric-Broyden):

$$B_{k+1} = B_k + \frac{(y_k - B_k s_k)s_k^T + s_k(y_k - B_k s_k)^T}{s_k^T s_k} - s_k^T(y_k - B_k s_k)\frac{s_k s_k^T}{(s_k^T s_k)^2}.$$

Another update in this context is the SR-1 update (Symmetric Rank-1):

$$B_{k+1} = B_k + \frac{(y_k - B_k s_k)(y_k - B_k s_k)^T}{(y_k - B_k s_k)^T s_k}.$$

3.1.2. Linear Convergence.

A major tool in the convergence proof of quasi-Newton methods is the *Bounded Deterioration Property (3.2)* .

THEOREM 3.2. *Let $D \subset I\!R_N$ be open and convex, $u, u_+ \in D, F \in C^1(I\!R_N)$, $F' \in Lip_\gamma(D)$, i.e.*

$$\|F'(x) - F'(y)\| \leq \gamma \|x - y\| \quad \forall\, x, y \in D.$$

Let $B \in I\!R_{N\times N}$ be given and B_+ the Broyden–Update

$$B_+ = B + \frac{(y - Bs)s^T}{\|s\|^2}$$

with $s := u_+ - u, \quad y := F(u_+) - F(u)$.
Then

$$\|B_+ - F'(u_*)\|_2 \leq \|B - F'(u_*)\|_2 + \frac{\gamma}{2}(\|u_+ - u_*\|_2 + \|u - u_*\|_2). \tag{3.2}$$

In the following we often replace the subscripts k and $k+1$ by no subscript and the subscript $+$, respectively.
If we set $E_k = B_k - F'(u_*)$ we obtain for the $\|\cdot\|_2$ -norm

$$\begin{aligned} \|E_+\| &= \|E + \frac{(y - Bs)s^T}{s^T s}\| = \|E(I - \frac{ss^T}{s^T s}) + \frac{(y - F'(u_*)s)s^T}{s^T s}\| \\ &\leq \|E\|\|I - \frac{ss^T}{s^T s}\| + \frac{1}{\|s\|^2}\|(y - F'(u_*)s)s^T\| \end{aligned}$$

and therefore

$$\|E_+\| \leq \|E\| + \frac{\gamma}{2}(\|u_+ - u_*\| + \|u - u_*\|). \tag{3.3}$$

The bounded deterioration property itself is sufficient for a locally linear rate of convergence. In particular it is true for Broyden's method.

THEOREM 3.3. *Let $D \subset I\!R_N$ be open and convex, $u, u_+ \in D, F \in C^1(I\!R_N)$, $F' \in Lip_\gamma(D)$. Let $u_* \in D$ with $F(u_*) = 0$ and let $[F'(u_*)]^{-1}$ exist. Then there exist $\epsilon > 0$ and $\delta > 0$, such that for all $u_0 \in I\!R_N$, $B_0 \in I\!R_{N\times N}$ with*

$$\|u_0 - u_*\| \leq \epsilon \quad \text{and} \quad \|B_0 - F'(u_0)\| \leq \delta$$

the iterates produced by $u_{k+1} = u_k + B_k^{-1}F(u_k)$ converge linearly to u_* and B_k remain bounded, if they satisfy the Bounded–Deterioration–Property.

Proof is carried out by induction. In all the previous proof we can work with the $\|\cdot\|_2$- norm.

3.1.3. Superlinear Convergence (in Finite Dimensions).

Let $\|\cdot\|_F$ denote the Frobenius norm. Then a simple calculation shows that for Broyden's method

$$\|E(I - \frac{ss^T}{s^Ts})\|_F = (\|E\|_F^2 - \frac{\|Es\|_2^2}{\|s\|_2^2})^{\frac{1}{2}} \leq \|E\|_F - \frac{1}{2\|E\|_F}\left(\frac{\|Es\|_2}{\|s\|_2}\right)^2 \tag{3.4}$$

for $E \in \mathbb{R}_{N\times N}, s \in \mathbb{R}_N$. This inequality allows for a much more refined estimate than (3.3).

THEOREM 3.4. *For Broyden's method we assume that for $E_k := B_k - F'(u_*)$, $e_k := u_k - u_*$ the following is true*

$$\|E_k\| \leq M,\ M > 0\ \forall\ k \in \mathbb{N},\quad \sum_{k=1}^{\infty} \|e_k\| < \infty. \tag{3.5}$$

Furthermore let $F' \in Lip_\gamma$. Then the Dennis–Moré condition holds:

$$\lim_{k\to\infty} \frac{\|(B_k - F'(u_*))s_k\|_2}{\|s_k\|_2} = 0.$$

The key inequality (3.3) can be improved with the Frobenius norm and (3.4):

$$\begin{aligned}
\|E_{k+1}\|_F &\leq \|E_k(I - \frac{s_ks_k^T}{s_k^Ts_k})\|_F + \frac{\|(y_k - F'(u_*)s_k)s_k^T\|_F}{s_k^Ts_k} \\
&\leq \|E_k\|_F - \frac{1}{2\|E_k\|_F}\left(\frac{\|E_ks_k\|_2}{\|s_k\|_2}\right)^2 + \frac{\|y_k - F'(u_*)s_k\|_2}{\|s_k\|_2}
\end{aligned}$$

The next theorem shows that the Dennis-Moré condition implies superlinear convergence for any quasi Newton method, in particular Broyden's method.

THEOREM 3.5. *Let the assumptions of Theorem 3.3 hold. Let $B_k \in \mathbb{R}_{N\times N}$ be a sequence of invertible matrices with:*

$$u_{k+1} = u_k + s_k,\quad B_ks_k = -F(u_k),\quad \lim_{k\to\infty} u_k = u_*.$$

If the Dennis–Moré condition holds, then the u_k converge superlinearly to u_* with $F(u_*) = 0$.

The Dennis-Moré condition is also necessary for superlinear convergence. We should point out that in the proofs of the theorems in this section we rely on estimates in the Frobenius norm.

3.2. BFGS-method for Control Problems.

3.2.1. Algorithm.

Consider optimal control problem as unconstrained minimization problem (1.6).

$$\text{Minimize } \phi(u) = \int_0^T L(S(u)(t), u(t), t)\, dt$$

- Given u_k, B_k
- Compute solution of $B_k s_k = -\nabla \phi(u_k)$
- Set $u_{k+1} = u_k + s_k$
- Compute $\nabla\phi(u_{k+1})$
- Set $y_k = \nabla\phi(u_{k+1}) - \nabla\phi(u_k)$
- Update B_k to B_{k+1} by a quasi-Newton update
- Return

The cost per iteration is the solution of a linear system and a gradient evaluation. Since B_k should be symmetric and positive definite we use a BFGS update which preserves these properties:

$$B_{k+1} = B_k + \frac{1}{< s_k, y_k >} < y_k, \cdot > y_k - \frac{1}{< s_k, B_k s_k >} < B_k s_k, \cdot > B_k s_k$$

Like Broyden's method also the BFGS method gives superlinear convergence in $\mathbb{R}_N$.

3.2.2. Example.

Consider

$$L(x,u,t) = e^{-t}\frac{1}{2}((x-1.5)^2 + (u-3)^2), \quad f(x,u,t) = u - t^2, \quad T = 0.3$$

Discretization: 4-th order (difference scheme, quadrature rule) $n = 20$. Initial data

$$x(0) = 1, \quad u_0 = 0, \quad (B_0 v)(t) = e^{-\rho t} v(t)$$

We tabulate $g_k = \| \nabla \phi(u_k)\|$ in the next tables.

TABLE 3.1
BFGS Method

k	g_k	g_k/g_{k-1}
0	0.5437e-01	
1	0.1149e-03	0.2113e-02
2	0.6442e-06	0.5608e-02
3	0.2525e-09	0.3919e-03

Perturb problem slightly $\bar{L}(x,u,t) = L(x,u,t) + 0.025u^4$.

TABLE 3.2
BFGS Method

k	g_k	g_k/g_{k-1}
0	0.2453e+01	
1	0.5098e+00	0.2078e+00
2	0.1452e+00	0.2849e+00
3	0.1269e+00	0.8922e+00
4	0.1600e-01	0.1235e+00
5	0.5337e-02	0.3335e+00
6	0.3750e-02	0.6689e+00
7	0.1021e-02	0.2860e+00
8	0.2548e-03	0.2495e+00
9	0.2142e-03	0.8409e+00
10	0.3267e-04	0.1525e+00
11	0.1325e-04	0.4055e+00
12	0.8974e-05	0.6772e+00
13	0.1794e-05	0.1999e+00
14	0.6576e-05	0.3666e+00
15	0.4980e-06	0.7573e+00
16	0.6896e-07	0.1385e+00

The numerical results show a substantially different convergence behavior for the two problems. Since for both examples the finite dimensional theory predicts a superlinear rate of convergence, we analyze in the next section the convergence for the underlying infinite dimensional problem. We will then be able to explain better the results of Tables 3.1 and 3.2.

3.3. Quasi Newton Methods in Infinite Dimensions.

3.3.1. Broyden's Method in Hilbert space.

As an example for quasi Newton methods we use again Broyden's method . Let X be a Hilbert space with inner product $< \cdot, \cdot >$, Y a Banach space, and $F : X \to Y$ a Fréchet-differentiable map with a Lipschitz continuous Fréchet-derivative in a neighborhood of u_*. The approximating operators for the derivative are denoted by

$$B_k \in L(X, Y),$$

where $L(X, Y)$ is the space of linear bounded operators from X into Y.

- Solve $B_k s_k = -F(u_k)$ for $s_k \in X$
- Set $u_{k+1} = u_k + s_k$
- Set $y_k = F(u_{k+1}) - F(u_k)$
- Set $B_{k+1} = B_k + \frac{1}{\|s_k\|^2}(y_k - B_k s_k) \otimes s_k$

For $\bar{y} \in Y$ and $\bar{x} \in X$ the operator $\bar{y} \otimes \bar{x} \in L(X, Y)$ is defined as

$$(\bar{y} \otimes \bar{x})(x) = < \bar{x}, x > \bar{y} \text{ for all } x \in X.$$

3.3.2. Convergence Analysis.

Linear convergence is true also in Hilbert space. The reason for this is that in the proof of Theorem 3.3 we use estimates in the $\| \cdot \|_2$-norm which can be carried over to the infinite dimensional case using the operator norm instead.

In order to show the superlinear rate of convergence we notice that the Hilbert space framework does not cause a problem to show that the Dennis-Moré condition is sufficient for superlinear convergence. Therefore it remains to check if e.g. for Broyden's method the Dennis-Moré condition holds. Notice that in Theorem 3.4 the Frobenius norm is used which does not allow for a straightforward extension to Hilbert space. We can show the following

THEOREM 3.6. *If the sequence of iterates* $u_{k+1} = u_k + s_k, \quad s_k = -B_k^{-1} F(u_k)$ *converges at a linear rate to* u_* *with* $F(u_*) = 0$ *and if the* B_k *remain bounded, then the Dennis-Moré condition holds in a weak sense:*

$$\lim_{k \to \infty} \frac{\lambda((B_k - F'(u_*))s_k)}{\|s_k\|} = 0$$

for all $\lambda \in Y'$*, the dual of* Y.

Since in the case $X = \mathbb{R}_N$ weak and strong convergence coincide, the previous theorem is an extension of Theorem 3.4 to the Hilbert space case. The proof uses $\|E^* \lambda\|$ instead of $\|E\|_F$.

3.3.3. Compactness Condition.

There are examples in ℓ^2 which show that Broyden's method does not converge superlinearly in general for infinite dimensional problems. Under an additional assumption , however, the superlinear rate holds also in Hilbert space.

DEFINITION 3.1. *A sequence of linear operators, $\{K_k\}$ is collectively compact if for every bounded set, $\mathcal{B}$, the set, $\cup_k K_k \mathcal{B}$, is relatively compact.*

LEMMA 3.7. *Let the sequence of iterates $u_{k+1} = u_k + s_k$, $\quad s_k = -B_k^{-1} F(u_k)$ converge at a linear rate to u_* with $F(u_*) = 0$ and let the B_k be bounded. If E_0 is compact, then the sequence $\{E_k\}$ is collectively compact.*

THEOREM 3.8. *Let F be Fréchet differentiable and $F'(\cdot)$ Lipschitz continuous in a neighborhood of u_* with $F(u_*) = 0$ and $F'(u_*)^{-1} \in L(Y, X)$. Then for each $c \in (0,1)$ there exists $\varepsilon > 0$ such that if*

$$\|u_0 - u_*\| \leq \varepsilon \text{ and } \|B_0 - F'(u_*)\| \leq \varepsilon,$$

then the Broyden iterates are well defined and converge linearly to u_*. If, in addition,

$$B_0 - F'(u_*) \in L(X, Y) \text{ is compact}$$

then the rate of convergenc is superlinear.

3.3.4. Example.

In general the Hessian looks as follows (1.16)

$$\begin{aligned} & < w, \nabla^2\phi(u)v > \\ = \; & < \xi(w), H_{xx}(x,u,\cdot)\xi(v) > + < w, H_{ux}(x,u,\cdot)\xi(v) > \\ + \; & < \xi(w), H_{xu}(x,u,\cdot)v > + < w, H_{uu}(x,u,\cdot)v > \end{aligned} \tag{3.6}$$

where x solves (1.3), p solves (1.10) and $\xi(w)$ solves (1.17). The noncompact part is $H_{uu}(x,u,\cdot)$. This explains the numerical results of the previous example because the term $0.025u^4$ in $H_{uu}(x,u,\cdot)$ cannot be approximated such that $B_0 - F'(u_*)$ is compact.

REFERENCES

[1] C. G. Broyden. A class of methods for solving simultaneous equations. *Math. Comp.*, 19:577–593, 1965.

[2] C. G. Broyden, J. E. Dennis, and J. J. Moré. On the local and superlinear convergence of quasi-Newton methods. *J. I. M. A.*, 12:223–246, 1973.

[3] J. E. Dennis. Toward a unified convergence theory for Newton-like methods. In L. B. Rall, editor, *Nonlinear Functional Analysis and Applications*, pages 425–472. Academic Press, 1971.

[4] A. Griewank. Rates of convergence for secant methods on nonlinear problems in Hilbert space. In J. P. Hennart, editor, *Numerical Analysis, Proceedings Guanajuato, Mexico 1984*, pages 138–157. Springer, 1985.

[5] A. Griewank. The local convergence of Broyden-like methods on Lipschitzian problems in Hilbert space. *SIAM J. Numer. Anal.*, 24:684–705, 1987.

[6] C. T. Kelley and E. W. Sachs. A new proof of superlinear convergence for Broyden's method in Hilbert space. *SIAM J. Optimization*, 1, 1991.

[7] C. T. Kelley and E. W. Sachs. Quasi-Newton methods and unconstrained optimal control problems. *SIAM J. Control and Optimization*, 25:1503–1517, 1987.

[8] J. Stoer. Two examples on the convergence of certain rank-2 minimization methods for quadratic functionals in Hilbert space. *Lin. Alg. and Appl.*, 28:37–52, 1984.

4. Quasi-Newton Methods with Structure for Control Problems.

4.1. Necessary Optimality Conditions.

4.1.1. Problem Formulation and Notation.

In this part we show how to overcome the difficulty to satisfy the compactness condition on the approximation of the Hessian by using more structure of the problem. Consider the unconstrained optimal control problem:

$$\text{Minimize} \int_0^T L(x(t),\, u(t),\, t)dt$$

over $u \in L^\infty_m[0,T]$ such that $x \in C_n[0,T]$ solves

$$\dot{x} = f(x,u,t) \quad , \quad x(0) = x_0 \quad , \quad t \in (0,T).$$

Here $f : I\!R_n \times I\!R_m \times [0,T] \to I\!R_n$ and $L : I\!R_n \times I\!R_m \times [0,T] \to I\!R$.
Define the Hamiltonian function

$$H(x,u,t) = f^T(x,u,t)\,p(t) + L(x,u,t),$$

where $p \in C_n[0,T]$ solves the adjoint equation

$$-\dot{p} = f_x^T p + L_x^T \quad , \quad p(T) = 0.$$

4.1.2. Nonlinear Equation.

The necessary optimality conditions yield that we have to solve the following system of nonlinear equations

$$(4.1) \qquad F(z) = F(p,x,u) \;=\; \begin{pmatrix} \dot{x} - f(x,u,t) \\ \dot{p} + H_x(x,u,t) \\ H_u(x,u,t) \end{pmatrix} \;=\; \begin{pmatrix} 0 \\ 0 \\ 0 \end{pmatrix}$$

for $z = (p,x,u) \in X^\infty$ and $F : X^\infty \to Y^\infty$ with

$$\begin{aligned} X^r &= W_n^{1,r}[0,T] \times W_n^{1,r}[0,T] \times L_m^r[0,T] \\ Y^r &= L_n^r[0,T] \times L_n^r[0,T] \times L_m^r[0,T] \end{aligned}$$

and boundary conditions.

4.2. Structure in the Jacobian.

4.2.1. Computation of the Jacobian.

ASSUMPTION 4.1. *f, L and their first and second partial derivatives with respect to x and u are continuous on $\mathbb{R}_n \times \mathbb{R}_m \times [0,T]$.*

Then F is Fréchet-differentiable and

$$F'(p,x,u)\begin{pmatrix}\pi\\ \xi\\ \nu\end{pmatrix} = \begin{pmatrix}0 & D-f_x & -f_u\\ D+f_x^T & H_{xx} & H_{xu}\\ f_u^T & H_{ux} & H_{uu}\end{pmatrix}\begin{pmatrix}\pi\\ \xi\\ \nu\end{pmatrix} \tag{4.2}$$

with $D = \frac{d}{dt}$ and all other components as multiplication operators.

4.2.2. Computed and Approximated Part.

We approximate $F'(p,x,u)$ by $J = A + C(z)$ with $C(z)$ containing first derivative information

$$C(z) = \begin{pmatrix}0 & D-f_x & -f_u\\ D+f_x^T & 0 & 0\\ f_u^T & 0 & 0\end{pmatrix}.$$

The operator A approximates the second order derivative terms,

$$A_* = \begin{pmatrix}0 & 0 & 0\\ 0 & H_{xx}(z_*) & H_{xu}(z_*)\\ 0 & H_{ux}(z_*) & H_{uu}(z_*)\end{pmatrix},$$

where $z_* = (p_*, x_*, u_*)$ is the solution of (4.1).

For example, let

$$A = \begin{pmatrix}0 & 0 & 0\\ 0 & A_{11} & A_{12}\\ 0 & A_{21} & A_{22}\end{pmatrix}$$

with

$$\begin{pmatrix}A_{11} & A_{12}\\ A_{21} & A_{22}\end{pmatrix} \in L^\infty_{(m+n)\times(m+n)}[0,T].$$

4.3. Pointwise BFGS Methods.

4.3.1. Pointwise BFGS Update.

In order to improve on the results from Table 3.2 we use more structure of the Jacobian. A is updated *pointwise* for each $t \in [0,T]$ by a BFGS-update

$$\begin{aligned} A_+ &= A + (y^{\#T} s)^+ y^\# y^{\#T} - (s^T A s)^+ A s s^T A \\ y^\# &= P_2(y - C(z_+)s) \quad , \quad y = F(z_+) - F(z) \end{aligned}$$

where $\alpha^+ = 1/\alpha$ unless $\alpha = 0$, where we set $\alpha^+ = 0$. Here s denotes the solution of $J s = -F(z)$ and

$$P_2 \begin{pmatrix} p \\ x \\ u \end{pmatrix} = \begin{pmatrix} 0 & 0 & 0 \\ 0 & 1 & 0 \\ 0 & 0 & 1 \end{pmatrix} \begin{pmatrix} p \\ x \\ u \end{pmatrix} = \begin{pmatrix} 0 \\ x \\ u \end{pmatrix}.$$

4.3.2. Computation of a Step.

For each J solve for the step $\zeta = (\pi, \xi, \nu)$

$$J \begin{pmatrix} \pi \\ \xi \\ \nu \end{pmatrix} = \begin{pmatrix} 0 & D - f_x & -f_u \\ D + f_x^T & A_{11} & A_{12} \\ f_u^T & A_{21} & A_{22} \end{pmatrix} \begin{pmatrix} \pi \\ \xi \\ \nu \end{pmatrix} = -F(z)$$

with boundary conditions on ξ, π. Since $F(z)$ is given by

$$F(z) = \begin{pmatrix} \dot{x} - f(x,u,t) \\ \dot{p} + H_x(x,u,t) \\ H_u(x,u,t) \end{pmatrix} \tag{4.3}$$

we can solve instead for

$$(p_+, x_+, u_+) = (p + \pi, x + \xi, u + \nu)$$

a two-point boundary value problem with an algebraic equation.

4.3.3. Numerical Example.

First we consider the example in Table 3.2 from Section 3.2.2 and use a pointwise BFGS-method. We obtain the results listed in Table 4.1 and see that by the use of a more sophisticated update we have recovered the superlinear rate.

Since the BFGS-update formula requires a division by a perhaps too small number we do not update for $t \in [0,T]$ if

$$\min\{|y^{\#T}(t)\hat{s}(t)|/\|y^\#(t)\|_2\|\hat{s}(t)\|_2, |\hat{s}(t)A(t)\hat{s}(t)|/\|\hat{s}(t)\|_2^2\} \leq h^4.$$

holds with $\hat{s} = P_2 s$.

TABLE 4.1
Pointwise BFGS update (Ex. 3.2.2)

k	$\|s_k\|_2$	$\|s_k\|_2/\|s_{k-1}\|_2$
0	0.2709D+01	
1	0.1040D+01	0.384
2	0.1865D+00	0.179
3	0.3976D-01	0.213
4	0.2192D-02	0.055
5	0.2386D-04	0.011
6	0.5984D-07	0.003

Let us consider a different example. Suppose that $n = 2, m = 1$ and

$$\dot{x}_1 = (1 - x_2^2)x_1 - 10x_2 + u, \quad ,\dot{x}_2 = x_1, \quad L(x_1, x_2, u) = \frac{1}{2}(x_1^2 + x_2^2 + u^2)$$

Furthermore, we choose $x_0 = (0,1)^T, \quad T = 5$ and initial data $A_0 = P_0 \in I\!R_{3\times 3}$ and $z_0 = (0,0,0)^T$.

We choose as discretization parameter $h = \frac{1}{121}$. The discretization of the two-point boundary value problem is performed by the trapezoid finite difference scheme used with a Richardson extrapolation to achieve 4th order accurate results.

TABLE 4.2
Pointwise BFGS update (Ex. 4.3.3)

k	$\|F(z_k)\|_2$	$\|F(z_k)\|_2/\|F(z_{k-1})\|_2$	No Upd %
0	0.94468D+01		0.0
1	0.23914D+01	0.253	0.0
2	0.24168D+00	0.101	4.1
3	0.24835D-01	0.103	4.1
4	0.26126D-02	0.105	7.4
5	0.11607D-02	0.444	4.1
6	0.17268D-02	1.488	0.8
7	0.44565D-03	0.258	3.3
8	0.37778D-02	8.477	1.7
9	0.25568D-02	0.677	0.8
10	0.23206D-02	0.908	0.0
11	0.14371D-02	0.619	6.6
12	0.14582D-01	10.147	6.6
13	0.13256D-01	0.909	0.0
14	0.70550D-02	0.532	1.7

Conclusion: *No* (superlinear) convergence.

4.4. Regularity of Derivative.

4.4.1. Second Order Sufficiency Conditions for Optimality.

In [6] it is assumed that along the optimal control and trajectory the matrix

$$\begin{pmatrix} H_{xx} & H_{xu} \\ H_{ux} & H_{uu} \end{pmatrix} \tag{4.4}$$

is positive definite for all $t \in [0,T]$. In [7] this condition was relaxed to

ASSUMPTION 4.2. *For the solution*

$$z_* \in X^\infty \text{ of } F(z_*) = 0$$

there exists $\alpha > 0$ *such that the matrix function (4.4) lies in* $L^\infty_{(m+n)\times(m+n)}[0,T]$ *and*

$$v^T H_{uu}(t) v \geq \alpha \|v\|^2 \qquad a.e. \text{ on } \quad [0,T] \tag{4.5}$$

for all $v \in \mathbb{R}_m$. *Furthermore let a solution* $Q \in W^{1,\infty}_{n\times n}[0,T]$ *of (4.6) exist.*

Here $Q \in W^{1,\infty}_{n\times n}$ is a solution of the matrix Riccati equation

$$\begin{aligned} Q' &= U_{12}^T Q + QU_{12} + QU_{11}Q - U_{22}, \qquad t \in [0,T] \\ Q(T) &= 0, \end{aligned} \tag{4.6}$$

where

$$\begin{aligned} U_{11} &= f_u H_{uu}^{-1} f_u^T \\ U_{12} &= -f_x + f_u H_{uu}^{-1} H_{ux} \\ U_{22} &= H_{xx} - H_{xu} H_{uu}^{-1} H_{ux} \end{aligned}$$

all evaluated along $(p_*(t), x_*(t), u_*(t))$, $t \in [0,T]$. This leads us to suspect that the use of a BFGS-update for the pointwise update matrix is inconsistent with the lack of a pointwise positive definiteness in (4.5). Therefore we will use in Section 4.5.1 updates which do not necessarily maintain the positive definiteness.

4.4.2. Regularity of Jacobian.

THEOREM 4.1. *Let Assumptions 4.1 and 4.2 hold and let* $z_* \in X^\infty$ *solve* $F(z_*) = 0$. *Then* $F'(z_*)$ *is Fréchet-differentiable as a mapping from* X^∞ *to* Y^∞ *and is regular in the following sense*

$$\|F'(z_*)\zeta\|_{Y^r} \geq \mu \|\zeta\|_{X^r} \quad \text{for all} \quad r \in [1,\infty], \zeta \in X_0^r \tag{4.7}$$

holds, and is a mapping onto $Y^r, r \in [1,\infty]$.

The implication for Newton's method where

$$z_+ = z - F'(z)^{-1}F(z)$$

is the following theorem.

THEOREM 4.2. *Let Assumptions 4.1 and 4.2 hold and let $z_* \in X^\infty$ solve $F(z_*) = 0$. Then for $z_0 \in X^\infty$ sufficiently close to z_* in the X^∞-norm, there is $\kappa > 0$ such that for the Newton iterates*

$$\|z_* - z_{k+1}\|_{X^\infty} \le \kappa \|z_* - z_k\|^2_{X^\infty}.$$

4.5. Pointwise SR1-Method.

4.5.1. Pointwise SR-1 Update.

Update formulas which do not require positive definiteness:
Pointwise Broyden-update

$$A_+ \;=\; A + (\hat{s}^T\hat{s})^+(y^\# - A\hat{s})\,\hat{s}^T \quad , \quad \hat{s} = P_2\, s,$$

Pointwise PSB-update

$$A_+ \;=\; A + (\hat{s}^T\hat{s})^+((y^\# - A\hat{s})\,\hat{s}^T + \hat{s}(y^\# - A\hat{s})^T) - \hat{s}^T(y^\# - A\hat{s})((\hat{s}^T\hat{s})^+)^2\,\hat{s}\,\hat{s}^T,$$

Pointwise SR-1-update

$$A_+ \;=\; A + ((y^\# - A\hat{s})^T\hat{s})^+(y^\# - A\hat{s})(y^\# - A\hat{s})^T.$$

4.5.2. Numerical Example.

We use the same data as for the results in Table 4.2.
No update if

$$|(y^\#(t) - A(t)\hat{s}(t))^T\hat{s}(t)| \le h^4\|y^\#(t) - A(t)\hat{s}(t)\|_2\|\hat{s}(t)\|_2$$

As we see in Table 4.3, superlinear convergence is recovered if we use a SR1-update instead of BFGS-update.

TABLE 4.3
Pointwise SR-1 update (Ex. 4.3.3)

k	$\|F(z_k)\|_2$	$\|F(z_k)\|_2/\|F(z_{k-1})\|_2$	No Upd %
0	0.94468D+01		0.0
1	0.23914D+01	0.253	1.7
2	0.18184D+00	0.076	1.7
3	0.29635D-01	0.163	1.7
4	0.37877D-03	0.013	1.7
5	0.15594D-06	0.000	1.7

Second Example

$$\dot{x}_1 = (1 - x_2^2)x_1 - x_2 + u, \quad \dot{x}_2 = x_1, \quad L(x_1, x_2, u) = \frac{1}{2}(x_1^2 + x_2^2 + u^2)$$

TABLE 4.4
Pointwise updates (Ex. 4.5.2)

k	$\|F(z_k)\|_2$	$\|F(z_k)\|_2/\|F(z_{k-1})\|_2$	No Upd %	Update
0	0.19680D+01		0.0	
1	0.21060D+00	0.107	0.0	BFGS
2	0.36148D-02	0.017	0.0	BFGS
3	0.27802D-03	0.077	0.0	BFGS
4	0.49937D-04	0.180	0.0	BFGS
5	0.42030D-05	0.084	0.0	BFGS
0	0.19680D+01		0.0	Broyden
1	0.21060D+00	0.107	0.0	Broyden
2	0.36539D-02	0.017	0.0	Broyden
3	0.22223D-03	0.061	0.0	Broyden
4	0.16664D-04	0.075	100.0	Broyden
5	0.19426D-06	0.012	100.0	Broyden
0	0.19680D+01		0.0	PSB
1	0.21060D+00	0.107	0.0	PSB
2	0.36200D-02	0.017	0.0	PSB
3	0.19240D-03	0.053	0.0	PSB
4	0.13755D-04	0.071	92.6	PSB
5	0.28245D-05	0.205	100.0	PSB
0	0.19680D+01		0.0	SR-1
1	0.21060D+00	0.107	1.7	SR-1
2	0.57159D-02	0.027	1.7	SR-1
3	0.10516D-01	1.840	1.7	SR-1
4	0.86692D-06	0.000	1.7	SR-1

4.6. Convergence Analysis .

4.6.1. Linear Convergence Rate.

THEOREM 4.3. *Let Assumptions 4.1 and 4.2 hold and let $z_* \in X^\infty$ solve $F(z_*) = 0$. There are $\varepsilon, \delta > 0$ such that if $z_0 \in X^\infty, A_0(t) \in \mathbb{R}_{(m+n)\times(m+n)}$, $t \in [0,T]$, satisfy*

$$\|z_0 - z_*\|_{X^\infty} < \varepsilon, \quad \|A_0 - A_*\|_\infty < \delta,$$

then the sequence of iterates produced by the pointwise PSB method or the Broyden method converges to z_* linearly in the X^∞−norm.

Proof via Bounded Deterioration Property in a pointwise Frobenius norm.

$$\|E\|_{\mathcal{F}} = \sup_{t\in[0,T]} \|E(t)\|_F$$

4.6.2. Superlinear Convergence Rate.

THEOREM 4.4. *Let Assumptions 4.1 and 4.2 hold and let $z_* \in X^\infty$ solve $F(z_*) = 0$. There are $\varepsilon, \delta > 0$ such that if $z_0 \in X^\infty, A_0(t) \in \mathbb{R}_{(m+n)\times(m+n)}$, $t \in [0,T]$, satisfy*

$$\|z_0 - z_*\|_{X^\infty} < \varepsilon, \quad \|A_0 - A_*\|_\infty < \delta,$$

then the sequence of iterates produced by the pointwise PSB method or the Broyden method satisfies for $r, s \in [1, \infty]$ with $r > s$

$$\lim_{k\to\infty} \frac{\|z_{k+1} - z_*\|_{X^s}}{\|z_k - z_*\|_{X^r}} = 0 .$$

Nonstandard superlinear convergence rate. Proof via pointwise Dennis-Moré-Condition.

REFERENCES

[1] A. Griewank. The solution of boundary value problems by broyden based secant methods. In J. Noye and R. May, editors, *Computational Techniques and Applications: CTAC 85, Proceedings of CTAC, Melbourne, August 1985*, pages 309–321. North Holland, 1986.

[2] W. E. Hart and S. O. W. Soul. Quasi-newton methods for discretized nonlinear boundary problems. *Journal of the Institute of Applied Mathematics*, 11:351–359, 1973.

[3] C. T. Kelley and E. W. Sachs. A quasi-Newton method for elliptic boundary value problems. *SIAM J. Numer. Anal.*, 24:516–531, 1987.

[4] C. T. Kelley and E. W. Sachs. A pointwise quasi-Newton method for unconstrained optimal control problems. *Numer. Math.*, 55:159–176, 1989.

[5] C. T. Kelley, E. W. Sachs, and B. Watson. A pointwise quasi-Newton method for unconstrained optimal control problems, II. *Journal of Optimization Th. and Applic.*, 71, 1991.

[6] E. B. Lee and L. Markus. *Foundations of Optimal Control Theory*. Wiley, New York-London-Sydney, 1967.

[7] H. Maurer. First and second order sufficient optimality conditions in mathematical programming and optimal control. *Math. Programming Study*, 14:163–177, 1981.

Lecture Notes in Economics and Mathematical Systems

For information about Vols. 1–210
please contact your bookseller or Springer-Verlag

Vol. 211: P. van den Heuvel, The Stability of a Macroeconomic System with Quantity Constraints. VII, 169 pages. 1983.

Vol. 212: R. Sato and T. Nôno, Invariance Principles and the Structure of Technology. V, 94 pages. 1983.

Vol. 213: Aspiration Levels in Bargaining and Economic Decision Making. Proceedings, 1982. Edited by R. Tietz. VIII, 406 pages. 1983.

Vol. 214: M. Faber, H. Niemes und G. Stephan, Entropie, Umweltschutz und Rohstoffverbrauch. IX, 181 Seiten. 1983.

Vol. 215: Semi-Infinite Programming and Applications. Proceedings, 1981. Edited by A.V. Fiacco and K.O. Kortanek. XI, 322 pages. 1983.

Vol. 216: H.H. Müller, Fiscal Policies in a General Equilibrium Model with Persistent Unemployment. VI, 92 pages. 1983.

Vol. 217: Ch. Grootaert, The Relation Between Final Demand and Income Distribution. XIV, 105 pages. 1983.

Vol 218: P. van Loon, A Dynamic Theory of the Firm: Production, Finance and Investment. VII, 191 pages. 1983.

Vol. 219: E. van Damme, Refinements of the Nash Equilibrium Concept. VI. 151 pages. 1983.

Vol. 220: M. Aoki, Notes on Economic Time Series Analysis: System Theoretic Perspectives. IX, 249 pages. 1983.

Vol. 221: S. Nakamura, An Inter-Industry Translog Model of Prices and Technical Change for the West German Economy. XIV, 290 pages. 1984.

Vol. 222: P. Meier, Energy Systems Analysis for Developing Countries. VI, 344 pages. 1984.

Vol. 223: W. Trockel, Market Demand. VIII, 205 pages. 1984.

Vol. 224: M. Kiy, Ein disaggregiertes Prognosesystem fur die Bundesrepublik Deutschland. XVIII, 276 Seiten. 1984.

Vol. 225: T.R. von Ungern-Sternberg, Zur Analyse von Märkten mit unvollständiger Nachfragerinformaton. IX, 125 Seiten. 1984.

Vol. 226: Selected Topics in Operations Research and Mathematical Economics. Proceedings, 1963. Edited by G. Hammer and D. Pallaschke IX, 478 pages. 1984.

Vol. 227: Risk and Capital. Proceedings, 1983. Edited by G. Bamberg and K. Spremann VII, 306 pages. 1984.

Vol. 228: Nonlinear Models of Fluctuating Growth. Proceedings, 1983. Edited by R.M. Goodwin, M. Krüger and A. Vercelli. XVII, 277 pages. 1984.

Vol. 229: Interactive Decision Analysis. Proceedings, 1983. Edited by M. Grauer and A.P. Wierzbicki. VIII, 269 pages. 1984.

Vol. 230: Macro-Economic Planning with Conflicting Goals. Proceedings, 1982. Edited by M. Despontin, P. Nijkamp and J. Spronk. VI, 297 pages. 1984.

Vol. 231: G.F. Newell, The M/M/8 Service System with Ranked Servers in Heavy Traffic. XI, 126 pages. 1984.

Vol. 232: L. Bauwens, Bayesian Full Information Analysis of Simultaneous Equation Models Using Integration by Monte Carlo. VI, 114 pages. 1984.

Vol. 233: G. Wagenhals, The World Copper Market. XI, 190 pages. 1984.

Vol. 234: B.C. Eaves, A Course in Triangulations for Solving Equations with Deformations. III, 302 pages. 1984.

Vol. 235: Stochastic Models in Reliability Theory Proceedings, 1984. Edited by S. Osaki and Y. Hatoyama. VII, 212 pages. 1984.

Vol. 236: G. Gandolfo, P.C. Padoan, A Disequilibrium Model of Real and Financial Accumulation in an Open Economy. VI, 172 pages. 1984.

Vol. 237: Misspecification Analysis. Proceedings, 1983. Edited by T.K. Dijkstra. V, 129 pages. 1984.

Vol. 238: W. Domschke, A. Drexl, Location and Layout Planning. IV, 134 pages. 1985.

Vol. 239: Microeconomic Models of Housing Markets. Edited by K. Stahl. VII, 197 pages. 1985.

Vol. 240: Contributions to Operations Research. Proceedings, 1984. Edited by K. Neumann and D. Pallaschke. V, 190 pages. 1985.

Vol. 241: U. Wittmann, Das Konzept rationaler Preiserwartungen. XI, 310 Seiten. 1985.

Vol. 242: Decision Making with Multiple Objectives. Proceedings, 1984. Edited by Y.Y. Haimes and V. Chankong. XI, 571 pages. 1985.

Vol. 243: Integer Programming and Related Areas. A Classified Bibliography 1981–1984. Edited by R. von Randow. XX, 386 pages. 1985.

Vol. 244: Advances in Equilibrium Theory. Proceedings, 1984. Edited by C.D. Aliprantis, O. Burkinshaw and N.J. Rothman. II, 235 pages. 1985.

Vol. 245: J.E.M. Wilhelm, Arbitrage Theory. VII, 114 pages. 1985.

Vol. 246: P.W. Otter, Dynamic Feature Space Modelling, Filtering and Self-Tuning Control of Stochastic Systems. XIV, 177 pages.1985.

Vol. 247: Optimization and Discrete Choice in Urban Systems. Proceedings, 1983. Edited by B.G. Hutchinson, P. Nijkamp and M. Batty VI, 371 pages. 1985.

Vol. 248: Pural Rationality and Interactive Decision Processes. Proceedings, 1984. Edited by M. Grauer, M. Thompson and A.P. Wierzbicki. VI, 354 pages. 1985.

Vol. 249: Spatial Price Equilibrium: Advances in Theory, Computation and Application. Proceedings, 1984. Edited by P.T. Harker. VII, 277 pages. 1985.

Vol. 250: M. Roubens, Ph. Vincke, Preference Modelling. VIII, 94 pages. 1985.

Vol. 251: Input-Output Modeling. Proceedings, 1984. Edited by A. Smyshlyaev. VI, 261 pages. 1985.

Vol. 252: A. Birolini, On the Use of Stochastic Processes in Modeling Reliability Problems. VI, 105 pages. 1985.

Vol. 253: C. Withagen, Economic Theory and International Trade in Natural Exhaustible Resources. VI, 172 pages. 1985.

Vol. 254: S. Müller, Arbitrage Pricing of Contingent Claims. VIII, 151 pages. 1985.

Vol. 255: Nondifferentiable Optimization: Motivations and Applications. Proceedings, 1984. Edited by V.F. Demyanov and D. Pallaschke. VI, 350 pages. 1985.

Vol. 256: Convexity and Duality in Optimization. Proceedings, 1984. Edited by J. Ponstein. V, 142 pages. 1985.

Vol. 257: Dynamics of Macrosystems. Proceedings, 1984. Edited by J.-P. Aubin, D. Saari and K. Sigmund. VI, 280 pages. 1985.

Vol. 258: H. Funke, Eine allgemeine Theorie der Polypol- und Oligopolpreisbildung. III, 237 pages. 1985.

Vol. 259: Infinite Programming. Proceedings, 1984. Edited by E.J. Anderson and A.B. Philpott. XIV, 244 pages. 1985.

Vol. 260: H.-J. Kruse, Degeneracy Graphs and the Neighbourhood Problem. VIII, 128 pages. 1986.

Vol. 261: Th.R. Gulledge, Jr., N.K. Womer, The Economics of Made-to-Order Production. VI, 134 pages. 1986.

Vol. 262: H.U. Buhl, A Neo-Classical Theory of Distribution and Wealth. V, 146 pages. 1986.

Vol. 263: M. Schäfer, Resource Extraction and Market Struucture. XI, 154 pages. 1986.

Vol. 264: Models of Economic Dynamics. Proceedings, 1983. Edited by H.F. Sonnenschein. VII, 212 pages. 1986.

Vol. 265: Dynamic Games and Applications in Economics. Edited by T. Basar. IX, 288 pages. 1986.

Vol. 266: Multi-Stage Production Planning and Inventory Control. Edited by S. Axsäter, Ch. Schneeweiss and E. Silver. V, 264 pages.1986.

Vol. 267: R. Bemelmans, The Capacity Aspect of Inventories. IX, 165 pages. 1986.

Vol. 268: V. Firchau, Information Evaluation in Capital Markets. VII, 103 pages. 1986.

Vol. 269: A. Borglin, H. Keiding, Optimality in Infinite Horizon Economies. VI, 180 pages. 1986.

Vol. 270: Technological Change, Employment and Spatial Dynamics. Proceedings, 1985. Edited by P. Nijkamp. VII, 466 pages. 1986.

Vol. 271: C. Hildreth, The Cowles Commission in Chicago, 1939–1955. V, 176 pages. 1986.

Vol. 272: G. Clemenz, Credit Markets with Asymmetric Information. VIII,212 pages. 1986.

Vol. 273: Large-Scale Modelling and Interactive Decision Analysis. Proceedings, 1985. Edited by G. Fandel, M. Grauer, A. Kurzhanski and A.P. Wierzbicki. VII, 363 pages. 1986.

Vol. 274: W.K. Klein Haneveld, Duality in Stochastic Linear and Dynamic Programming. VII, 295 pages. 1986.

Vol. 275: Competition, Instability, and Nonlinear Cycles. Proceedings, 1985. Edited by W. Semmler. XII, 340 pages. 1986.

Vol. 276: M.R. Baye, D.A. Black, Consumer Behavior, Cost of Living Measures, and the Income Tax. VII, 119 pages. 1986.

Vol. 277: Studies in Austrian Capital Theory, Investment and Time. Edited by M. Faber. VI, 317 pages. 1986.

Vol. 278: W.E. Diewert, The Measurement of the Economic Benefits of Infrastructure Services. V, 202 pages. 1986.

Vol. 279: H.-J. Büttler, G. Frei and B. Schips, Estimation of Disequilibrium Modes. VI, 114 pages. 1986.

Vol. 280: H.T. Lau, Combinatorial Heuristic Algorithms with FORTRAN. VII, 126 pages. 1986.

Vol. 281: Ch.-L. Hwang, M.-J. Lin, Group Decision Making under Multiple Criteria. XI, 400 pages. 1987.

Vol. 282: K. Schittkowski, More Test Examples for Nonlinear Programming Codes. V, 261 pages. 1987.

Vol. 283: G. Gabisch, H.-W. Lorenz, Business Cycle Theory. VII, 229 pages. 1987.

Vol. 284: H. Lütkepohl, Forecasting Aggregated Vector ARMA Processes. X, 323 pages. 1987.

Vol. 285: Toward Interactive and Intelligent Decision Support Systems. Volume 1. Proceedings, 1986. Edited by Y. Sawaragi, K. Inoue and H. Nakayama. XII, 445 pages. 1987.

Vol. 286: Toward Interactive and Intelligent Decision Support Systems. Volume 2. Proceedings, 1986. Edited by Y. Sawaragi, K. Inoue and H. Nakayama. XII, 450 pages. 1987.

Vol. 287: Dynamical Systems. Proceedings, 1985. Edited by A.B. Kurzhanski and K. Sigmund. VI, 215 pages. 1987.

Vol. 288: G.D. Rudebusch, The Estimation of Macroeconomic Disequilibrium Models with Regime Classification Information. VII,128 pages. 1987.

Vol. 289: B.R. Meijboom, Planning in Decentralized Firms. X, 168 pages. 1987.

Vol. 290: D.A. Carlson, A. Haurie, Infinite Horizon Optimal Control. XI, 254 pages. 1987.

Vol. 291: N. Takahashi, Design of Adaptive Organizations. VI, 140 pages. 1987.

Vol. 292: I. Tchijov, L. Tomaszewicz (Eds.), Input-Output Modeling. Proceedings, 1985. VI, 195 pages. 1987.

Vol. 293: D. Batten, J. Casti, B. Johansson (Eds.), Economic Evolution and Structural Adjustment. Proceedings, 1985. VI, 382 pages.

Vol. 294: J. Jahn, W. Knabs (Eds.), Recent Advances and Historical Development of Vector Optimization. VII, 405 pages. 1987.

Vol. 295. H. Meister, The Purification Problem for Constrained Games with Incomplete Information. X, 127 pages. 1987.

Vol. 296: A. Börsch-Supan, Econometric Analysis of Discrete Choice. VIII, 211 pages. 1987.

Vol. 297: V. Fedorov, H. Läuter (Eds.), Model-Oriented Data Analysis. Proceedings, 1987. VI, 239 pages. 1988.

Vol. 298: S.H. Chew, Q. Zheng, Integral Global Optimization. VII, 179 pages. 1988.

Vol. 299: K. Marti, Descent Directions and Efficient Solutions in Discretely Distributed Stochastic Programs. XIV, 178 pages. 1988.

Vol. 300: U. Derigs, Programming in Networks and Graphs. XI, 315 pages. 1988.

Vol. 301: J. Kacprzyk, M. Roubens (Eds.), Non-Conventional Preference Relations in Decision Making. VII, 155 pages. 1988.

Vol. 302: H.A. Eiselt, G. Pederzoli (Eds.), Advances in Optimization and Control. Proceedings, 1986. VIII, 372 pages. 1988.

Vol. 303: F.X. Diebold, Empirical Modeling of Exchange Rate Dynamics. VII, 143 pages. 1988.

Vol. 304: A. Kurzhanski, K. Neumann, D. Pallaschke (Eds.), Optimization, Parallel Processing and Applications. Proceedings, 1987. VI, 292 pages. 1988.

Vol. 305: G.-J.C.Th. van Schijndel, Dynamic Firm and Investor Behaviour under Progressive Personal Taxation. X, 215 pages.1988.

Vol. 306: Ch. Klein, A Static Microeconomic Model of Pure Competition. VIII, 139 pages. 1988.

Vol. 307: T.K. Dijkstra (Ed.), On Model Uncertainty and its Statistical Implications. VII, 138 pages. 1988.

Vol. 308: J.R. Daduna, A. Wren (Eds.), Computer-Aided Transit Scheduling. VIII, 339 pages. 1988.

Vol. 309: G. Ricci, K. Velupillai (Eds.), Growth Cycles and Multisectoral Economics: the Goodwin Tradition. III, 126 pages. 1988.

Vol. 310: J. Kacprzyk, M. Fedrizzi (Eds.), Combining Fuzzy Imprecision with Probabilistic Uncertainty in Decision Making. IX, 399 pages. 1988.

Vol. 311: R. Färe, Fundamentals of Production Theory. IX, 163 pages. 1988.

Vol. 312: J. Krishnakumar, Estimation of Simultaneous Equation Models with Error Components Structure. X, 357 pages. 1988.

Vol. 313: W. Jammernegg, Sequential Binary Investment Decisions. VI, 156 pages. 1988.

Vol. 314: R. Tietz, W. Albers, R. Selten (Eds.), Bounded Rational Behavior in Experimental Games and Markets. VI, 368 pages. 1988.

Vol. 315: I. Orishimo, G.J.D. Hewings, P. Nijkamp (Eds), Information Technology: Social and Spatial Perspectives. Proceedings 1986. VI, 268 pages. 1988.

Vol. 316: R.L. Basmann, D.J. Slottje, K. Hayes, J.D. Johnson, D.J. Molina, The Generalized Fechner-Thurstone Direct Utility Function and Some of its Uses. VIII, 159 pages. 1988.

Vol. 317: L. Bianco, A. La Bella (Eds.), Freight Transport Planning and Logistics. Proceedings, 1987. X, 568 pages. 1988.

Vol. 318: T. Doup, Simplicial Algorithms on the Simplotope. VIII, 262 pages. 1988.

Vol. 319: D.T. Luc, Theory of Vector Optimization. VIII, 173 pages. 1989.

Vol. 320: D. van der Wijst, Financial Structure in Small Business. VII, 181 pages. 1989.

Vol. 321: M. Di Matteo, R.M. Goodwin, A. Vercelli (Eds.), Technological and Social Factors in Long Term Fluctuations. Proceedings. IX, 442 pages. 1989.

Vol. 322: T. Kollintzas (Ed.), The Rational Expectations Equilibrium Inventory Model. XI, 269 pages. 1989.

Vol. 323: M.B.M. de Koster, Capacity Oriented Analysis and Design of Production Systems. XII, 245 pages. 1989.

Vol. 324: I.M. Bomze, B.M. Pötscher, Game Theoretical Foundations of Evolutionary Stability. VI, 145 pages. 1989.

Vol. 325: P. Ferri, E. Greenberg, The Labor Market and Business Cycle Theories. X, 183 pages. 1989.

Vol. 326: Ch. Sauer, Alternative Theories of Output, Unemployment, and Inflation in Germany: 1960–1985. XIII, 206 pages. 1989.

Vol. 327: M. Tawada, Production Structure and International Trade. V, 132 pages. 1989.

Vol. 328: W. Güth, B. Kalkofen, Unique Solutions for Strategic Games. VII, 200 pages. 1989.

Vol. 329: G. Tillmann, Equity, Incentives, and Taxation. VI, 132 pages. 1989.

Vol. 330: P.M. Kort, Optimal Dynamic Investment Policies of a Value Maximizing Firm. VII, 185 pages. 1989.

Vol. 331: A. Lewandowski, A.P. Wierzbicki (Eds.), Aspiration Based Decision Support Systems. X, 400 pages. 1989.

Vol. 332: T.R. Gulledge, Jr., L.A. Litteral (Eds.), Cost Analysis Applications of Economics and Operations Research. Proceedings. VII, 422 pages. 1989.

Vol. 333: N. Dellaert, Production to Order. VII, 158 pages. 1989.

Vol. 334: H.-W. Lorenz, Nonlinear Dynamical Economics and Chaotic Motion. XI, 248 pages. 1989.

Vol. 335: A.G. Lockett, G. Islei (Eds.), Improving Decision Making in Organisations. Proceedings. IX, 606 pages. 1989.

Vol. 336: T. Puu, Nonlinear Economic Dynamics. VII, 119 pages. 1989.

Vol. 337: A. Lewandowski, I. Stanchev (Eds.), Methodology and Software for Interactive Decision Support. VIII, 309 pages. 1989.

Vol. 338: J.K. Ho, R.P. Sundarraj, DECOMP: an Implementation of Dantzig-Wolfe Decomposition for Linear Programming. VI, 206 pages.

Vol. 339: J. Terceiro Lomba, Estimation of Dynamic Econometric Models with Errors in Variables. VIII, 116 pages. 1990.

Vol. 340: T. Vasko, R. Ayres, L. Fontvieille (Eds.), Life Cycles and Long Waves. XIV, 293 pages. 1990.

Vol. 341: G.R. Uhlich, Descriptive Theories of Bargaining. IX, 165 pages. 1990.

Vol. 342: K. Okuguchi, F. Szidarovszky, The Theory of Oligopoly with Multi-Product Firms. V, 167 pages. 1990.

Vol. 343: C. Chiarella, The Elements of a Nonlinear Theory of Economic Dynamics. IX, 149 pages. 1990.

Vol. 344: K. Neumann, Stochastic Project Networks. XI, 237 pages. 1990.

Vol. 345: A. Cambini, E. Castagnoli, L. Martein, P Mazzoleni, S. Schaible (Eds.), Generalized Convexity and Fractional Programming with Economic Applications. Proceedings, 1988. VII, 361 pages. 1990.

Vol. 346: R. von Randow (Ed.), Integer Programming and Related Areas. A Classified Bibliography 1984–1987. XIII, 514 pages. 1990.

Vol. 347: D. Ríos Insua, Sensitivity Analysis in Multi-objective Decision Making. XI, 193 pages. 1990.

Vol. 348: H. Störmer. Binary Functions and their Applications. VIII, 151 pages. 1990.

Vol. 349: G.A. Pfann, Dynamic Modelling of Stochastic Demand for Manufacturing Employment. VI, 158 pages. 1990.

Vol. 350: W.-B. Zhang, Economic Dynamics. X, 232 pages. 1990.

Vol. 351: A. Lewandowski, V. Volkovich (Eds.), Multiobjective Problems of Mathematical Programming. Proceedings, 1988. VII, 315 pages. 1991.

Vol. 352: O. van Hilten, Optimal Firm Behaviour in the Context of Technological Progress and a Business Cycle. XII, 229 pages. 1991.

Vol. 353: G. Riccil (Ed.), Declslon Processes In Economics. Proceedings, 1989. III, 209 pages 1991.

Vol. 354: M. Ivaldi, A Structural Analysis of Expectation Formation. XII, 230 pages. 1991.

Vol. 355: M. Salomon. Deterministic Lotsizlng Models for Production Planning. VII, 158 pages. 1991.

Vol. 356: P. Korhonen, A. Lewandowski, J . Wallenius (Eds.), Multlple Crltena Decision Supporl. Proceedings, 1989. XII, 393 pages. 1991.

Vol. 358: P. Knottnerus, Linear Models with Correlaled Disturbances. VIII, 196 pages. 1991.

Vol. 359: E. de Jong, Exchange Rate Determination and Optimal Economlc Policy Under Various Exchange Rate Regimes. VII, 270 pages. 1991.

Vol. 360: P. Stalder, Regime Translations, Spillovers and Buffer Stocks. VI, 193 pages . 1991.

Vol. 361: C. F. Daganzo, Logistics Systems Analysis. X, 321 pages. 1991.

Vol. 362: F. Gehreis, Essays In Macroeconomics of an Open Economy. VII, 183 pages. 1991.

Vol. 363: C. Puppe, Distorted Probabilities and Choice under Risk. VIII, 100 pages . 1991

Vol. 364: B. Horvath, Are Policy Variables Exogenous? XII, 162 pages. 1991.

Vol. 365: G. A Heuer, U. Leopold-Wildburger. Balanced Silverman Games on General Discrete Sets. V, 140 pages. 1991.

Vol. 366: J. Gruber (Ed.), Econometric Decision Models. Proceedings, 1989. VIII, 636 pages. 1991.

Vol. 367: M. Grauer, D. B. Pressmar (Eds.), Parallel Computing and Mathematical Optimization. Proceedings. V, 208 pages. 1991.

Vol. 368: M. Fedrizzi, J. Kacprzyk, M. Roubens (Eds.), Interactive Fuzzy Optimization. VII, 216 pages. 1991.

Vol. 369: R. Koblo, The Visible Hand. VIII, 131 pages.1991.

Vol. 370: M. J. Beckmann, M. N. Gopalan, R. Subramanian (Eds.), Stochastic Processes and their Applications. Proceedings, 1990. XLI, 292 pages. 1991.

Vol. 371: A. Schmutzler, Flexibility and Adjustment to Information in Sequential Decision Problems. VIII, 198 pages. 1991.

Vol. 372: J. Esteban, The Social Viability of Money. X, 202 pages. 1991.

Vol. 373: A. Billot, Economic Theory of Fuzzy Equilibria. XIII, 164 pages. 1992.

Vol. 374: G. Pflug, U. Dieter (Eds.), Simulation and Optimization. Proceedings, 1990. X, 162 pages. 1992.

Vol. 375: S.-J. Chen, Ch.-L. Hwang, Fuzzy Multiple Attribute Decision Making. XII, 536 pages. 1992.

Vol. 376: K.-H. Jöckel, G. Rothe, W. Sendler (Eds.), Bootstrapping and Related Techniques. Proceedings, 1990. VIII, 247 pages. 1992.

Vol. 377: A. Villar, Operator Theorems with Applications to Distributive Problems and Equilibrium Models. XVI, 160 pages. 1992.

Vol. 378: W. Krabs, J. Zowe (Eds.), Modern Methods of Optimization. Proceedings, 1990. VIII, 348 pages. 1992.